World in Transition

Members of the German Advisory Council on Global Change (WBGU)
(as on 21 March 2003)

Prof Dr Hartmut Graßl (chair)
Director of the Max Planck Institute for Meteorology, Hamburg

Prof Dr Dr Juliane Kokott (vice chair)
Director of the Institute of European and International Business Law at the University of
St. Gallen, Switzerland

Prof Dr Margareta E Kulessa
Professor at the University of Applied Sciences Mainz, Section Business Studies

Prof Dr Joachim Luther
Director of the Fraunhofer Institute for Solar Energy Systems, Freiburg

Prof Dr Franz Nuscheler
Director of the Institute for Development and Peace, Duisburg

Prof Dr Dr Rainer Sauerborn
Medical Director of the Department of Tropical Hygiene and Public Health at the
University of Heidelberg

Prof Dr Hans-Joachim Schellnhuber
Director of the Potsdam Institute for Climate Impact Research (PIK) and Research Director
of the Tyndall Centre for Climate Change Research in Norwich, United Kingdom

Prof Dr Renate Schubert
Director of the Center for Economic Research at the ETH Zurich, Switzerland

Prof Dr Ernst-Detlef Schulze
Director at the Max Planck Institute of Biogeochemistry in Jena

**German Advisory Council
on Global Change**

World in Transition

Towards Sustainable Energy Systems

Earthscan

London and Sterling, VA

GERMAN ADVISORY COUNCIL ON GLOBAL CHANGE (WBGU)
Secretariat
Reichpietschufer 60-62, 8th Floor
D-10785 Berlin
Germany

http://www.wbgu.de

Time of going to press, German version: 21.3.2003, entitled
Welt im Wandel: Energiewende zur Nachhaltigkeit. Springer-Verlag, Berlin Heidelberg New York, 2003
ISBN 3-540-40160-1

First published by Earthscan in the UK and USA in 2004

ISBN: 1-84407-882-9

Printed and bound in the UK by Cromwell Press Ltd
Translation by Christopher Hay, Darmstadt
Cover design by Meinhard Schulz-Baldes using the following illustrations:
Wind mills (M. Schulz-Baldes), solar thermal power plant (Plataforma solar de Almeria), 3-stone hearth (R. Sauerborn),
petrol pump, oil pump, dam, smokestack (Pure Vision Photo Disc Deutschland)

For a full list of publications please contact:

Earthscan
8–12 Camden High Street
London, NW1 0JH, UK
Tel: +44 (0)20 7387 8558
Fax: +44 (0)20 7387 8998
Email: earthinfo@earthscan.co.uk
Web: **www.earthscan.co.uk**

22883 Quicksilver Drive, Sterling, VA 20166-2012, USA

Earthscan publishes in association with WWF-UK and the International Institute for Environment and Development

A catalogue record for this book is available from the British Library

Library of Congress Cataloging-in-Publication Data

Wissenschaftlicher Beirat der Bundesregierung Globale Umweltveränderungen (Germany)
   World in transition : conservation and sustainable use of the biosphere / German Advisory Council on Global Change.
     p. cm.
   Includes bibliographical references (p.).
   ISBN 1-85383-802-0 (cloth)
     1. Biological diversity conservation--Government policy--Germany. 2. Nature conservation--Government policy--Germany.
3. Sustainable development--Government policy--Germany. I. Title

   QH77.G3 W57 2001
   333.95'16'0943--dc21

                                                                                    2001023313

This book is printed on elemental chlorine-free paper

Council Staff and Acknowledgements

Secretariat

Scientific Staff

Prof Dr Meinhard Schulz-Baldes
(Secretary-General)

Dr Carsten Loose
(Deputy Secretary-General)

Dietrich Brockhagen (DEA oek)

Dr Martin Cassel-Gintz (until 30.06.2002)

Dipl-Pol Judith C Enders (01.05. to 31.07.2002)

Dr Ursula Fuentes Hutfilter

Dipl Umweltwiss Tim Hasler (from 01.09.2002)

Dipl Pol Lena Kempmann (from 01.10.2002)

Dr Angela Oels (until 06.08.2002)

Dr Thilo Pahl (until 31.01.2003)

Dr Benno Pilardeaux
(Media and Public Relations)

Administration, Editorial work and Secretariat

Vesna Karic-Fazlic (Accountant)

Martina Schneider-Kremer, MA (Editorial work)

Margot Weiß (Secretariat)

Scientific Staff to the Council Members

Dr Carsten Agert (Fraunhofer Institute for Solar Energy Systems, Freiburg, from 01.08.2002)

Referendar jur Tim Bäuerle (Heidelberg, until 31.12.2002)

Cand rer pol Markus Dolder (ETH Zürich, Institute for Research in Economics, until 31.08.2002)

Lic rer pol Stefanie Fankhauser (ETH Zürich, Institute for Research in Economics, until 31.07.2002)

Dr Thomas Fues (Institute for Development and Peace, Duisburg)

Dr Jürgen Kropp (Potsdam Institute for Climate Impact Research)

Dr Jacques Léonardi (Max Planck Institute for Meteorology, Hamburg)

Referendar jur Christian Lutze (Heidelberg, from 01.01.2003)

Dr Franziska Matthies (Tyndall Centre for Climate Change Research, Norwich, UK)

Dr Tim Meyer (Fraunhofer Institute for Solar Energy Systems, Freiburg, until 31.07.2002)

Dipl Volksw Kristina Nienhaus (ETH Zürich/Akademie für Technikfolgenabschätzung in Baden-Württemberg, Stuttgart, from 09.09.2002)

Dipl Volksw Marc Ringel (University Mainz)

Dipl Biol Angelika Thuille (Max Planck Institute for Biogeochemistry, Jena)

The Council ows its gratitude to the important contributions and support by other members of the research community. This report builds on the following expert's studies:

- Dr Maritta von Bieberstein Koch-Weser (Earth 3000, Bieberstein) (2002): Nachhaltigkeit von Wasserkraft.
- Dr Ottmar Edenhofer, Dipl Volksw Nicolas Bauer and Dipl Phys Elmar Kriegler (Gesellschaft für Sozio-ökonomische Forschung – GSF, Potsdam) (2002): Szenarien zum Umbau des Energiesystems.
- Prof Dr Ing habil Hans-Burkhard Horlacher (TU Dresden) (2002): Globale Potenziale der Wasserkraft.
- Dr Ing Martin Kaltschmitt, Dr oec Dipl Ing Dieter Merten, Dipl Ing Nicolle Fröhlich and Dipl-Phys Moritz Nill (Institut für Energetik und Umwelt GmbH, Leipzig) (2002): Energiegewinnung aus Biomasse.
- Crescencia Maurer (Senior Associate in the Institutions and Governance Program of the World Resources Institute – WRI, Washington, DC) (2002): The Transition from Fossil to Renewable Energy Systems: What Role for Export Credit Agencies?
- Dr Joachim Nitsch (DLR, Institut für Technische Thermodynamik, Stuttgart) (2002): Potenziale der Wasserstoffwirtschaft.
- Dipl Geoökol Christiane Ploetz (VDI-Technologiezentrum, Abteilung Zukünftige Technologien Consulting, Düsseldorf) (2002): Sequestrierung von CO_2: Technologien, Potenziale, Kosten und Umweltauswirkungen.
- Dr Fritz Reusswig, Dipl Oec Katrin Gerlinger and Dr Ottmar Edenhofer (Gesellschaft für Sozioökonomische Forschung – GSF, Potsdam) (2002): Lebensstile und globaler Energieverbrauch. Analyse und Strategieansätze zu einer nachhaltigen Energiestruktur.
- Keywan Riahi (Institute for Applied Systems Analysis – IIASA, Laxenburg) (2002): Data From Model Runs With MESSAGE.
- Dr Franz Trieb and Dipl Systemwiss Stefan Kronshage (DLR, Institut für Technische Thermodynamik, Stuttgart) (2002): Berechnung von Weltpotenzialkarten.

Valuable support was provided during an in-depth discussion with scientific experts. The WBGU thanks the participants Prof Nakicenovic (IIASA, Laxenburg), Dr Nitsch (DLR, Stuttgart) and Prof Dr von Weizsäcker (MdB – Enquete Commission on Globalisation, Berlin).

The Council also wishes to thank all those who, in numerous instances, promoted the progress of this report through their comments and advice:

Jan Christoph Goldschmidt (Fraunhofer Institute for Solar Energy Systems, Freiburg), Dr Thomas Hamacher (Max Planck Institute of Plasma Physics, Garching), Dr Klaus Hassmann (Siemens AG), Prof Dr Klaus Heinloth (University Bonn), Prof Dr Dieter Holm (former University Pretoria), Prof Dr Eberhard Jochem (Fraunhofer Institute for Systems and Innovation Research, Karlsruhe), Prof Dr Wolfgang Kröger (Paul Scherrer Institute, Villingen), Prof Dr Matheos Santamouris (University Athens).

For their substantial assistance on issues relating to rural electrification and energy supply in developing countries, provided in connection with the preparation of the World Energy Outlook 2002, we thank Dr Fatih Birol, Chief Economist and Head of Economic Analysis Division of the International Energy Agency (IEA, Paris) and Marianne Haug, Director of the Energy Efficiency Department, Technology and R&D of IEA as well as Laura Cozzi, Energy Analyst of the Economic Analysis Division of IEA.

The Council is much indebted to the persons who received the WBGU delegation visiting the PR of China from March 10 to 22, 2002. Many experts from politics, administration and science offered guided tours, prepared presentations and were available for in-depth discussions and conversations. In particular the Council wishes to thank Ambassador Joachim Broudré-Gröger (German Embassy Beijing) and Wilfried Wolf (Leader of the Economic Division of the German Embassy Bejing), without whose support the substantive and organizational preparation and performance of the study tour would not have been possible, and the experts at Tsinghua University, Beijing, and at the University of Shanghai, who participated in highly informative energy expert panels with the Council.

The Council thanks Christopher Hay (Übersetzungsbüro für Umweltwissenschaften, Darmstadt) for his expert translation of this report into English from the German original.

Finally, we wish to thank Bernd Killinger, who, as an intern, carried out research and assembled literature, as well as Sabina Rolle, who, as student assistant, supported our work.

Outline of Contents

Contents

Boxes

Tables

Figures

Acronyms and Abbreviations

ACP	African, Caribbean and Pacific Group of States
AFREC	African Energy Commission
ARD	Monitoring and Measuring Afforestation – Reforestation – Deforestation (Kyoto Protocol to the UNFCCC)
ASEAN	Association of South East Asian Nations
BMBF	Bundesministerium für Bildung und Forschung [Federal Ministry of Education and Research (Germany)]
BMU	Bundesministerium für Umwelt, Naturschutz und Reaktorsicherheit [Federal Ministry for Environment, Nature Conservation and Reactor Safety (Germany)]
BMWA	Bundesministerium für Wirtschaft und Arbeit [Federal Ministry of Economics and Labour (Germany)]
BMZ	Bundesministerium für wirtschaftliche Zusammenarbeit und Entwicklung [Federal Ministry for Economic Cooperation and Development (Germany)]
BWR	Boiling Water Reactor
CCGT	Combined Cycle Gas Turbine
CDF	Comprehensive Development Framework (World Bank)
CDM	Clean Development Mechanism (UNFCCC)
CERUPT	Certified Emission Reduction Unit Procurement Tender, The Netherlands
CHP	Combined Heat and Power
CIS	Commonwealth of Independent States
COPD	Chronic Obstructive Pulmonary Disease
CSD	Commission on Sustainable Development (UN)
CTI	Climate Technology Initiative (IEA)
DA	Development Assistance
DAC	Development Assistance Committee (OECD)
DALYs	Disability Adjusted Life Years
DENA	Deutsche Energie Agentur [German Energy Agency]
DNI	Direct Normal Incidence
DSM	Demand Side Management
DTIE	Division for Technology, Industry and Economy (UNEP)
ECAs	Export Credit and Investment Insurance Agencies (OECD)
ECT	Energy Charter Treaty
EDF	European Development Fund
EEA	European Economic Area
EGR	Enhanced Gas Recovery
EIA	Environmental Impact Assessment
EIB	European Investment Bank
EOLE	Programme Français de Développement de Centrales Éoliennes Raccordées au Réseau Électrique
EOR	Enhanced Oil Recovery
ERUPT	Emission Reduction Unit Procurement Tender Programme, The Netherlands
ESF	European Science Foundation

ESMAP	Energy Sector Management Assistance Programme (World Bank, UN)
EU	European Union
FAO	Food and Agriculture Organization (UN)
FDI	Foreign Direct Investment
FETC	Federal Energy Technology Center (USA)
GATT	General Agreement on Tariffs and Trade
GATS	General Agreement on Trade in Services
GDP	Gross Domestic Product
GEF	Global Environment Facility (UNDP, UNEP, World Bank)
GTZ	Gesellschaft für Technische Zusammenarbeit [German Society on Development Cooperation]
GREET	Global Renewable Energy Education and Training (UNESCO)
HDI	Human Development Index
HIPC Initiative	Heavily Indebted Poor Countries Initiative
HPI	Human Poverty Index
HTR	High Temperature Reactor
HVDC	High-Voltage Direct Current
IAEA	International Atomic Energy Agency
IBRD	International Bank for Reconstruction and Development (World Bank)
ICID	International Commission on Irrigation and Drainage
ICIMOD	International Centre for Integrated Mountain Development (Nepal)
ICOLD	International Commission on Large Dams
IEA	International Energy Agency
IFAD	International Fund for Agricultural Development (FAO)
IFC	International Finance Corporation (World Bank Group)
IfE	Lehrstuhl für Energiewirtschaft und Anwendungstechnik der TU München [Institute for Energy Economy and Application Technology, University Munich]
IHA	International Hydropower Association (UNESCO)
IIASA	International Institute for Applied Systems Analysis (Laxenburg, Austria)
IMF	International Monetary Fund
INEF	Institute for Development and Peace, University Duisburg
INPA	Instituto Nacional de Pesca y Agricultura (Columbia) [National Institute for Fisheries and Agriculture, Columbia]
IPCC	Intergovernmental Panel on Climate Change (WMO, UNEP)
IPS	Institute for Policy Studies
IPSE	Intergovernmental Panel on Sustainable Energy (recommended by the Council)
IREICS	International Renewable Energy Information and Communication System (WSP)
ISE	Fraunhofer Institute for Solar Energy Systems, Freiburg/Br. (Germany)
ISEA	International Sustainable Energy Agency (recommended by the Council)
ITER	International Experimental Fusion Reactor
JBIC	Japan Bank for International Cooperation
JI	Joint Implementation (Kyoto Protocol to the UNFCCC)
KfW	Kreditanstalt für Wiederaufbau [The German Development Bank]
LLDC	Least Developed Countries
LPG	Liquified Petroleum Gas
LWR	Light Water Reactor
MACRO	Top-down Macroeconomic Model (IIASA)
MCFC	Molton Carbonat Fuel Cell
MESA	Multilateral Energy Subsidization Agreement (recommended by the Council)
MESSAGE	Model for Energy Supply Strategy Alternatives and their General Environmental Impact (IIASA)
MIND	Model of Investment and Technological Development (PIK)

MOX	Mixed Oxide
NAFTA	North American Free Trade Agreement
NATO	North Atlantic Treaty Organisation
NEXI	Nippon Export and Investment Insurance
NICs	Newly Industrializing Countries
ODA	Official Development Assistance
OECD	Organisation for Economic Co-operation and Development
OLADE	Organización Latinoamericana de Energia (Central America)
OPEC	Organization of Petroleum Exporting Countries
OPIC	Overseas Private Investment Corporation
OSPAR	Convention for the Protection of the Marine Environment of the North-East Atlantic
PAA	Parts of Assigned Amounts
PAFC	Phosphoric Acid Fuel Cell
PEEREA	Energy Charter Protocol on Energy Efficiency and Related Environmental Aspects
PEMFC	Proton Exchanger Membrane Fuel Cell
PIK	Potsdam Institute for Climate Impact Research (Germany)
POP	Persistent Organic Pollutant
PRSP	Poverty Reduction Strategy Papers (IWF, World Bank)
PV	Photovoltaics
PWR	Pressurized Water Reactor
RBMK	Reactor Bolsoi Mochnosti Kipyashiy – Large Power Boiling Reactor
RECS	Renewable Energy Certification System
RNE	Rat für Nachhaltige Entwicklung [German Council for Sustainable Development]
SGP	Small Grant Programme (GEF)
SMEs	Small and Medium-sized Enterprises
SOFC	Solid Oxide Fuel Cell
SRES	Special Report on Emission Szenarios (IPCC)
SRU	Rat von Sachverständigen für Umweltfragen [Council of Environmental Experts (Germany)]
TERI	Tata Energy Research Institute, India
TRIPS	Trade-Related Aspects of Intellectual Property Rights
UN	United Nations
UNDESA	UN Department of Economic and Social Affairs
UNEP	United Nations Environment Programme
UNEP CCEE	Collaborating Centre on Energy and Environment (UNEP)
UNESCO	United Nations Educational, Scientific and Cultural Organization
UNFCCC	United Nations Framework Convention on Climate Change
UNFfD	Monterrey Conference on Financing for Development
UNFPA	United Nations Fund for Population Activities
UNIDO	United Nations Industrial Development Organisation
WBGU	Wissenschaftlicher Beirat der Bundesregierung Globale Umweltveränderungen [German Advisory Council on Global Change]
WCD	World Commission on Dams
WEC	World Energy Council
WERCP	World Energy Research Coordination Programme (UN, recommended by the Council)
WHO	World Health Organization (UN)
WSP	World Solar Programme
WSSD	World Summit on Sustainable Development
WTO	World Trade Organization
YLD	Years Lived With Disability
YLLS	Years of Life Lost

Summary for policy-makers

The first section of this summary for policy-makers presents in brief the prime concerns surrounding today's energy systems, while the second proposes the criteria that need to be met to turn energy systems towards sustainability. The third section, building upon an exemplary scenario, sets out a possible path for transforming the global energy system within the 21st century; this will require a substantial redirection of energy policies over the coming decades. On that basis, the fourth section proposes a roadmap with concrete goals and policy options for action by which to implement this global transformation.

1
Why it is essential to transform energy systems worldwide

The German Advisory Council on Global Change (WBGU) illustrates in the present report that it is essential to turn energy systems towards sustainability worldwide – both in order to protect the natural life-support systems on which humanity depends, and to eradicate energy poverty in developing countries. Nothing less than a fundamental transformation of energy systems will be needed to return development trajectories to sustainable corridors. A further important aspect is that such a global reconfiguration of energy systems would promote peace by reducing dependency upon regionally concentrated oil reserves.

1.1
The use of fossil energy sources jeopardizes natural life-support systems

Today, 80 per cent of worldwide energy use is based on fossil energy sources, and this share is rising. Burning these fuels releases emissions to the environment, where they cause climatic changes, air pollution and human disease. The effects of emissions can be local (in the case of grit, benzene or soot), regional (aerosols, short-lived gases) or global (persistent greenhouse gases). Global climate protection is the supreme challenge presenting an urgent need to transform energy systems.

Emissions of persistent greenhouse gases – above all carbon dioxide, but also methane and nitrous oxide – contributed substantially over the past 100 years to a 0.6°C increase in the mean ground-level air temperature. For the next 100 years, the Intergovernmental Panel on Climate Change (IPCC) forecasts a rise in mean temperature ranging between 1.4 and 5.8°C, depending upon humanity's behaviour and without taking climate protection measures into consideration. The Council considers a mean global temperature change of more than 2°C compared to pre-industrial levels to be intolerable. The predicted shift in climatic regions, in combination with more frequent weather extremes such as floods and drought, has the potential to impair severely, for millions of people, the natural basis of human existence. Developing countries are particularly threatened. Damage to sensitive ecosystems is already evident today. The risk of irreversible ecosystem damage grows in line with the level and rate of warming.

Besides carbon dioxide, the burning of fossil fuels generates benzene and soot emissions with numerous damaging effects on health and ecosystems. It also generates nitrogen oxides, hydrocarbons and carbon monoxide, which promote the formation of ground-level ozone and reduce the self-purifying capacity of the atmosphere. Nitrogen and sulphur oxides, as well as ammonia, are converted chemically in the atmosphere and enter soils through acid deposition. Present energy systems damage the natural environment in many and diverse ways, jeopardize human health and exert massive influences upon biogeochemical cycles.

1.2
Two thousand million people lack access to modern forms of energy

Improving access to advanced energy in developing countries is a fundamental contribution to poverty reduction and key to attaining the United Nations Millennium Development Goals. For some 2.4 thousand million people, notably in rural parts of Asia and Africa, energy supply depends largely or entirely upon biomass use (firewood, charcoal or dung) for cooking and heating. On average, 35 per cent of energy consumed in developing countries derives from biomass; in parts of Africa this share reaches 90 per cent. According to the World Health Organization, emissions from the burning of biomass and coal indoors cause the death of 1.6 million people every year. This is substantially more than the one million deaths caused by malaria. A transformation of energy systems towards sustainability is therefore essential in order to overcome development problems.

2
A corridor for sustainable energy policy: Guard rails for a global transformation

Sustainable transformation paths are bounded by so called 'guard rails'. The Council defines these guard rails as those levels of damage which can only be crossed at intolerable cost, so that even short-term utility gains cannot compensate for such damage

---

Box 1

Guard rails for sustainable energy policy

Ecological guard rails

CLIMATE PROTECTION
A rate of temperature change exceeding 0.2°C per decade and a mean global temperature rise of more than 2°C compared to pre-industrial levels are intolerable parameters of global climate change.

SUSTAINABLE LAND USE
10–20 per cent of the global land surface should be reserved for nature conservation. Not more than 3 per cent should be used for bioenergy crops or terrestrial CO_2 sequestration. As a fundamental matter of principle, natural ecosystems should not be converted to bioenergy cultivation. Where conflicts arise between different types of land use, food security must have priority.

PROTECTION OF RIVERS AND THEIR CATCHMENT AREAS
In the same vein as terrestrial areas, about 10–20 per cent of riverine ecosystems, including their catchment areas, should be reserved for nature conservation. This is one reason why hydroelectricity – after necessary framework conditions have been met (investment in research, institutions, capacity building, etc.) – can only be expanded to a limited extent.

PROTECTION OF MARINE ECOSYSTEMS
It is the view of the Council that the use of the oceans to sequester carbon is not tolerable, because the ecological damage can be major and knowledge about biological consequences is too fragmentary.

PREVENTION OF ATMOSPHERIC AIR POLLUTION
Critical levels of air pollution are not tolerable. As a preliminary quantitative guard rail, it could be determined that pollution levels should nowhere be higher than they are today in the European Union, even though the situation there is not yet satisfactory for all types of pollutants. A final guard rail would need to be defined and implemented by national environmental standards and multilateral environmental agreements.

Socio-economic guard rails

ACCESS TO ADVANCED ENERGY FOR ALL
It is essential to ensure that everyone has access to advanced energy. This involves ensuring access to electricity, and substituting health-endangering biomass use by advanced fuels.

MEETING THE INDIVIDUAL MINIMUM REQUIREMENT FOR ADVANCED ENERGY
The Council considers the following final energy quantities to be the minimum requirement for elementary individual needs: By the year 2020 at the latest, everyone should have at least 500kWh final energy per person and year and by 2050 at least 700kWh. By 2100 the level should reach 1,000kWh.

LIMITING THE PROPORTION OF INCOME EXPENDED FOR ENERGY
Poor households should not need to spend more than one tenth of their income to meet elementary individual energy requirements.

MINIMUM MACROECONOMIC DEVELOPMENT
To meet the macroeconomic minimum per-capita energy requirement (for energy services utilized indirectly) all countries should be able to deploy a per-capita gross domestic product of at least about US$3,000, in 1999 values.

KEEPING RISKS WITHIN A NORMAL RANGE
A sustainable energy system needs to build upon technologies whose operation remains within the 'normal range' of environmental risk. Nuclear energy fails to meet this requirement, particularly because of its intolerable accident risks and unresolved waste management, but also because of the risks of proliferation and terrorism.

PREVENTING DISEASE CAUSED BY ENERGY USE
Indoor air pollution resulting from the burning of biomass and air pollution in towns and cities resulting from the use of fossil energy sources causes severe health damage worldwide. The overall health impact caused by this should, in all WHO regions, not exceed 0.5 per cent of the total health impact in each region (measured in DALYs, disability adjusted life years).

(Box 1). For instance, if, in the interests of short-term economic gains, the energy sector is transformed too late, global warming will be driven to the point at which the costs of inaction would be much higher over the long term due to the economic and social upheaval that is then to be expected. Guard rails are not goals: They are not desirable values or states, but minimum requirements that need to be met if the principle of sustainability is to be adhered to.

3
Turning energy systems towards sustainability is feasible: A test run for system transformation

The sustainability of scenarios for energy futures can be tested against the guard rails set out in the previous section. In principle, many developments are conceivable that would turn today's worldwide energy systems towards sustainability. Insofar, the scenario created in this report should be viewed as one example (Fig. 1). Building upon scenarios for the stabilization of CO_2 concentrations in the atmosphere at a maximum of 450ppm, this report shows that the global transformation of energy systems over the next 100 years is in principle technologically and economically feasible.

The exemplary path charted by the Council embraces four key components:
1. Major reduction in the use of fossil energy sources;
2. Phase-out of the use of nuclear energy;
3. Substantial development and expansion of new renewable energy sources, notably solar;
4. Improvement of energy productivity far beyond historical rates.

Analysis of this path yields the following key findings:
- Worldwide cooperation and approximation of living conditions facilitate rapid technology development and dissemination. High economic growth can then, in conjunction with a strong increase in energy productivity, lead to sustainable energy supply.
- It will only be possible to meet minimum climate protection requirements if binding CO_2 reduction requirements are in place.
- Energy policy activities need to be supported by further measures to reduce greenhouse gas emissions from other sectors (for instance nitrous oxide and methane from agriculture) and to preserve natural carbon stocks.

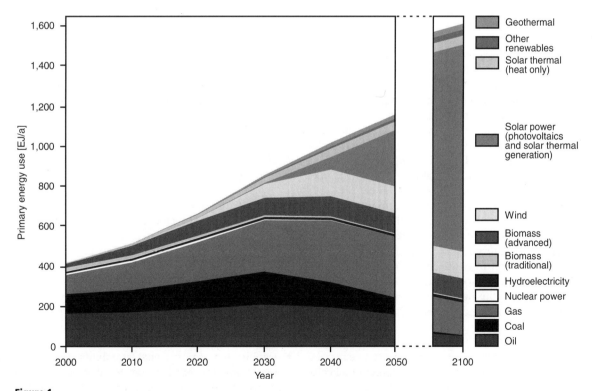

Figure 1
Transforming the global energy mix: The exemplary path until 2050/2100.
Source: WBGU

- While the exemplary path developed here is based upon a stabilization of atmospheric CO_2 concentrations at 450ppm, due to uncertainties attaching to climate system behaviour this can by no means be taken as a safe stabilization level. The Council recommends retaining options by which to achieve lower stabilization concentrations.
- Even if climate protection goals are met, a fossil-nuclear path entails substantially larger risks, as well as much higher environmental impacts. Moreover, it is significantly more expensive over the medium to long term than a path relying upon promoting renewables and improving energy efficiency, mainly due to the costs of CO_2 sequestration.
- Due to the long time lags, the next 10–20 years are the decisive window of opportunity for transforming energy systems. If this transformation is initiated later, disproportionately high costs must be expected.
- The transformation will only succeed if the transfer of capital and technology from industrialized to developing countries is intensified. To this end, industrialized countries will need to strengthen technology development significantly in the fields of energy efficiency and renewable energy sources, for instance by raising and redirecting research and development expenditure, implementing market penetration strategies, providing price incentives and developing appropriate infrastructure. This can reduce the initially high costs of the new technologies and can accelerate attainment of market maturity, thus in turn facilitating transfer to developing countries.
- Over the short and medium term, it is essential to swiftly tap those renewable energy sources which are already technologically manageable and relatively cost-effective today. These are in particular wind and biomass. Over the long term, the rising primary energy requirement can only be met through vigorous utilization of solar energy – this holds by far the largest sustainable potential. To tap this potential in time, installed capacity will need to grow ten-fold every decade – now and over the long term.
- The utilization of fossil energy sources will continue to be necessary over the next decades. Wherever possible, this needs to be done in such a fashion that the efficiency potential is tapped and both the infrastructure and generating technology can be converted readily to renewable sources. In particular, the efficient use of gas, for instance in combined heat and power generation and in fuel cells, can perform an important bridging function on the path towards a hydrogen economy.

- A certain volume of carbon sequestration in appropriate geological formations (e.g. depleted oil and gas caverns) will be necessary as a transitional technology during this century in order to remain within the climate guard rails. For ecological reasons, the Council rejects use of the oceans for carbon sequestration.

4
Milestones on the WBGU transformation roadmap: Targets, time tables and policies

4.1
Protecting natural life-support systems

To keep global warming within tolerable limits, global carbon dioxide emissions need to be reduced by at least 30 per cent from 1990 levels by the year 2050 (overview: Fig. 2). For industrialized countries, this means a reduction by some 80 per cent, while the emissions of developing and newly industrializing countries are allowed to rise by at most 30 per cent. Without a fundamental transformation of energy systems, emissions must be expected to double or even quadruple in developing and newly industrializing countries over that period. This is why in these countries, too, a rapid redirection of energy production and utilization is essential. The focus of such activities needs to be placed on promoting renewables and enhancing efficiency. In view of the considerable uncertainties, e.g. regarding the behaviour of the climate system, these emissions reduction goals are minimum requirements.

4.1.1
Improving energy productivity

In order to minimize resource consumption, global energy productivity (the ratio of gross domestic product to energy input) needs to be improved by 1.4 per cent every year initially, and then by at least 1.6 per cent as soon as possible. At that rate, energy productivity would treble by 2050 from 1990 levels. Moreover, minimum efficiencies of more than 60 per cent should be aimed at by 2050 for large fossil-fuelled power plants. To this end, the Council recommends
- establishing international standards prescribing minimum efficiencies for fossil-fuelled power plants in a stepwise process from 2005 onwards, based on the corresponding European Union (EU) directive.

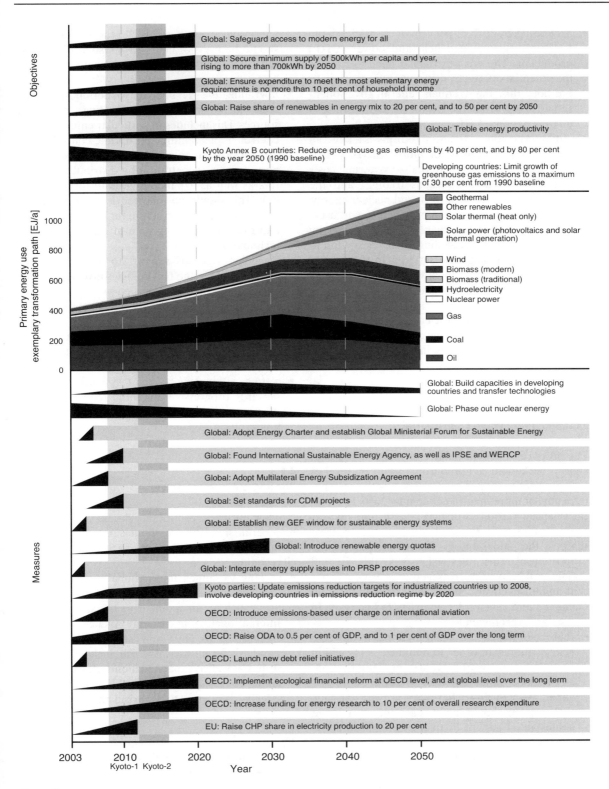

Figure 2
Overview of the transformation roadmap proposed by the German Advisory Council on Global Change (WBGU).
CDM=Clean Development Mechanism, CHP=combined heat and power, GDP=gross domestic product, GEF=Global
Environment Facility, IPSE=Intergovernmental Panel on Sustainable Energy, ODA=Official Development Assistance,
OECD=Organisation for Economic Co-operation and Development, PRSP=Poverty Reduction Strategy Papers,
WERCP=World Energy Research Coordination Programme
Source: WBGU

- generating, by 2012, 20 per cent of electricity in the EU through combined heat and power (CHP) production. There is a particular need to harness the potential offered by distributed production. To promote this, the German federal government should argue within the EU for the swift setting of binding national CHP quotas.
- initiating ecological financial reforms as a key tool for creating incentives for more efficiency. This includes measures to internalize external costs (e.g. CO_2 taxation, emissions trading) and the removal of subsidies for fossil and nuclear energy.
- improving the information provided to end users in order to promote energy efficiency, e.g. by means of mandatory labelling for all energy-intensive goods, buildings and services. In the case of goods traded internationally, cross-national harmonization of efficiency standards and labels is recommendable.
- exploiting the major efficiency potentials in the use of energy for heating and cooling through instruments of regulatory law targeting the primary energy requirement of buildings.

4.1.2
Expanding renewables substantially

The proportion of renewable energies in the global energy mix should be raised from its current level of 12.7 per cent to 20 per cent by 2020, with the long-term goal of more than 50 per cent by 2050. Ecological financial reforms will make fossil and nuclear sources more expensive and will thus reduce their share in the global energy mix. Consequently, the proportion of renewables will rise. As this rise will remain well below the envisaged increase to 20 per cent and, respectively, 50 per cent, the Council recommends that renewables be expanded actively. In particular, it recommends

- that countries agree upon national renewable energy quotas. In order to minimize costs within such a scheme, a worldwide system of internationally tradable renewable energy credits should be aimed at by 2030. Its flexibility notwithstanding, such a system should commit each country to meet a substantial part of its quota through domestic generation.
- continuing and broadening market penetration strategies (e.g. subsidy schemes over limited periods, guaranteed feed-in tariffs, renewable energy quota schemes). Until significant market volume has been achieved, guaranteed feed-in tariffs under which payments decline over time are amongst the particularly expedient options. When a sufficiently large market volume of individual

energy sources has been reached, assistance should be transformed into a system of tradable renewable energy credits or green energy certificates.
- upgrading energy systems to permit the large-scale deployment of fluctuating renewable sources. This includes in particular enhancing grid control, implementing appropriate control strategies for distributed generators, upgrading grids to permit strong penetration by distributed generators as well as expanding grids to form a global link. This should be followed later by the establishment of an infrastructure for hydrogen storage and distribution, using natural gas as a bridging technology.
- providing vigorous support to disseminate and further develop the technologies involved in solar and energy-efficient construction.
- building and strengthening human-resource and institutional capacities in developing countries and intensifying technology transfer in order to improve the framework conditions for the establishment of sustainable energy systems.
- setting within export credit systems, from 2005 onwards, progressive minimum requirements for the permissible carbon intensity of energy production projects.

4.1.3
Phasing out nuclear power

No new nuclear power plants should be given planning permission. The use of nuclear power should be terminated worldwide by 2050. To this end, the Council recommends
- seeking to launch international negotiations on the phase-out of nuclear power. This process could begin with an amendment to the statutes of the International Atomic Energy Agency (IAEA).
- establishing by 2005 new, stricter IAEA safety standards for all sites at which nuclear material is stored, as well as expanded monitoring and action-taking competencies of the IAEA in the field of safeguards relating to terrorism and proliferation.

4.2
Eradicating energy poverty and seeking minimum levels of supply worldwide

Access to advanced energy is a vital element for poverty reduction and development. The Council therefore recommends adopting as an international target that access to advanced energy is safeguarded for the entire world population from 2020, and that, from that time onwards, all individuals have access to at least 500kWh per person and year to meet elementary final energy requirements (Fig. 2). In this endeavour, care needs to be taken that socio-economic disparities are reduced in connection with all measures seeking to transform energy systems. The proportion of household income spent on energy should not exceed 10 per cent. Access to advanced energy is also a key contribution to achieving the United Nations Millennium Development Goals.

4.2.1
Focussing international cooperation on sustainable development

IMPLEMENTING NEW WORLD BANK POLICY IN ASSISTANCE DELIVERY PRACTICE

The Council takes the view that the World Bank, which supports countries in expanding their energy systems, should also promote sustainable energy in order to facilitate the leapfrogging of unsustainable development stages. In efforts to promote the transformation of energy systems, the World Bank has not yet moved sufficiently from the conceptual to the operational level. An urgent need thus remains to redirect its assistance delivery procedures, which until now have predominantly financed fossil fuels according to the least-cost principle. The Council recommends that

- the new assistance delivery approach of the World Bank is implemented in practice, starting immediately. The German federal government should use its membership on the Board of Governors of the World Bank to work towards this.

INTEGRATING SUSTAINABLE ENERGY SUPPLY WITHIN POVERTY REDUCTION STRATEGIES

In late 1999, the International Monetary Fund (IMF) and the World Bank began focussing their policies vis-à-vis Least Developed Countries primarily on poverty reduction. Poverty Reduction Strategy Papers (PRSPs) serve to steer the medium-term development of countries and provide a basis for eliciting international support. The Council recommends

- integrating sustainable energy supply within PRSP processes in order to raise the profile of energy-related issues in development cooperation.

STRENGTHENING THE ROLE OF REGIONAL DEVELOPMENT BANKS

The role of regional development banks should be strengthened. These have good regional connections and more intimate knowledge of local problems than global institutions do. The Council recommends that

- Germany, in connection with its involvement in these banks and within the EU context, works towards the promotion of energy supply in developing countries through the regional development funds;
- the EU makes targeted use of the European Development Fund to promote renewables in the ACP (African, Caribbean, Pacific) states.

4.2.2
Strengthening the capabilities of developing countries

PROMOTING ECONOMIC AND SOCIAL DEVELOPMENT IN LOW-INCOME COUNTRIES

To turn energy systems towards sustainability, a minimum degree of economic development is a precondition. Many countries fall far short of the per-capita income required for this. The Council therefore recommends not only intensifying development cooperation in the field of basic services and sustainable energy supply, but also intensifying cooperation with low-income countries in particular, in both quantitative and qualitative terms. Furthermore, within the context of the WTO 'Development Round', improved access for goods from all low-income countries to the markets of industrialized and newly industrializing countries should be urged.

LAUNCHING NEW DEBT RELIEF INITIATIVES

In general, heavily indebted developing countries have little scope to cope with price fluctuations on world energy markets. Their ability to finance improvements to the efficiency of their energy supply systems and to advance the deployment of renewable energy technologies is similarly limited. To embark on transformation, wide-ranging debt relief is needed. The Council recommends that

- the German federal government argues for new debt relief initiatives within the G7/G8 context.

4.2.3
Combining regulatory and private-sector elements

It is essential to take measures on both the supply and demand side in order to improve access to advanced low-emission energy forms and to renewable energy sources, and to improve the energy efficiency in developing, newly industrializing and transition countries.

SUPPLY SIDE: COMBINING LIBERALIZATION AND
PRIVATIZATION WITH REGULATORY INTERVENTIONS
On the supply side, privatization and liberalization
need to be combined with regulatory interventions
undertaken by the state. The mix of these three
spheres will need to vary depending upon the specific
circumstances of a region. Liberalization and privatization require an attractive environment for private-
sector investors and the tapping of international
sources of capital. Stronger state intervention
requires the setting of standards, and also an expansion of public-private partnerships, possibly supported by bilateral and multilateral development
cooperation activities.

DEMAND SIDE: INCREASING THE PURCHASING
POWER OF THE POOR
On the demand side, the aim must be to increase purchasing power in relation to energy, particularly of
the poor. This can be done by target-group specific
subsidies, or by expanding micro-finance systems. To
also increase the willingness to use energy more sustainably, measures taken on the demand side need to
give consideration to culture-specific and gender-
specific framework conditions.

4.3
Mobilizing financial resources for the global transformation of energy systems

To finance the global transformation of energy systems towards sustainability, there is an urgent need to
mobilize additional financial resources, as well as to
create new transfer mechanisms or strengthen existing ones in order to support economically weaker
countries in this transformation process. The Council
welcomes the programme on 'Sustainable energy for
development' geared to establishing strategic partnerships which the German government announced
at the World Summit on Sustainable Development.
Over the next five years, a total of €1,000 million will
be budgeted for this programme.

MOBILIZING PRIVATE-SECTOR CAPITAL
To mobilize private-sector capital for the global
transformation of energy systems, the Council recommends
• facilitating access to developing country markets
  for small and medium-sized suppliers of renewable energy technologies within the context of
  public-private partnerships;
• establishing by 2010 a German and, if possible, EU
  standard for the Clean Development Mechanism.
  This standard should permit exclusively, with
  exceptions to be substantiated in each case, pro-

jects that promote renewables (excluding large
hydroelectric dams due to currently unresolved
sustainability problems), improve the energy efficiency of existing facilities or engage in demand-
side management.

BOOSTING DEVELOPMENT COOPERATION FUNDING
At 0.27 per cent of gross domestic product (GDP) in
2001, German official development assistance
(ODA) funding is far from the internationally agreed
target of 0.7 per cent. However, Germany has committed itself to increasing ODA funding to a level of
0.33 per cent of GDP by 2006. Even an increase to
some 1 per cent of GDP would be commensurate to
the severity of the problems prevailing. The Council
recommends
• as a matter of urgency, raising ODA funding
  beyond the level of 0.33 per cent announced for
  2006, and proposes allocating, as a first step, at
  least 0.5 per cent of GDP by 2010.

HARNESSING INNOVATIVE FINANCING TOOLS
To implement the global transformation of energy
systems, it will be essential to tap new sources of
finance. Specially, the potential of raising charges for
the use of global commons deserves examination.
The Council recommends
• raising from 2008 onwards an emissions-based
  user charge on international aviation, provided
  that this sector is not yet subject by then to international emissions reduction commitments.

STRENGTHENING THE GLOBAL ENVIRONMENT
FACILITY AS AN INTERNATIONAL FINANCING
INSTITUTION
The Global Environment Facility (GEF), operated
jointly by UNDP, UNEP and the World Bank, should
be used as a catalyst for global environmental protection measures. The Council recommends
• concentrating by 2005 the financial assistance provided for efficiency technologies and renewable
  resources in a newly created GEF 'window for sustainable energy systems'. In order to be able to
  give greater consideration to development policy
  aspects in the deployment of funds, a simplification of the incremental costs approach should be
  considered. With a view to the high levels of funding required to promote the global transformation
  of energy systems, GEF resources need to be
  expanded considerably.

4.4
Using model projects for strategic leverage, and engaging in energy partnerships

SENDING OUT SIGNALS THROUGH MODEL PROJECTS
The Council argues in favour of using model projects for the introduction of new renewables on a large scale to deliver strategic leverage for a global transformation of energy systems towards sustainability. Such model projects could have global knock-on effects. They would showcase how technology leaps can be implemented in energy projects. The Council recommends initiating the following model projects:

- A strategic energy partnership between the European Union and North Africa, integrating into European power supply the potential of solar energy use in a manner profitable for both sides;
- Developing the infrastructure needed to substitute traditional biomass use by biogenic bottled gas;
- Energy-efficient buildings in the low-cost sector, piloted by South African townships;
- Improving the power quality in weak electric grids in rural African regions;
- '1 million huts electrification programme' for developing countries, generating the necessary internal dynamics for off-grid rural electrification.

FORMING STRATEGIC PARTNERSHIPS TO TURN
ENERGY SYSTEMS TOWARDS SUSTAINABILITY
Existing or emergent policy initiatives promoting a global transformation of energy systems towards sustainability provide a framework for action. The Council recommends that, in addition to the World Conference for Renewable Energy due to take place in 2004, the following policy processes in particular are used as catalysts to promote this transformation:

- The international initiatives adopted at the World Summit on Sustainable Development
  - Energy Initiative for Poverty Eradication and Sustainable Development,
  - Global Village Energy Partnership,
  - Global Network on Energy for Sustainable Development.
- The economic partnership agreement currently being negotiated between the EU and the ACP states.

4.5
Advancing research and development

Turning energy systems towards sustainability is a major technological and social challenge on a scale comparable to that of a new industrial revolution. For it to succeed, a major research and development effort is necessary. This concerns renewable energy sources, infrastructure, end-use efficiency technologies as well as the provision of knowledge on the conservation and expansion of natural carbon stocks and sinks. The social sciences also need to contribute, by analysing the individual and institutional barriers to this transformation process and developing strategies to overcome these barriers.

However, for many years now expenditure for research and development in the energy sector has been declining. At present, across the OECD only some 0.5 per cent of turnover in the energy sector is devoted to research and development activities, and the percentage is dropping. Only if there is sustained, high investment in research and development can there be a prospect of renewable-energy technologies and efficiency-enhancing measures coming into widespread use over the medium and long term at low cost. The Council recommends

- increasing at least ten-fold, above all through re-allocation of resources from other areas, by 2020 the direct state expenditure in industrialized countries for research and development in the energy sector from its current level of about US$1,300 million annually (average across the OECD for the 1990–1995 period). The focus needs to be shifted rapidly away from fossil and nuclear energy towards renewables and efficiency.
- establishing within the UN system a World Energy Research Coordination Programme (WERCP) to draw together the various strands of national-level energy research activities, in analogy to the World Climate Research Programme.

4.6
Drawing together and strengthening global energy policy institutions

ESTABLISHING COORDINATING BODIES AND
NEGOTIATING A WORLD ENERGY CHARTER
To promote a global transformation of energy systems towards sustainability, it is essential to coordinate activities at global level and consequently to draw together international institutions and actors. The Council recommends strengthening and expanding the institutional architecture of global energy policy in a stepwise process, building upon existing organizations:

- As a first step, a World Energy Charter should be negotiated at the planned World Conference for Renewable Energy to be held in Germany in 2004. This should contain the key elements of sustainable, global energy policy and provide a joint basis for action at global level.

- Moreover, this conference should decide upon – or better still establish – a Global Ministerial Forum for Sustainable Energy responsible for coordinating and determining the strategic direction of the relevant actors and programmes.
- In parallel, a Multilateral Energy Subsidies Agreement (MESA) should be negotiated by 2008. This agreement could provide for the stepwise removal of subsidies for fossil and nuclear energy, and could establish rules for subsidizing renewable energy and energy efficiency technologies.
- At least the OECD states should commit themselves to national renewable energy quotas of at least 20 per cent by 2015. It would be important in this context to agree to negotiate the globalization and flexibilization of this system, such negotiations leading by 2030 at the latest to a worldwide system of tradable renewable energy credits.
- In support of these activities, a group of like-minded, advanced states should adopt a pioneering role on the path towards sustainable energy policies. The European Union would be a suitable candidate for such a leadership role.
- Building upon the steps above, the institutional foundations of sustainable energy policy could be further strengthened by concentrating competencies at global level. To this end, the role of the Ministerial Forum could be further expanded.
- Using the experience gained until that date, by about 2010 the establishment of an International Sustainable Energy Agency (ISEA) should be examined.

ENHANCING POLICY ADVICE AT THE
INTERNATIONAL LEVEL

It is important that the political implementation of a global transformation of energy systems towards sustainability receives continuous support through independent scientific input, as is currently the case in climate protection policy. To this end, the Council recommends

- establishing an Intergovernmental Panel on Sustainable Energy (IPSE) charged with analysing and evaluating global energy trends and identifying options for action.

5 Conclusion: Political action is needed now

To protect natural life-support systems and eradicate energy poverty alike, there is an urgent need to transform energy systems. This transformation will be feasible without severe adverse effects upon societal and economic systems if policy-makers grasp the opportunity to shape this process over the next two decades. The intended effects can only be expected to

emerge after a certain time lag. This lag makes swift action all the more important. The costs of inaction would be much higher over the long term than the costs of initiating this transformation. Every delay will make it more difficult to change course.

The direction of transformation is clear: The energy efficiency must be increased, and massive support for renewables must be provided. It will be particularly important in this endeavour to reduce dependency on fossil fuels. The long-term objective is to break the ground for a solar age.

In the view of the Council, the transformation is feasible. It is also financeable if, in addition to intensified use of existing mechanisms (e.g. GEF, ODA, World Bank and regional development bank loans) and enhanced incentives for private-sector investors (e.g. through public-private partnerships), innovative financing avenues (such as user charges for global commons) are pursued. The present report highlights the key opportunities for steering energy systems towards sustainability, guided by a transformation roadmap.

For the worldwide transformation of energy systems to succeed, it will need to be shaped in a step-wise and dynamic manner, for no one can predict today with sufficient certainty the technological, economic and social developments over the next 50–100 years. Long-term energy policy is thus a searching process. It is the task of policy-makers to rise to this challenge. The World Conference for Renewable Energy announced by the German chancellor at the Johannesburg World Summit on Sustainable Development offers an excellent opportunity to take action.

Worldwide, energy demand is rising swiftly. This has been the case above all in the industrialized and transition countries since the onset of industrialization, despite massive efficiency improvements achieved in some fields. Since the end of the Second World War, the energy hunger of the developing and newly industrializing countries has also been on the rise. In the Least Developed Countries, however, energy poverty prevails. Some 2.4 thousand million people need access to modern forms of energy to catch up with the economic development of the industrialized countries. Energy is a precondition to economic growth, and thus world consumption is set to rise sharply in the 21st century. Present structures of energy use pose severe environmental risks and raise major barriers to development in many countries. Moreover, these global structures are a source of security risks. Steering energy systems towards sustainability is thus a key task of global environment and development policy in the 21st century.

PROBLEMS ASSOCIATED WITH PRESENT PATTERNS OF ENERGY USE

Although mineral oil resources extractable at low cost are expected to be exhausted during the 21st century, rising energy demand is not primarily a problem of limited resources, as was commonly feared in the 1970s. The problems associated with present patterns of energy use stem rather from the emission of gases and particles to the atmosphere. This is because energy use is based largely upon fossil fuels. Human-induced global climate change is the most severe consequence, joined by a host of further environmental and health problems: The extraction, processing and use of fossil fuels destroys landscapes, generates acid rain and eutrophicates marginal seas. It also causes respiratory diseases attributable to both ambient air pollution in conurbations and indoor air pollution. Use of wood or charcoal as fuel leads to deforestation of entire landscapes in many developing countries. Not least, many inter-state conflicts result from a desire to control resources, including oil.

CONFLICTING GOALS IN GLOBAL ENERGY POLICY

The task of supplying the world's population with energy thus harbours a goal conflict: On the one hand, the right of developing countries to develop must be observed while, on the other hand, global climate change must be held within acceptable bounds. It needs to be kept in mind that the global distribution of energy use is highly uneven. Just one-fifth of the world's population uses some three-quarters of the world's energy supply. This is the great challenge in moving energy systems onto a sustainable path, and the starting point of the present report. The following guiding questions arise:

- What bearing does global warming have upon global energy policy?
- Which energy sources need to be harnessed more vigorously in the future?
- What are the present and prospective technological options?
- How can energy systems in industrialized and transition countries be transformed in an environmentally benign fashion?
- How can it be ensured that all people have access to a basic supply of modern energy forms?
- How can energy supply be ensured and expanded in developing countries in a manner that is both cost-effective and environmentally sound?
- How can forms of energy use harmful to human health be overcome?
- How should a global energy policy be shaped – structurally, institutionally and financially?
- Which challenges does the scientific community face?
- Which concrete policies need to be adopted over the next two decades?

In this report, the German Advisory Council on Global Change (WBGU) has taken up these guiding questions, has sought answers and has identified ways to overcome goal conflicts. The time horizon for political action extends over the next 50 years. Some scenarios, such as on the long-term structure of the global energy system, probe the future even further, to the year 2100. It follows that the catalogue of instruments developed by the Council must not be

considered static. The measures recommended here
are rather intended as a basic pattern of transformation, in the full knowledge that a search process
extending over such a long period cannot be predicted. Nonetheless, search processes, too, need to be
initiated and shaped.

THIS REPORT'S INNOVATIVE CONTRIBUTION
Opinion papers and reports on sustainable energy
policy abound. Only recently, in mid-2002, the Study
Commission of the German Bundestag on 'Sustainable Energy Supplies in View of Globalization and
Liberalization' presented its final report. The present
report – which concentrates explicitly on the global
level – breaks new ground in four respects:
1. The two overarching objectives of protecting the
   world's climate and overcoming energy poverty
   are given equal standing, and avenues are sought
   by which to resolve goal conflicts. The discussion
   presented here of ways to move global energy systems onto a sustainable path does justice to the
   right of developing countries to catch up with the
   industrialized world.
2. To signpost this transformation path, the Council
   defines, in this report, the attributes required by
   such a path if it is to be sustainable, using a 'guard
   rail' approach. Guard rails define maximum limits
   of damage; if these limits are crossed today or in
   the future, this would entail intolerable consequences. Guard rails are not goals, but minimum
   requirements that must be met if the principle of
   sustainability is to be adhered to. This is the Council's understanding of sustainability: The guard
   rails mark out the boundaries of the scope for sustainable action.
3. Building on the guard rail approach, the report
   sets out a corridor for sustainable energy policy
   and draws up a roadmap for a global transition of
   energy systems to sustainability by the year 2050.
   The Council understands such a transformation as
   a search process, for no one is able to predict
   future developments with sufficient accuracy.
   Nonetheless, with its concrete time schedules and
   substantive goals, the transformation roadmap
   proposes key elements of policy redirection at
   both the national and international levels.
4. The report's recommendations tie in with ongoing
   policy processes, highlighting concrete interventions and alternatives. These recommendations
   are summarized for policy-makers.
The transformation of energy systems is a Herculean
task, comparable to a new industrial revolution. The
report shows why this is so and what needs to be
done.

Social and economic energy system linkages

2.1
Introduction

Energy is an essential precondition for human development. From the first time wood was used for light and heat thousands of years ago to the latest energy technologies, greater quality and efficiency in energy use has been an aim of and driving force behind innovation and progress. Three major transitions in the development of energy systems led to greater quality in the types of energy: the use of coal-fired steam engines allowed for new, more efficient production processes and simultaneously reduced dependence on ever scarcer 'traditional' fuels (wood, manure). The second transition, from coal to oil, increased mobility as the combustion engine was developed. Finally, the use of electricity (light, computers) led humankind into the age of information.

These developments, especially industrialization and urbanization, have brought about great structural changes in the economy and in society. Fluid fuels and grid-based types of energy, which are cleaner and can be used more flexibly, also increased the quality of energy use. However, the amount of energy used grew several fold along with the technological innovations and the concomitant structural

changes in society and the economy. At the same time, energy systems moved from dependency on traditional fuels to dependency on fossil energy sources. In the 1960s, oil surpassed coal as the most important fossil energy carrier, a position coal had held for around half a century (Fig. 2.1-1). The transport sector, in particular, is almost completely dependent on oil as an energy source.

2.2
The global setting

2.2.1
Rising energy and carbon productivity – trends up to 2020

Today, 80 per cent of the energy we use worldwide comes from fossil sources (Table 2.2-1). For the next few decades, the fossil resources currently available will suffice to meet demand. However, it is probable that energy prices will rise during this period as the extraction of fossil resources becomes more complicated, and hence more expensive. Traditional biomass continues to play a dominant role in many developing nations, especially in rural areas (UNDP

Figure 2.1-1
Share of various energy sources in world primary energy consumption. In 100 years, drastic changes in the energy mix are possible, as the example of coal illustrates.
Source: Nakicenovic et al., 1998

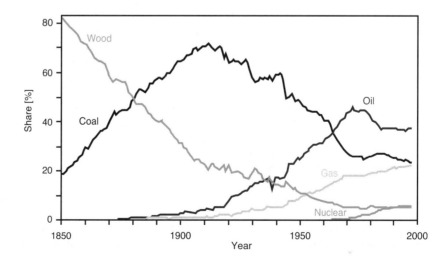

Energy source	Primary energy [EJ]	Share [%]	Static range of reserves [Years]	Static range of resources [Years]	Dynamic range of resources [Years]
Oil	142	35.3	45	~200	95
Natural gas	85	21.1	69	~400	230
Coal	93	23.1	452	~1,500	1,000
Sum of fossil energy sources	*320*	*79.6*			
Hydropower	9	2.2		renewable	
Traditional biomass	38	9.5		renewable	
New renewables	9	2.2		renewable	
Sum of renewables	*56*	*13.9*			
Nuclear power	26	6.5	50	>>300	
Total sum	*402*	*100.0*			

Table 2.2-1
World consumption of primary energy in 1998 by energy source with an indication of range. The static range is the number of years the known reserves/resources will last at current rates of annual production. In other words, it describes how long a resource will be available if consumption remains at current levels. In contrast, the dynamic range is based on the expected annual increase in production. Source: UNDP et al., 2000

et al., 2000). But globally, it only makes up around 10 per cent of all energy generated. The fastest growing shares are of natural gas and 'new' renewable energy sources such as wind, photovoltaics, and solar thermal, though they only make up around 2 per cent of energy generation worldwide. The International Energy Agency (IEA) expects new renewables to grow by 3.3 per cent annually and natural gas by 2.4 per cent up to 2030. The growing share of gas, which is mostly due to the development of inexpensive combined-cycle (gas and steam) turbines, is eating away at the share of coal and nuclear energy. Nevertheless, coal is still the most commonly used energy source for the generation of electricity. Nuclear energy has been stagnant or falling somewhat; the IEA expects its share to drop to around 5 per cent by 2030. The use of nuclear energy is only growing in a few (mostly Asian) countries (IEA, 2002c).

Population growth and economic and technological development are the main factors that determine the world's energy needs. Per capita energy consumption increases – with considerable variance – when incomes increase, as a comparison of numerous countries reveals (Fig. 2.2-1). And yet, it also becomes clear that the same level of energy consumption can bring about quite different levels of material prosperity: Japan's per capita energy use is roughly the same as South Korea's, but Japan's per capita income is seven times greater. In the last two centuries, the global gross national product increased by 3 per cent annually, while global energy demand only rose by 2 per cent per year (IPCC, 1996). Hence, economic energy productivity has increased by 1 per cent annually. This is not only due to technological progress (greater efficiency), but also to changing patterns in energy services (such as shifts between sectors) and the replacement of fuels with more modern types of energy (such as gas replacing wood for cooking). In addition, changing consumption patterns and lifestyles can affect energy productivity (Nakicenovic et al., 1998; Section 2.2.3).

Pollution usually goes hand in hand with the increasing use of energy, though not proportionally: global emissions of carbon dioxide are not keeping pace with the increase in energy consumption. The use of less carbon-intensive fossil energy carriers, such as natural gas, as well as the use of nuclear energy or renewable energy carriers instead of carbon-intensive fossil energy carriers like coal are changing the mix of energy carriers and leading to decarbonization. Per unit energy consumed, carbon dioxide emissions are falling globally by 0.3 per cent each year.

2.2.2
Energy use by sector

Today, the world's main energy consumer is industry with some two-fifths of the world's primary energy consumption. Private households and commercial buildings consume somewhat less, while the transport sector consumes around a fifth (Table 2.2-2; IPCC, 2000b). In Asia, the share of industry is greater (59 per cent), while industry only takes up around one-third in the OECD countries, where transport makes up one-fourth, compared to only 15 per cent in Asia. The agricultural sector only uses some 3 per cent of commercial energy globally. Between 1970 and 1990 annual global growth rates were greatest in the buildings sector (heating, cooling, lighting, etc.) at 2.9 per cent, followed by the transport sector at 2.8 per cent (IPCC, 2000b). In the first half of the 1990s, the use of primary energy worldwide only grew by 0.7

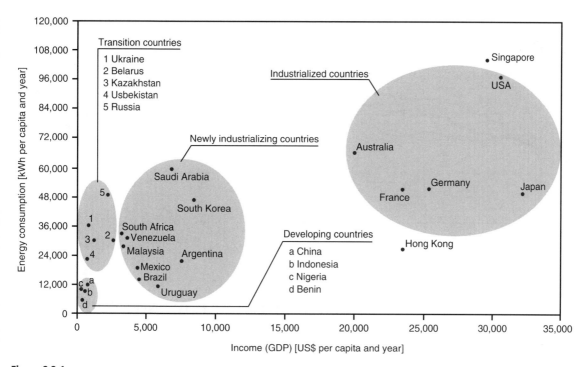

Figure 2.2-1
The relation between mean income (GDP per capita) and energy consumption (per capita demand in kWh) in 1997 for various country groups. The primary energy consumption of a country is shown – i.e. its industry and transport are also included – divided by the number of inhabitants. The clusters for developing, newly industrializing, transition, and industrialized countries are clearly disparate. Energy demand increases as income rises, but energy consumption reaches a plateau once income has reached a very high level.
Source: modified after WRI, 2001 and World Bank, 2001c

per cent, but the growth of consumption in the transport sector was greater at 1.7 per cent, especially in the developing world (Table 2.2-2).

In industry, energy is mainly used to manufacture a few energy-intensive goods such as steel, paper, cement, aluminium, and chemicals. The demand for these goods is growing in quickly developing countries – for instance, due to the expansion of infrastructure – while demand for these goods, except paper, is falling or stable in industrialized countries.

Some of the manufacturing plants for these goods have moved to newly industrializing countries (NICs).

The important factors in energy consumption in buildings include population density, urbanization, the number of housing units, per capita floor area, the number of people per household, age distribution, income per household, and the floor space used commercially. In general, the greater the degree of urbanization, the more energy is consumed per household

Table 2.2-2
Share of sectors in primary energy consumption according to various country groups, and growth rates between 1990 and 1995. Households and commercial buildings are taken together.
Source: IPCC, 2000b

	OECD		Transition countries		Asia		Africa and Latin America		World	
	Total	Rate 90–95	Total	Rate 90–95	Total	Rate 90–95	Total	Rate 90–95	Total	Rate 90–95
	[%]	[%/a]	[%]	[%/a]	[%]	[%/a]	[%]	[%/a]	[%]	[%/a]
Industry	33	0.9	51	-7.3	59	5.9	36	3.5	41	0.2
Households/buildings	40	1.9	32	-6.8	22	4.8	33	3.8	34	0.8
Transport	25	1.6	14	-6.0	15	7.6	26	4.2	22	1.7
Agriculture	2	1.6	3	-10.6	5	5.6	4	12.6	3	0.8
Total	*100*	*1.6*	*100*	*-7.1*	*100*	*5.9*	*100*	*4.1*	*100*	*0.7*

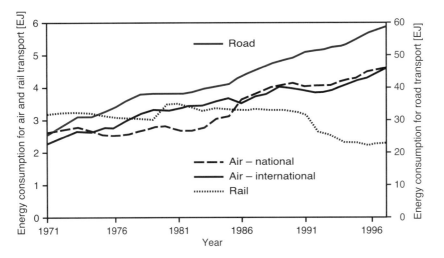

Figure 2.2-2
Worldwide energy
consumption in the
transport sector from 1971
to 1996. While road
transport more than
doubled over the 25-year
period, rail transport
declined, particularly in the
1990s.
Source: WRI et al., 2002

– mainly due to the greater income in cities (Naki-cenovic et al., 1998). In industrialized countries, space heating and air-conditioning takes up a large part of energy consumption in buildings. Per capita energy consumption in buildings is not only increasing in industrialized countries, but also in NICs. In developing countries, cooking and heating are the largest factors.

The volume of traffic and the technologies used determine the amount of energy consumed in the transport sector. In the past few decades, passenger traffic in private cars and aircraft as well as freight transport on roads have increased greatly. The comparatively small share of transport by railway continues to fall, while transport by road (73 per cent) and air (12 per cent) dominate energy consumption in the transport sector (Fig. 2.2-2).

The use of energy for transport depends on economic activities, infrastructure, residential patterns, and the prices for fuels and vehicles (IPCC, 2000b). The greater the population density, the lower the energy use for transport (Newman and Kenworthy, 1990). In industrialized countries, including Germany, new cars are becoming more energy-efficient (IPCC, 2001c). However, in passenger transport, the transport services delivered relative to the energy used have decreased in most European countries and Japan since 1970: the growing number and lower capacity utilization (fewer passengers) of cars outweigh the lower fuel consumption of the vehicle fleets, and the trend towards larger cars and more powerful motors has worsened the situation even further (IPCC, 2000b). Global emissions from the transport sector can also be expected to increase as traffic increases in the developing countries.

2.2.3
Lifestyles and energy consumption

In modern consumer societies, lifestyles have often replaced old class distinctions. Today, differences in income and value-orientations are the major factors that determine a person's lifestyle. Lifestyles in industrialized countries have become greatly differentiated. People use their lifestyle to express a personal and group-specific identity: they say who they are or who they want to be. Individuals may choose their lifestyle, but the lifestyles to choose from come about within social structures and trends in the course of social interaction: people compare themselves with others, look for role models, or set themselves apart from others. Unsustainable consumption thus cannot be reduced to individual consumer characteristics as laziness or egoism, but rather has to be seen and assessed in a societal context.

Mobility is a part of self-fulfilment for many people. Ecological criteria are often felt to be obstacles towards this goal. Lifestyles and consumption potential also affect social prestige. Ecological behaviour – such as the use of public transport or taking vacations in one's own country instead of going abroad – still often has a negative image. Differences in lifestyles manifest themselves in patterns of energy use and CO_2 emissions. Often, there is a connection with the available income of a household: emissions increase along with income.

At the same time, a number of other factors influence energy consumption:
- Individual characteristics (such as value-orientations, environmental awareness, age, gender, profession, education, origin, religion);
- The social environment (e.g. culture, social values, role models);

- Structures and institutions (infrastructure, residential environment, income, media, market transparency, access to information and consulting).

Greater prosperity and an increase in energy consumption have long gone hand in hand in western industrialized countries and were seen as interdependent prerequisites in the first 25 years after the Second World War. Under the pressure of the oil crises, however, the equation 'more prosperity = greater energy consumption' began to be called into question. In the meantime, the thesis that economic development and a high standard of living can be partially decoupled from an increase in energy consumption has been empirically proven for many OECD countries. A comparison of energy consumption in countries with a similar economic state of development also reveals that there are various ways to attain the same level of prosperity (Reusswig et al., 2002). The great variance of income in countries with the same level of energy consumption illustrates this fact clearly (Fig. 2.2-1).

2.3
Energy in industrialized countries

2.3.1
Energy supply structures

Within industrialized countries, a distinction can be made between two groups in terms of energy and carbon productivity: the US, Canada and Australia on the one hand, and the OECD countries of western Europe (mostly the member states of the EU) and Japan on the other. The OECD countries in North America have the highest per capita consumption of primary energy in the world, more than twice as much as the western European OECD countries: the energy productivity of the US and Canada, which depend heavily on the use of fossil energy, is 42 per cent lower than that of the OECD countries of western Europe and 100 per cent below that of Japan. The western European industrialized countries and Japan consume energy much more efficiently and are gradually lowering their carbon intensities.

ENERGY SOURCES AND ENERGY NEEDS
Domestic reserves of conventional energy sources largely determine the structure of primary energy consumption in industrialized countries. In the US, oil made up 39 per cent of the primary energy in 1997, followed by gas at 24 per cent and coal at 23 per cent. Nuclear energy made up 8 per cent, renewable

energy sources 4 per cent, and hydropower 2 per cent. Primary energy consumption rose steadily during the 1990s. By 2020 an annual increase of 0.9 per cent is forecast, compared to a 1.3 per cent increase between 1971 and 1997 (IEA, 2001b). According to this forecast, of all fossil energy carriers natural gas will grow the fastest at 1.3 per cent annually. The share of mineral oil will increase from 39 per cent to 41 per cent due to greater demand in the transport sector. Overall, the renewable sources of energy (excluding hydropower) will grow the fastest at 1.6 per cent annually, but given the low levels at which they are starting, their share in overall primary energy consumption will not increase considerably under the expected conditions.

In the OECD countries of western Europe, the consumption of primary energy will probably grow similar to the trend in the US at around 1 per cent annually, which is only slightly lower than the 1.2 per cent average annual increase between 1971 and 1997. But the structure of primary energy carriers will change considerably, especially in comparison with North America. According to the estimates of the IEA, the shares of coal and nuclear power will continue to drop (from 20 per cent to 14 per cent and 14 per cent to 9 per cent, respectively). In contrast, natural gas will grow by 3 per cent annually, increasing its share of primary energy consumption from 20 per cent to 31 per cent. Though the consumption of renewable energy will also continue to increase, its share will only rise from 4 to 5 per cent (IEA, 2001b).

TRENDS IN ENERGY DEMAND BY SECTOR
In the US, the increase in transport of 1.6 per cent annually by 2020 will determine energy demand most of all. The growth in traffic will far outpace increases in the efficiency of fuel consumption. In contrast, the energy demand of the industrial sector will only increase moderately at 0.5 per cent annually. Continuing structural change – the GDP share of the service sector, which is less energy-intensive, will continue to increase – will mean that industry's share in overall energy demand will decrease (Fig. 2.3-1).

In the EU, the transport sector presently consumes more than 32 per cent of final energy (over 80 per cent of which is in road transport) (EEA, 2001), thus making it an important factor in the increase of primary energy consumption in western Europe. In contrast, energy demand in the industrial sector has remained constant over the past 30 years in the OECD countries of western Europe.

DEPENDENCY ON IMPORTS
In industrialized countries, the security of energy supply is a key policy objective. The National Energy Policy Development Group in the United States esti-

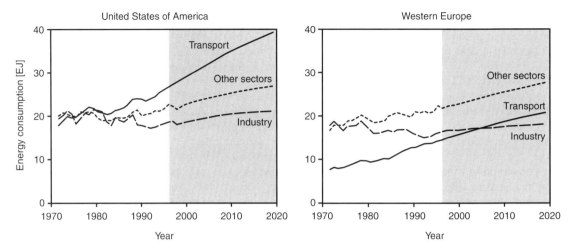

Figure 2.3-1
Previous development and IEA forecast for future energy consumption of the economic sectors of industrialized countries up to 2020.
Source: IEA, 2001b

mates that over the next 20 years the consumption of mineral oil will increase by 33 per cent, natural gas by 50 per cent, and electricity by 45 per cent (National Energy Policy Development Group, 2001). Hence, the gap between domestic production and demand will widen further. At current levels of consumption, US coal reserves will last another 250 years, given that 24 per cent of the coal consumed in the US is imported. The share of imports for other fossil energy sources will rise more steeply; by 2020, the US will probably have to cover some 70 per cent of its mineral oil consumption with imports. This trend will have important geopolitical consequences (Section 2.6.2).

The EU's dependency on imports is greater still. In the next 20–30 years, imports will increase from 50 per cent to 70 per cent of overall consumption, thus almost reaching the level of dependency of Japan, which currently imports 80 per cent of its energy. The EU's imports could reach 90 per cent for mineral oil, 70 per cent for natural gas, and even 100 per cent for coal. In light of its increasing dependency on imports, the EU elaborated a strategy in its Green Paper to ensure the security of energy supply. The core recommendations in the Green Paper include greater promotion of renewable energy by means of financial and tax incentives and resolute policies to influence energy demand (EU Commission, 2000a).

SUBSIDY AND RESEARCH POLICY IN THE ENERGY SECTOR

Subsidies are a key tool in energy policy. They are used to lower extraction and production costs, increase profits for producers, or lower prices for consumers. To ensure the security of energy supply, subsidies aim to guarantee a certain share of domes-

tic production and the greatest possible variety of energy carriers (IEA, 1999). While Germany mostly subsidized nuclear energy in the 1960s and 1970s, most of its energy subsidies now go to hard coal (UBA, 1997). Germany's coal subsidies are by far the greatest in all of Europe (Fig. 2.3-2). In addition to direct subsidies, fossil energy carriers also benefit from forms of indirect tax relief, such as the tax exemption for aviation kerosene and the distinction made between diesel and petrol.

In the US, too, subsidies especially promote fossil energy: 50 per cent of the energy subsidies in the US – a total of US$6,200 million – go to fossil energy, 18 per cent to renewables, and 10 per cent to nuclear power generation (EIA, 2000).

State expenditures for research and development represent a special kind of subsidy. Though they do not directly affect the current extraction and supply of energy or energy prices, they do influence the future development of energy markets and are thus crucial for the transformation of energy systems. A small number of countries dominate state research and development expenditures; all of them are industrialized countries. In 1995, only 10 of the 26 member states of the IEA accounted for 98 per cent of all of the research expenditures in the energy sector (IEA, 1997). In the last two decades, research budgets in the energy sector have been drastically cut in almost all industrialized countries with the exception of Japan (Fig. 2.3-3). The cuts in the R&D budgets affected all energy sources. Between 1980 and 1995, global expenditures for fossil energy dropped by 58 per cent, by 56 per cent for renewables, and by 40 per cent for nuclear (Margolis and Kammen, 1999). Public R&D expenditures, averaged across all industrialized countries, focus on promoting fossil and nuclear

Figure 2.3-2
Comparison of state
subsidies for hard coal
mining in four EU member
states in 1994 and 2001.
Source: EU Commission,
2001a

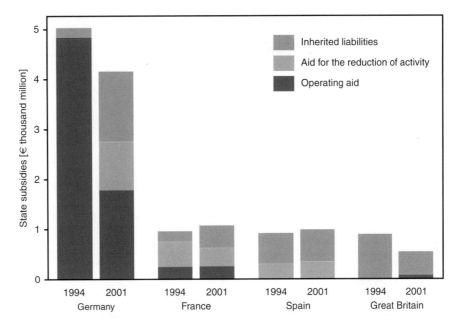

energy (55 per cent). Renewables and energy efficiency make up 40 per cent (UNDP et al., 2000).

Parallel to the drop in public research and development expenditures, the private research expenditures in industrialized countries have also fallen,
especially in the US, Italy, Spain and Great Britain
(Erdmann, 2001). But the different definitions and
methods make it hard to compare industrialized
countries. However, it can be stated that the energy
sector is one of the industries with the lowest
research and development expenditures in terms of
worldwide revenue. For example, in 1995 the US
energy sector only invested 0.5 per cent of its revenue
in research and development. In comparison, the
pharmaceuticals industry and the telecommunications industry each spent more than 10 per cent of
their revenue on research (Margolis and Kammen,
1999).

2.3.2
Principles and objectives of energy policy

The energy policies of the industrialized countries
pursue three objectives: energy supply security, low
prices resp. cost-effectiveness, and low environmental impact. The most important objective is to provide
and maintain the security of supply. This objective is
often the reason given for the considerable state
intervention on the markets for grid-based energy
(electricity, gas) in the form of monopolies protected
and regulated by the government. Domestic energy
reserves – usually fossil fuels – have been extracted
to ensure a large degree of energy independence,
especially during crises – hence the heterogeneous

mix of primary energy sources in the industrialized
countries. The objective of becoming independent of
energy imports became especially important after
the oil crises in the 1970s.

Many countries aim to attain the second goal of
energy policy – providing energy at the lowest possible price – by, among other things, establishing security reserves. The primary energy reserves (especially
oil and coal) not only serve to ensure supply, but are
also used to stabilize world market prices. For grid-
based energy supply, state investment and price regulations were used for a long time to prevent uneco-
nomic investments and protect consumers from
excessive (monopoly) prices of utility companies.
However, in many instances this policy did not produce the desired economic efficiency. After the comparatively positive experience with liberalization in
the US and Great Britain, the EU and other industrialized countries also launched the liberalization of
their electricity and gas markets in the 1990s to attain
greater economic efficiency and lower prices through
competition (Section 2.3.3).

Environmental protection is the third objective of
the energy policies of the industrialized countries. In
the 1970s and 1980s, the finiteness of fossil primary
energy resources and the adoption of local clean air
policies were the main issues in this field. In the
meantime, climate change has become the focal
theme in many countries. However, national governments vary greatly in the importance they attach to
this objective. Furthermore, the EU and the US differ on the status that expanding renewable energy
should have.

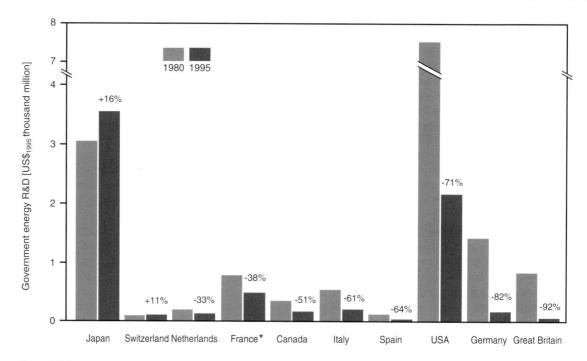

Figure 2.3-3
Comparison of public research and development (R&D) expenditures of selected OECD countries in the energy sector in the years 1980 and 1995. *Data for France are of 1985.
Source: Margolis and Kammen, 1999 (after IEA, 1997)

2.3.3
Liberalization of markets for grid-based energy supply

STARTING POINT
For the grid-based supply of electricity and gas, there have long been exceptional areas that were not subject to competition, i.e. where the state directly handled the supply of electricity or regulated it comprehensively. The economic justification for this is that the value chain of the generation of electricity is grid-based for long-distance transport and regional distribution to consumers. As provisioning is cheaper via one line than via multiple lines, these grids constitute what is called a 'natural' monopoly.

Debate on the regulatory setting governing the supply of electricity and gas began at the end of the 1970s with the liberalization measures in the US and Great Britain. The reasons given for the structural reforms launched for greater competition in the energy sector are primarily the goal of lower prices and the decentralization of the energy sector (distributed power). Deregulation and/or changes in the regulations ('re-regulation') are two ways of reaching these goals. To some extent, environmental protection interests are also given as further reasons.

The majority of the industrialized countries are deregulating (especially regional monopolies, price controls, and investment controls) to ensure the greatest level of competition possible on the electricity markets. The monopolies are to be reduced to a minimum of grid-based types of energy by breaking up energy provision, transit transport, local distribution, and the sale of electricity into separate sectors. In the areas where the market cannot provide any competition, there is to be competition for the market: tenders for temporary licenses.

LIBERALIZATION OF ELECTRICITY AND GAS MARKETS IN THE EU
The EU's liberalization efforts are based on the Single Market Directives for electricity and gas adopted in 1997 and 1998. They are to be implemented on the electricity market first. Liberalization is to take place in increments to facilitate adaptation for the power companies. According to the Single Market Directive and the resolutions of EU government heads, starting in 2004 industrial customers will be able to choose their power company, followed by private consumers in mid-2007. The efforts to liberalize the electricity markets have advanced so far in some member states of the EU – such as Great Britain, Sweden, Finland and Germany – that all customers can already choose their power company. Often, the power companies are vertically integrated firms, i.e. they serve all of the links in the value chain from primary to final energy. The Single Market Directive for electricity requires

this vertical integration to be broken up: the sectors of provisioning, grid-bound transport, and sales have to be separated, at least in terms of accounting. The transit grids have to be separate organizations within the company. All energy generators have to have access to the grids.

The gas markets in the EU are also undergoing a liberalization process, but one that is less advanced than the electricity markets. The Single Market Directive for natural gas obliges EU states to open their gas markets in increments. First, 20 per cent of the total annual gas consumption of a member state has to be open to competition by 2000, then 28 per cent by 2003, and finally 33 per cent by 2008. Great Britain and Germany have already completely opened their gas markets on paper, but competition is very slow in coming, especially in Germany (IEA, 2001a).

LIBERALIZATION OF GRID-BASED ENERGY SUPPLY IN THE USA
The United States pioneered the liberalization of the energy sector among industrialized countries with its National Energy Act of 1978. However, the implementation of this basic law was left up to the energy authorities of the various states, resulting in a heterogeneous mix of institutional designs and primary energy sources used. California was considered a forerunner of future energy policy due to the liberalization of its markets for grid-based energy supply and the use of renewable energy sources. But California's energy crisis has led to a much more differentiated assessment of the state's deregulation strategy. The goal of ensuring power supply moved clearly back into the spotlight once the liberalized electricity markets began to present a risk of blackouts (both controlled, i.e. 'rolling', and uncontrolled) on the east coast of the US. The old tenet of previous US administrations that energy shortages are regional, temporary phenomena has been abandoned. Under the direction of the US Vice-President, a task force has developed proposals for a future national energy policy. The 'National Energy Policy Report' (National Energy Policy Development Group, 2001) focuses on expanding domestic oil and gas production and mining coal to reduce the country's dependency on energy imports. In addition, nuclear power is to be expanded as a non-polluting alternative to coal, oil and gas.

2.3.4
Renewable energies in industrialized countries

EUROPEAN UNION
In accordance with an EU Directive of September 2001 (EU Commission, 2001b), the share of renewable energy in the EU's gross domestic consumption is to reach 12 per cent by 2010, with the share of renewables in total electricity produced rising to 22.1 per cent. The mix of renewable energy sources in the EU differs from one member state to another. Hydropower is used most often, covering an especially large part of the electricity generation in Austria and Sweden due to the local geographic conditions there. In Germany, a melange of policy tools (the rates for power fed into the grid, market penetration programmes, voluntary measures, etc.) have been promoting the 'new' sources of renewable energy since the early 1990s. As a result, the share of renewable energy has increased quickly in the past 10 years, both in terms of the primary energy consumption and power consumption (Fig. 2.3-4). Similar developments – especially for wind power – are found in many other EU countries.

USA
In the US, the current share of renewable energy sources (including hydropower) in electricity is also relatively slight at 6–7 per cent (IEA, 2002b). Today, new renewable energy sources like biomass, geothermal power, wind power, and solar energy only make up 2 per cent of electricity generation, a figure expected to rise to 2.8 per cent by 2020 (National Energy Policy Development Group, 2001). Accord-

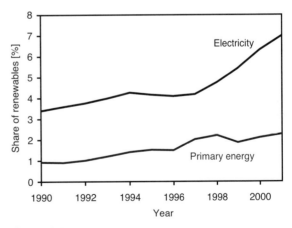

Figure 2.3-4
Development of renewables as a share of primary energy and electricity (without waste incineration plants) in Germany. Both shares doubled in ten years.
Source: BMU, 2002a

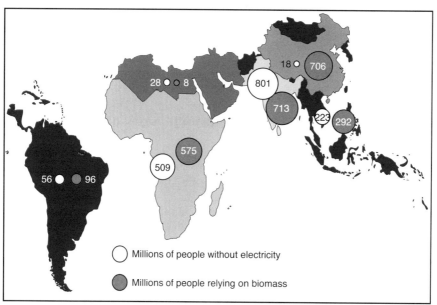

Figure 2.4-1
Regional distribution of people without access to electricity and those dependent on biomass for their energy supply. The different colours indicate regions and countries, which relate to the cited data.
Source: IEA, 2002c

○ Millions of people without electricity

● Millions of people relying on biomass

ing to the National Energy Policy Group, the main reason for the comparatively low consumption of renewable energy in relation to its technical potential is the high cost compared to conventional resources. Hence, greater efforts to promote renewables are proposed. They include budget increases for research and development in renewable energy and expanded tax rebates for power generated from renewables (wind, biomass). The US promotes renewables much less than western Europe, usually limiting its work to research and development programmes. Although the US is the third largest user of wind power after Germany and Denmark, few states in the US can match the growth trends for the consumption of wind power in Germany, for instance. Hence, renewable energy's share of primary energy is expected to grow more slowly in the US than in the EU in the next few years (IEA, 2001b).

2.4
Energy in developing and newly industrializing countries

2.4.1
Energy supply structures

DEVELOPING COUNTRIES
Access to modern energy is an essential part of the fight against poverty and a prerequisite for reaching the Millennium Development Goals (DFID, 2002). Energy fosters income, education, social involvement, and health, particularly freeing women of such

time-consuming activities as collecting firewood and fetching water.

Today, 1,640 million people – some 27 per cent of the world's population – have no access to electricity. 99 per cent of these people live in developing countries, 80 per cent of them in rural areas (IEA, 2002c; Fig. 2.4-1). This energy poverty goes hand in hand with a low index of human development (Fig. 2.4-2). China is an important exception here: 90 per cent of the population has access to electricity. Per capita income will grow the fastest in developing countries over the next few decades. In a business-as-usual scenario, more than 60 per cent of the growth in demand for primary energy will come from developing countries between 2000 and 2030 (IEA, 2002c).

Another problem specific to developing countries is drawing the attention of the world's population: the considerable health risks, especially to women and children, resulting from the use of wood and manure for cooking and heating (Box 2.4-1). The WHO estimates that 1.6 million people die each year from indoor air pollution – more than twice as many as from the effects of air pollution in cities and almost twice as many as die from malaria annually (WHO, 2002b). The development of income and technology will not solve this problem alone. The IEA (2002c) estimates that the number of people who use traditional biomass for cooking and heating will increase from 2,400 million at present to 2,600 million by 2030.

The developing countries are not pursuing any unified energy policy. Nevertheless, some patterns can be identified:
• The demand for commercial energy is growing faster than the GDP, except in the poorest developing countries. A 10 per cent increase in GDP is

Figure 2.4-2
Per capita energy
consumption and Human
Development Index for
1991/1992, based on data for
100 countries.
Source: Reddy, 2002

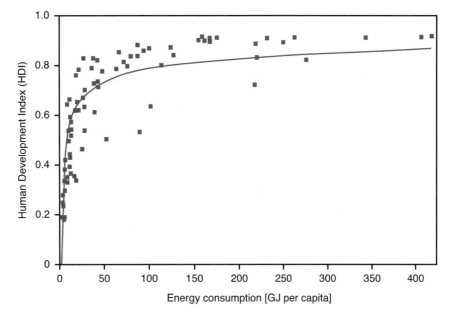

accompanied by a 12 per cent increase in demand for commercial energy (Leach, 1986). Above all, population growth is responsible for the above-average growth of the commercial energy sector (OTA, 1991): 90 per cent of the global population growth currently stems from developing countries. At GDP growth rates of 2–3 per cent per year, the growth rates in the consumption of commercial energy would be greater than the GDP even if the energy mix remained unchanged.

- Many developing countries are setting up infra-structures, for instance for transport. In the process, a lot of materials whose manufacture requires a great amount of energy – such as steel and concrete – are consumed, which will lead to great increases in the use of commercial energy in the mid-term.
- Rising rates of urbanization are leading to an increase in the share of commercial energy. Bio-mass is mostly used in rural areas. Nevertheless, many of the poor in urban slums will continue to be dependent on biomass and coal for heating and cooking.
- Modern methods of production are making elec-tric appliances such as refrigerators, televisions, radios and computers more affordable for con-sumers. This will, in turn, increase demand both among the consumers connected to the grid and the businesses that manufacture these goods, some of which are located in developing countries.
- The energy sector in developing countries suffers from inefficiency and poor controls. In 1992, state energy subsidies in developing countries amounted to a total of US$50,000 million, more than the official development assistance (ODA)

for these countries (DFID, 2002). What is worse: these subsidies often do not reach the proper tar-get groups or promote sustainable technologies. For instance, in Ethiopia 86 per cent of the subsi-dies for petroleum do not go to the poor (Kebede and Kedir, 2001).

NEWLY INDUSTRIALIZING COUNTRIES
Newly industrializing countries are closing the gap to industrialized countries and will be able to move beyond the characteristics of developing countries through their own dynamics in the foreseeable future. Structurally, they have adapted the modern-ization patterns of the industrialized countries and are imitating their models of economic growth and development. According to the World Bank, these countries include: the 'Tiger States' of South Korea and Taiwan; OPEC countries such as Saudi Arabia and Iran; South American countries rich in resources such as Brazil and Argentina; South Africa; and a few small (and rich) tourism island states such as the Bahamas and Mauritius.

The newly industrializing countries lie between the industrialized and the developing countries both in terms of per capita GDP and per capita energy consumption. But there are considerable differences within the newly industrializing countries: While Uruguay consumes some 12,000kWh per capita, South Korea consumes some 50,000kWh per capita. Newly industrializing countries generally place great store on economic development, with less attention paid to environmental or social problems. Environ-mentally friendly energy sources thus play a subordi-nate role in these countries. For example, during a prosperous period of investment in South Korea in

Box 2.4-1

Energy carrier usage as a function of household income in developing countries

Figure 2.4-3 illustrates the connection observed between household income, the energy services in demand, and the energy sources. The supply of energy required for basic needs is listed in the bottom row. As households become more prosperous, they use liquefied petroleum gas or fossil fuels instead of traditional biomass. Energy services such as refrigerators, etc. are not available to the poorest households. Where they are in demand, fossil fuels and – to a lesser extent – electricity are used. Electricity is only used

for the 'advanced' energy services when income reaches a certain level. The chart does not reflect the differences between urban and rural areas.

But there are exceptions to this diagram: in southern and south-east Asia, no connection was observed between income and the use of traditional biomass (Hulscher, 1997), i.e. the wealthy continue to cook and heat in traditional ways.

The development towards modern types of energy will take a long time and never be truly complete if energy policies rely simply upon economic development rather than strongly supporting change.

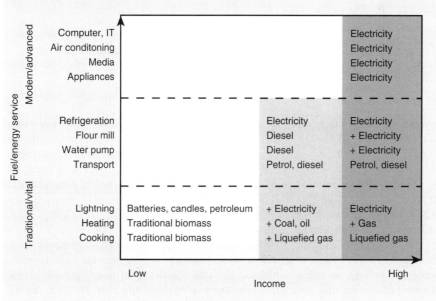

Figure 2.4-3
The mix of energy sources and energy services for households in developing countries relative to the income of the household. The plus symbols indicate that households use the type of energy concerned in addition to the types of energy already used.
Source: modified after IEA, 2002c

1995 the share of hydro, wind and solar only made up 0.3 per cent of the primary energy supply, while imported oil accounted for more than 60 per cent (Brauch, 1998).

The energy policies of most newly industrializing countries do not include targets for efficiency strategies or investments in renewable energy sources, except for hydropower (EIA, 2001). Though some countries like Brazil use their great potential for hydropower, they nonetheless rely on fossil fuels to cover the largest part of their energy demand (EIA, 2002).

The restrictions on the economic options of newly industrializing countries caused by oil imports have been exacerbated by the devaluation of the currencies of many Asian countries. This trend has led to a political refocus in ASEAN member states; now, for the first time, strategies for greater energy efficiency are being formulated. Nonetheless, no shift in investments towards renewable energy sources can be identified to date (Luukkanen and Kaivo, 2002). Box

2.4-2 describes reform efforts for the representative case of India.

2.4.2
Trends in energy demand by sector

In developing countries, private households are the largest group of consumers, followed by industry and transport. Households make up some 25 per cent of energy demand worldwide. But in China, they make up 37 per cent, in India and Indonesia 54 per cent, and in Nigeria even 80 per cent (Fig. 2.4-4, WRI, 2002). As the economy becomes stronger, the share of energy used in households drops noticeably as more energy is used for industry, transport or agriculture.

The types and amounts of energy used in developing countries vary widely due to the prevailing differences in income distribution as well as in institutions and infrastructures. In most developing countries, at

Box 2.4-2

The case of India: Development patterns, reforms and institutional design in the energy sector

THE SUPPLY/DEMAND GAP IN POWER SUPPLY

India makes up 16 per cent of the world's population but only consumes 2 per cent of the world's electricity. Electricity is growing faster than all other energy sectors in India at 8 per cent. With the industry sector growing at 9 per cent annually, the demand for energy is also growing at 9 per cent and thus far above the average GDP growth of 4.5 per cent per annum in the last 50 years. The power supply is insufficient: supply has worsened further since 1991, and in 1997 12 per cent of households – and even 18 per cent during peak times – did not have power. This gap entailed costs in the amount of 1.5–2 per cent of GDP. The unreliability of the power supply – power outages and frequent fluctuations in voltage – has led to the implementation of incentives to replace electricity with kerosene lamps and install diesel generators and voltage regulators. At the same time, the introduction of energy-efficient technologies faces obstacles; for example, the service life of energy-saving light bulbs is shorter due to the voltage fluctuations.

70 per cent of Indians live in rural areas, half of them below the poverty line of US$1 per capita per day. Thus, only a third of the electricity is used in rural areas, though 86 per cent of the villages already have access to electricity. Here, biomass is used to meet most energy needs.

INSTITUTIONAL REFORM IN THE ENERGY SECTOR IS SLOW IN COMING

The partial liberalization of the energy sector in the 1990s led to greater imports of coal, oil and technology, a modernization of power plants, and an incremental reduction of subsidies. But even in 1998, 64 per cent of private power consumption was still subsidized. And energy is subsidized even more in agriculture, where 20 per cent of electricity is consumed. India's electricity market is still a long way from complete liberalization and deregulation. For instance, power companies still cannot choose their coal provider; 64 per cent of power plants are coal-fired. Hence, operators of coal power plants cannot choose to use coal with little sulphur and heavy metal. The inflexibility of state monopolies and the unclear distribution of responsibilities among state organizations have delayed liberalization and deregulation. The reforms, which have gradually reduced subsidies and done away with adjusted prices, have raised power prices for consumers. Although the supply of energy has increased, 75 per cent of the greater supply has not been paid for because the power is illegally tapped due to the higher prices.
Sources: Lookman and Rubin, 1998; IEA, 1999; Gupta et al., 2001; World Bank, 2001a; Ghosh, 2002

least half of the overall energy demand comes from rural areas. In these regions, the energy pattern of the extremely poor are very similar: some 80 per cent of the low per capita energy consumption (<30GJ of primary energy per year) arises at home, almost exclusively for cooking (World Bank, 2001a).

In the newly industrializing countries, the demand for final energy has risen continually in the past few decades, especially in industry and transport. A comparison of the demand structure of industrialized, developing and transition countries reveals that energy demand varies according to the size of the industrial sector. One salient aspect is the extent to which the newly industrializing countries have approached the sector pattern of industrialized countries. The share of the transport sector in the newly industrializing countries is growing, though the size of this share still varies greatly. In South Korea, for instance, it is lower than in the OECD at 20 per cent vs. 32 per cent, respectively, for 1992 (Brauch, 1998).

Figure 2.4-4
Energy demand by sector in the four most populated developing countries – China, India, Indonesia and Nigeria – as well as the newly industrializing country of Brazil. The global mean is provided for comparison. All of the values are for 1997.
Source: WRI, 2002

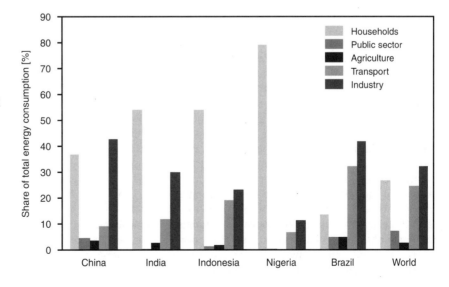

2.5
Energy in transition countries

2.5.1
Energy use

The transition countries are those in eastern Central Europe, eastern and south-eastern Europe, the Baltic states, and the Commonwealth of Independent States (CIS). In the former Soviet Union, the energy sector was expanded on the basis of centrally-planned targets. These targets left no room for free enterprise and environmental criteria as they focused primarily on such political goals as the spatial integration of the economy or a plentiful energy supply for industry and the population. In the process, the centralized, hierarchic planning of all energy sources led to the creation of unified energy complexes that ensured the supply of energy to the whole of the Soviet Union (UN-ECE, 2001). As the production, transport and distribution of energy did not depend on considerations of costs or efficiency, a situation was created in which it became possible to tap remote reserves of mineral oil and gas in the permafrost regions of Siberia (von Hirschhausen and Engerer, 1998). On the other hand, practically the whole energy cycle lacked incentives for energy efficiency, especially as generally no reliable records were kept of the volume of energy extracted, transported and consumed. The result was an energy system characterized by a very great extraction volume, excessive energy consumption, and great losses during transport and conversion.

The collapse of the socialist system in the early 1990s opened the floodgates for a comprehensive process of transition in these states. The transformation of the former socialist planned economies into market economies led to partial de-industrialization in all of the transition countries. Today, the CIS faces the challenge of ensuring its energy supply with the resources available domestically and the infrastructures inherited. Most of the transition countries have to pay a high price for imports of primary energy. The current infrastructure entails a great reliance on imports from other CIS states, notably Russia. In contrast, countries that have extensive energy resources – such as Russia, Azerbaijan, Kazakhstan, Turkmenistan and Uzbekistan – primarily face the problem of mobilizing capital for the development, maintenance and modernization of their mineral oil, gas and electricity industry. Not only does domestic energy supply have to be ensured, but also enough energy resources made available for exports, as energy exports generally make up a large part of the export revenue of these countries. In Russia, for example, oil and gas exports make up some 50 per cent of the total export revenue and some 20 per cent of GDP (EBRD, 2001).

In the CIS, energy demand fell by more than 20 per cent between 1990 and 1997 and has since only risen slightly (UN-ECE, 2001). The economic collapse and the widespread inability of both industry and private households to pay their power bills are primarily responsible for this downturn. The energy sector is suffering from considerable payment defaults as the political order prevents utility companies from cutting off the power supply to industrial or private customers who have not paid. The lost revenue is then lacking when investments are desperately needed to service and modernize the energy sector. This development has produced different outcomes for the various energy sources (EBRD, 2001; UN-ECE, 2001):

- Oil production in the CIS fell by 31 per cent between 1990 and 2000, from 571 to 395 million tonnes. The main reason is the difference between world market prices and fixed domestic prices, which makes the sale of oil to domestic refineries unattractive, especially in light of their liquidity problems.
- During the same period, production of coal dropped by 56 per cent from 703 million tonnes to less than 300 million tonnes. The main reasons are the closing of uneconomical mines and the use of more environmentally friendly and often less expensive energy sources, especially natural gas.
- In the same period, gas production fell by 14 per cent from 815,000 million m^3 to around 700,000 million m^3, with the drop in domestic demand partially compensated for by a 12 per cent increase in exports.
- A total of 28 per cent less electricity was generated, with thermal power plants – which account for some 70 per cent of the electricity generated – most directly affected. In contrast, the share of electricity generated by nuclear power plants (in Russia, Ukraine and Armenia) and hydropower dams remained basically stable; the share of each figured just above 15 per cent of the total power supply in 1997.

Renewable energy plays a minor role in the CIS, only attaining some 6 per cent, almost all of which comes from hydropower. Geothermal power makes up a very small percentage. Renewable energy is expected to grow more slowly than energy demand. In the mid to long term, however, the situation could change if unsafe nuclear power plants are shut down and prices for fossil energy rise as further reforms take hold on the energy markets.

2.5.2
Trends in energy demand by sector

In the CIS, the service sector's share of GDP grew from 35 per cent to 57 per cent from 1990 to 1998 due to partial de-industrialization and is expected to continue to grow. This development has slowed down energy demand, however, due to the high energy productivity of the service sector. In other sectors (households, retail, agriculture, and public service providers), energy demand has fallen much less than in industry, presumably due to the energy supply policy and the state guarantees that ensure power supply (Fig. 2.5-1). The IEA estimates that energy demand in these sectors will increase by 2.2 per cent annually up to 2010. While the energy demand of industry will also increase, growth rates there will be lower (IEA, 2001b).

The energy demand of the transport sector in the CIS is expected to grow especially quickly, with annual growth rates of 3.1 per cent up to 2020. In 2020, transport will make up some 53 per cent of the overall consumption of mineral oil (IEA, 2001b). In the Soviet Union, environmentally friendly transport modes such as trains and local public transport made up a much larger share of overall passenger and freight transport than in western industrialized countries. The share of transport in overall energy consumption was thus far smaller. But since the mid-1990s, the share of road traffic has been progressively rising: insufficient investment in railway infrastructure and public local transport is making them much less attractive than road transport. While there are currently only 100 cars for every 1,000 people in Russia (compared to 510 cars for every 1,000 people in Germany; IEA, 2001b), the anticipated rise in income is expected to lead to a larger number of cars. The effects of the EU's eastward expansion are discussed in Box 2.5-1.

2.5.3
Subsidies as a cause of inefficient energy consumption

In light of the energy sector's predicament in the CIS, one would expect that measures to increase energy productivity would be the highest priority both for the government and businesses. And yet, productivity gains have hardly been realized despite the great potential for savings. In 1997, the energy productivity of the CIS countries was around US$100 per megawatt hour (purchasing-power parity) and thus almost 7 times lower than the average for OECD countries (UN-ECE, 2001). The potential energy sav-

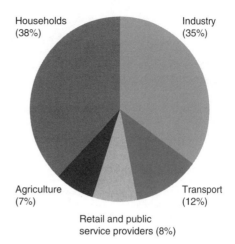

Figure 2.5-1
Sectorial pattern of energy demand in Russia, Ukraine and Uzbekistan.
Source: modified after WRI, 2001

ings that could be taken advantage of in the CIS countries amount to around 15–18EJ annually, or almost 40 per cent of the energy used. 90 per cent of this potential is in Russia and Ukraine alone. Roughly a third of this potential is in the energy and fuel sector itself, but the share of the industrial sector in potential savings is even greater (Russia: 30–37 per cent, Ukraine: 55–59 per cent). The third major area for energy savings is the building sector. Generally, buildings in the CIS states are not equipped with heating and gas meters. In addition, material for insulation is often too expensive. The savings potential in the building sector is estimated at 3EJ, or 16–18 per cent of the overall potential (UN-ECE, 2001).

The continuing subsidies for energy consumption are probably the main reason why this potential to save energy is not being taken advantage of in most CIS countries. Subsidy policies are characterized by the following:
- Energy prices are kept below the cost of generation by legal or political means as large sections of the industry and the population would not be able to pay higher energy prices.
- There is a cross-subsidy of households by industry: energy prices in the industrial sector are kept roughly twice as high as for private consumption, but not high enough to provide incentives for increases in efficiency.
- For political reasons, payment defaults may not be countered with the stoppage of power supply. There are either no insolvency proceedings, or they do not work. The energy industry has played an important role in the toleration of poor payment practices in industry and thus prevented both a restructuring of the industrial sector and an increase in energy productivity (EBRD, 2001).

Box 2.5-1

The effects of eastward EU expansion upon European energy supply

The energy sector of the central and eastern European states is still in a restructuring phase. The key outcome of EU accession will be the liberalization of grid-based energy industries. This will lead to a mix of energy sources in the accession states that is determined more by the market than by state targets. Liberalization is also expected to result in a large drop in the share of coal (from 55 per cent in 1990 to 38 per cent in 2020 in a business-as-usual scenario) and an increase in the share of natural gas (from 15 to 30 per cent over the same period). The concomitant modernization of power plants and heating systems will lead to a considerable drop in pollutant emissions. It is also expected to reverse the upward greenhouse gas emissions trend, levels of which have recently been rising again after several years of economic recession. Enhancing energy efficiency will remain a priority after EU accession, especially as the available financial resources can generate the greatest effects in this sphere of action. However, the goal of the energy policies of many new member states continues to be to expand exports to western Europe, which are limited at the moment due to technical constraints.

The future of the 22 nuclear reactors in central and eastern Europe remains problematic. On average, they provide some 30 per cent of the power supply or 6 per cent of total energy supply. Up to now, the EU has arranged for the stepwise decommissioning of a total of 6 nuclear generating units of the first generation of Soviet design in Bulgaria, Lithuania and Slovakia. The EU aims to shut down all reactors of the first generation (another 2 units) and increase the safety standards of later reactors. But there is some doubt about whether the funds provided by the EU (a total of €850 million since 1990) are a sufficient contribution towards financing the considerable costs of decommissioning and safety improvements. An estimated €5,000 million is required just for safety improvements in the new member states over the next 10 years.

At the same time, the question arises of the effects on the CO_2 emissions inventories of central and eastern European states. If the nuclear power plants are replaced by thermal power plants, CO_2 emissions are expected to increase by the same amount as they would have to decrease in accordance with the targets agreed in the Kyoto Protocol (generally 8 per cent below the levels of 1990). Nonetheless, in light of the drop in gross national product and the greater energy productivity, the Kyoto targets are not necessarily endangered. However, the potential volume of carbon trading under the Kyoto Protocol and the resulting revenue would plummet.

But the greatest challenge for the EU's climate policies and those of its (future) member states may not stem from developments in the energy sector, but rather from the growth in traffic driven by the expansion of the EU. Expansion is expected to double the annual growth rate of the volume of transport between new members states and the current EU to 10 per cent, as well as that of exports from the new member states to 6 per cent, while the share of railway transport is expected to continue to decrease as it has done in the past years. In addition, the entrance of the new member states will probably only further accelerate their growth in the number of private cars, which already skyrocketed over the past decade. The share of old, heavily polluting cars will also be much larger than in the current EU.

Sources: EU Commission, 1999b; Matthes, 1999; EU Commission, 2000b; Jantzen et al., 2000; IPTS, 2001.

The subsidies not only make energy saving uneconomical for industrial and private consumers, but also detrimentally affect the liquidity of power companies, which are no longer able to come up with the funds to invest in reducing losses in transport and conversion. At the same time, there is a lack of predictable market conditions and long-term prospects for profit for foreign investors, whose capital will probably be indispensable for the modernization of energy systems in the foreseeable future.

2.5.4
Privatization, liberalization and (re-)regulation of energy industries

Efforts to liberalize the energy industries of the CIS are mostly based on the US/UK model. This model provides for a separation of energy provision, sales, and grid operation, the splitting up and privatization of state power companies, and an independent regulatory authority. Liberalization efforts have been implemented to various extents: Only a few countries, such as Azerbaijan, have completely done without reform, while many countries in the CIS have converted the various components of their energy industry into private companies. However, control of these companies has largely remained in the hands of the political elite (in the form of state holdings or through the transfer of shares to former state actors in the energy industry complex). At the same time, the possibilities for foreign investors to have holdings have been severely limited.

The restructuring process in the energy industry could be accelerated if Russia joins the World Trade Organization (WTO) in the next few years. Russian legislation and practices in the fields of industrial subsidies, taxation, and customs policy do not currently meet WTO requirements. Adaptation to the WTO's conditions can be expected to lead to greater competition in the energy sector – and hence to greater efficiency. At the same time, the region's attractiveness for foreign direct investors would increase (EBRD, 2001; CEFIR and Club 2015, 2001).

2.6
Economic and geopolitical framework conditions

The growing interdependencies in a globalized world and the transformation of world politics since the early 1990s have fundamentally altered the framework conditions of global energy policy. Firstly, a more integrated global economy has led to a rise in energy consumption in the transportation and exchange of goods, services and personnel. Secondly, the end of the Cold War has facilitated the frontier-free tapping of energy resources – even in regions to which the West's transnational corporations previously had little or no access.

2.6.1
Globalization as a new framework condition for energy policy action

For the energy sector, globalization is nothing new. The internationalization of markets and market players occurred first, and is most advanced, in the energy industry (Enquete-Commission, 2001). Due to the often vast geographical distances between energy carrier production, on the one hand, and energy conversion and use, on the other, this sector has always been a driving force in the deepening of international trade relations. The liberalization of grid based energy in many countries has also contributed to the growth of the international energy trade.

When considering the impact of world trade on energy supply and use, it is important to bear in mind that global economic integration facilitates the transfer of standards in energy efficiency, products, technologies, production processes and management systems. In this aspect of globalization, the industrialized countries' foreign direct investment (FDI) in other regions of the world plays a key role. FDI rose significantly in the 1990s: Global FDI (inflow and outflow) amounted to 2.7 per cent of global GNP in 1990, and to 4.3 per cent eleven years later (UNCTAD, 2002).

On the one hand, the globalization of world trade fosters opportunities to export technologies for the generation of renewables. On the other, there is a danger that poor-quality, energy-inefficient technologies and products – such as obsolete machinery, vehicles and plant – will be exported to developing and newly industrializing countries, negatively impacting on energy efficiency (Enquete Commission, 2001).

Alongside the rise in commodity flows, passenger transport services – especially air travel – have a direct impact on the energy sector. Since 1968, the number of tourists travelling by air has increased more than five-fold, from 131 million (1968) to 693 million in 2001 (World Tourism Organization, 2002), and tourist numbers are expected to climb to 1,600 million by 2020. World airline passenger traffic is currently growing by 4–6 per cent annually (Lee et al., 2001), and air traffic's share of total passenger transport volume is expected to quadruple to 36 per cent by 2050 compared with a figure of 9 per cent for 1990 (WBGU, 2002).

Another energy-relevant effect of globalization is the transfer of Western lifestyles to the less industrialized regions of the world. The global media corporations and the entertainment industry are driving forces here, among others. This transfer typically leads to changing consumption patterns, as is evident from – among other things – the rise in private households' energy consumption, especially for domestic purposes (living space, refrigeration/cooling) and the acquisition of electrical domestic appliances and communications equipment. It also has an impact on mobility and leisure.

2.6.2
Geopolitics

The link between global energy supplies and geopolitics relates primarily to oil and gas, as they are more regionally concentrated in the Earth's crust than hard coal and lignite. The spatial gap between energy use, on the one hand, and the extraction of energy sources, on the other, means that the importing countries are heavily dependent on a single geographical zone, the resource and energy 'ellipse', comprising the Middle East region and the countries of the Caucasus-Caspian region (Fig. 2.6-1). These regions, which are among the most politically unstable in the world, harbour 70 per cent of the world's oil and 65 per cent of its gas reserves (Table 2.6-1). Their significance for global energy supplies varies, however: At the current rate of output, the oil reserves in the Persian Gulf are likely to last for another 90 years, compared with just 20 years for the Caucasus-Caspian region, while the figures for the gas reserves are 270 and 80 years respectively (Scholz, 2002). The transportation conditions also differ. Oil has been produced in the Persian Gulf for a long time and can be distributed worldwide via pipelines and tankers. Gas, on the other hand, with the exception of liquefied petroleum gas (LPG) tankers, needs a networked system of pipelines extending all the way to the end user; until now, this has only been profitable for distances up to 6,000km (Müller, 2002).

Since the industrialized countries' own oil and gas reserves – where they exist – are rapidly dwindling, demand must now be met by an increase in imports.

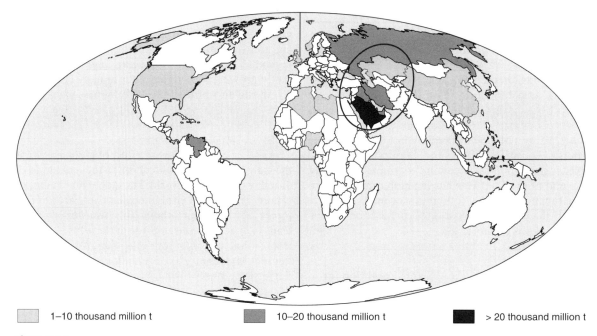

| | 1–10 thousand million t | | 10–20 thousand million t | | > 20 thousand million t |

Figure 2.6-1
Countries with oil reserves exceeding thousand million tonnes. The regional distribution of reserves within the countries is not shown. The resource and energy 'ellipse' harbours around 70 per cent of the world's oil reserves and around 40 per cent of its gas reserves.
Source: BGR, 2000

For 2020, Germany's dependency on energy imports is likely to be 75 per cent, with the corresponding figures for the EU and the USA being 70 per cent and 62 per cent respectively (Enquete Commission, 2001). The petroleum exporting countries have joined together to form OPEC, thus establishing a strong negotiating position (Box 2.6-1).

During the Cold War, securing the oil supply from the Middle East was a major priority for the USA and its NATO allies. This geostrategic goal was underpinned by the Carter Doctrine: "An attempt by an outside force to gain control of the Persian Gulf region will be regarded as an assault on the vital

Table 2.6-1
Regional distribution of fossil energy reserves in 2000.
Source: Enquete Commission, 2001

Region	Mineral oil [%]	Gas [%]	Coal [%]
Middle East	65	35	0
CIS	6	38	23
North America	6	5	26
Central and South America	9	5	2
Europe	2	4	12
Africa	7	6	7
Asia/Pacific	5	7	30
Total	*100*	*100*	*100*

interests of the United States of America, and such an assault will be repelled by any means necessary, including military force." (US President Jimmy Carter in his State of the Union Address, 23 January 1980).

The demise of the Soviet Union led to a shift in the global political landscape and geopolitical options. The security interest of the Newly Independent States in the Caucasus and Central Asia in reducing their dependency on Russia gave the USA the opportunity to gain a foothold in the region – they pumped in massive amounts of economic and military aid, and, in the wake of 11 September 2001, established military bases within the framework of the 'International Coalition against Terrorism'. At the same time, the USA has also focussed increasingly on the oil-rich regions of Africa: West Africa already produces 15 per cent of the USA's mineral oil imports, and this figure is set to increase to 25 per cent within the next ten years through the expansion of production plants and the construction of a pipeline between southern Chad and the Atlantic ports (The Economist, 14.09.2002). In the mid to long term, the Caucasus and West Africa could form an important supplementary source of energy supplies alongside the Gulf region. The ongoing crises in the Middle East and Iraq, the political unpredictability of Iran, the growing domestic problems in Saudi Arabia, and the terrorist threat posed by Islamic funda-

Box 2.6-1

OPEC's role as an energy policy actor

The Organization of the Petroleum Exporting Countries (OPEC) was founded in 1960 by Saudi Arabia, Venezuela, Iraq, Iran and Kuwait. Qatar (1961), Indonesia (1962), Libya (1962), the United Arab Emirates (1967), Algeria (1969) and Nigeria (1971) joined the organization later. Today, OPEC is a powerful alliance of newly industrializing countries operating in the international energy market. As the amount of energy carriers exported by the OPEC countries meets around one-third of the global demand for primary energy, OPEC has considerable influence over the development of global energy systems. The 11 member states define themselves as developing countries whose goal is to safeguard oil revenue and economic growth in the long term. In economic and structural terms, oil revenue is of key importance to the OPEC states: Oil exports by the 11 states totalled US$254,000 million in 2000, amounting to around 30 per cent of their GDP (US$860,000 million in 2000). There is a substantial prosperity gap within OPEC. At one end of the scale is Nigeria, whose per capita GDP is US$319; at the other is Qatar, whose per capita GDP is US$24,000. Within the OPEC countries, the energy sector is completely dominated by oil and gas.

In 1998, OPEC represented just 40 per cent of the international mineral oil market, but it controls 78.5 per cent of the known oil reserves worldwide. By comparison, its global market share of refined products is just 10 per cent. OPEC operates as a cartel in the international oil market. Its influence has declined noticeably since the non-OPEC states – Mexico, Russia, Norway, Great Britain and China – entered the market, but it remains a powerful force worldwide nonetheless. OPEC's internal cohesion is being challenged, above all, by the conflict between densely-populated countries with few oil reserves, on the one hand, and the less populous countries with substantial oil reserves, on the other. In future, the OPEC states are likely to face several economic challenges simultaneously:

- An increase in oil extraction from non-OPEC sources,
- new discoveries by small producers,
- an expansion in production capacities within its member states, especially in Venezuela and Nigeria,
- fluctuations in demand in the petroleum importing countries caused by economic cycles,
- greater competition within the energy market due to the expected greater impact of gas supplies and prices,
- better drilling and exploration technologies, leading to falling operational costs. If non-OPEC countries increase their oil output as a result, competition will intensify.

OPEC's influence on the world's oil market and its ability to enforce high oil prices are therefore likely to decline.

Sources: Salameh, 2000; IEA, 2001b; OPEC, 2001; Jabir, 2001; Odell, 2001

mentalists make the Gulf region increasingly unattractive as a source of supply. Nonetheless, the Gulf is likely to remain the main supplier of oil to the USA for the foreseeable future.

Alongside the geopolitically motivated diversification of oil and gas sources, transport routes are also being diversified for similar reasons. A pipeline from Kazakhstan through Russia to the Black Sea port of Novorossiysk could be vulnerable to pressure from Russia. A Mediterranean route from the Caspian Sea through Azerbaijan, Armenia, Georgia and Turkey or, alternatively, through Iran to Turkey's Mediterranean port of Ceyhan would run through extremely fragile states if the first route were chosen, or, if the second were chosen, through Iran. For security policy reasons, the US energy corporations are therefore targeting their current strategic planning towards securing a transportation route from the Caspian Sea through Turkmenistan, Afghanistan and Pakistan to the Indian Ocean – on condition that the USA can guarantee political and military security. In fact, the World Bank recently announced plans to build a gas pipeline from Turkmenistan to Pakistan which will run through Afghan territory (Agence France Press, 2002).

Through its energy and geopolitical strategy on the 'energy ellipse', the USA appears to be pursuing several objectives:

- Safeguarding energy supplies through the diversification of sources and transportation routes;
- Preventing political and military control of production areas and transportation routes from falling into the hands of rival powers (Russia/China), potentially hostile states (Iran) or local warlords who could disrupt the highly vulnerable transportation routes through terrorist acts;
- Finally, developing a position of strength vis-à-vis potential economic rivals.

There is significant potential for inter-state conflict in the Caucasus-Caspian region, which is a geostrategically important and sensitive region where US, Russian and Chinese interests collide. China is seeking to gain access to energy sources in Kazakhstan, with the construction of a pipeline from Kazakhstan to China being a strategic objective (Morse, 1999). There is assumed to be a link between the military conflict over Chechnya and Russia's strategic plans for oil pipelines.

A further potential source of conflict is the major oil and gas reserves in the seabed, as the rights to these reserves are disputed (Klare, 2001). The five Caspian seaboard countries, for example, have so far failed to agree on the distribution of the oil and gas rights. Seven different states are engaged in a dispute over the oil and gas rights in the South China Sea. Similar conflicts over rights of ownership have arisen in the offshore areas of the Persian Gulf, the Red Sea, the Timor Sea, and the Gulf of Guinea.

2.7
The institutional foundation of global energy policy

In the past, energy policy was viewed primarily as a national government function, with security of supply being the main objective. Over the last decade, three factors which encourage more internationalization in the energy policy field have increased in significance:

• The recognition that global climate protection needs international cooperation,
• The liberalization of the energy sector in major industrialized countries, but also in transition and developing countries, which intensifies trade in energy goods and services,
• The need for the global expansion of energy supply, especially in the developing countries, both for economic and political reasons and in the interests of climate protection.

Partly as a result of this development, numerous actors are now involved in energy policy issues at international level (Fig. 2.7-1).

A coherent global energy policy requires a coordinated approach and linkage with various other policy fields (including transport, environmental and development policy). This can only take place through effective and coordinated institutions at international level. The following sections provide an overview of the existing legal and institutional bases of international energy policy in core areas, i.e. knowledge base, organization and financing. It is intended to demonstrate whether the preconditions for an effective energy policy at global level are already in place, and if not, where there is a need for action. The text confines itself to the key institutions and their main functions.

2.7.1
Knowledge base

The range of scientific positions on energy and climate policy is broad, with studies often arriving at conflicting conclusions. This further enhances the importance of institutions whose agenda is to supply an international scientific assessment as a basis for policy-relevant recommendations. The findings of the Intergovernmental Panel on Climate Change (IPCC) are viewed as important for the development of global energy policy. The IPCC was established by the World Meteorological Organization (WMO) and

UNEP in 1988, and its Secretariat is hosted by the WMO in Geneva. The UNEP division responsible for implementing the Energy Programme works closely with the independent UNEP Collaborating Centre on Energy and Environment (UCCEE), which plays an active role in research and analysis as well as supporting the implementation of national and regional programme activities.

Together with the UN Department of Social and Economic Affairs (UNDESA) and the World Energy Council (WEC), UNDP prepared a global energy report, entitled 'World Energy Assessment', for the 9th Session of the UN Commission on Sustainable Development (CSD) in 2000. The report's key demands include a basic supply of commercial energy services worldwide and the provision of advice to developing countries on developing and implementing energy projects.

The World Energy Outlook, published on a regular basis by the International Energy Agency (IEA), is the most important source of energy statistics and energy sector analyses worldwide, and also tracks global energy supply trends. The latest report explores the link between energy and poverty (IEA, 2002c).

2.7.2
Organization

As well as a sound scientific basis, a global energy policy also needs institutions which define objectives and are responsible for adopting and implementing appropriate measures. Political declarations, international treaties and the work of relevant UN agencies are especially important in this context.

2.7.2.1
Political declarations

In recent decades, progress has been made not only with the definition of problems by the scientific community, but also in addressing these problems at international conferences and in declarations and treaties.

At the United Nations Conference on Environment and Development (UNCED) in Rio de Janeiro in 1992, the international community committed itself, for the first time, to comprehensive and ambitious targets in the twin areas of human development and environmental protection, set out in Agenda 21 and the Rio Declaration. Although none of Agenda 21's 40 chapters focusses explicitly on the issues of energy or transport, energy does feature as a subsection of Chapter 9 (Protection of the atmosphere) and

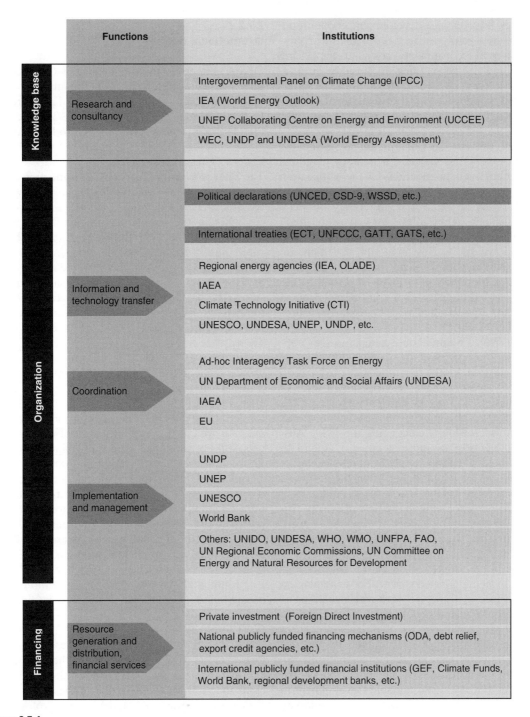

Functions	Institutions

Knowledge base

Research and consultancy
- Intergovernmental Panel on Climate Change (IPCC)
- IEA (World Energy Outlook)
- UNEP Collaborating Centre on Energy and Environment (UCCEE)
- WEC, UNDP and UNDESA (World Energy Assessment)

Organization

- Political declarations (UNCED, CSD-9, WSSD, etc.)

- International treaties (ECT, UNFCCC, GATT, GATS, etc.)

Information and technology transfer
- Regional energy agencies (IEA, OLADE)
- IAEA
- Climate Technology Initiative (CTI)
- UNESCO, UNDESA, UNEP, UNDP, etc.

Coordination
- Ad-hoc Interagency Task Force on Energy
- UN Department of Economic and Social Affairs (UNDESA)
- IAEA
- EU

Implementation and management
- UNDP
- UNEP
- UNESCO
- World Bank
- Others: UNIDO, UNDESA, WHO, WMO, UNFPA, FAO, UN Regional Economic Commissions, UN Committee on Energy and Natural Resources for Development

Financing

Resource generation and distribution, financial services
- Private investment (Foreign Direct Investment)
- National publicly funded financing mechanisms (ODA, debt relief, export credit agencies, etc.)
- International publicly funded financial institutions (GEF, Climate Funds, World Bank, regional development banks, etc.)

Figure 2.7-1
Global energy policy today: The key institutions and their main functions. For abbreviations, see List of Abbreviations.
Source: WBGU

Chapter 14 (Promoting sustainable agriculture and rural development). Transport is mentioned in Chapter 7 (Promoting sustainable human settlement development) and also in Chapter 9.

The Nineteenth Special Session of the United Nations General Assembly in New York five years after UNCED (Earth Summit +5) identified a specific need to support the developing countries in building a sustainable energy supply, e.g. through technology transfer and development cooperation. The meeting urgently recommended the incorporation of external environmental costs into the pricing structure and the dismantling of subsidies on non-sustainable energy carriers. It also identified a significant need to improve the coordination of energy-related activities within the United Nations (UN, 1997).

In April 2001, the UN Commission on Sustainable Development (CSD) adopted a number of global energy policy recommendations in the publication 'Energy for Sustainable Development: A Policy Agenda'. They cover access to energy services, energy efficiency, renewables, next-generation fossil fuel technologies, nuclear power, rural energy systems, and transport. Research and development, capacity building, technology transfer, access to information, mobilization of financial resources, the removal of market distortions, and the inclusion of stakeholders are identified as cross-cutting measures (UN-ECOSOC, 2001).

At the World Summit on Sustainable Development (WSSD) in 2002, energy was included as a separate agenda item for the first time, with the focus on renewables, access to energy services, organization of energy markets, and energy efficiency. However, the balance sheet is sobering: Due to the blocking tactics by the anti-target lobby comprising the USA, Australia and the OPEC countries, a target of at least 15 per cent of global energy supply from renewables by 2010 could not be adopted. For other targets, too – such as the removal of market distortions or boosting research and development in the field of energy efficiency – no success indicators or timeframes were adopted, and nor were these targets established on a legally binding basis. Nonetheless, numerous 'type 2' initiatives were adopted, with states, corporations and NGOs pledging to cooperate. For example, the EU announced that it would work with a coalition of like-minded states and regions to establish quantifiable, timebound targets for the development of renewables and pledged €700 million to improve access to reliable and affordable energies in the developing countries. UNEP launched the Global Network on Energy for Sustainable Development to promote 'clean' energy technologies in developing countries. The German Chancellor announced that Germany would host the International Conference for Renewable Energies, to take place in June 2004 in Bonn.

2.7.2.2
International treaties

Of the numerous treaties dealing with aspects of international energy policy, only the most important agreements are examined below. They are the Energy Charter Treaty, the WTO/GATT rules, and the UNFCCC together with the Kyoto Protocol.

ENERGY CHARTER TREATY
The Energy Charter Treaty (ECT) is the most significant international treaty dealing explicitly with cross-border cooperation among industrialized countries in the energy sector. The Treaty, which evolved from the 1991 European Energy Charter, came into force in 1998. 46 states, mainly from Europe and Central Asia, have ratified the Treaty (as at 11.09.2002). However, several major signatory states (Russia, Japan, Norway and Australia) have yet to ratify the ECT, and various other states – notably the USA and Canada – have not acceded to the Treaty.

The aim of the ECT is to promote economic growth through the liberalization of investment and trade. To this end, it extends the GATT rules to the energy sector. Minimum standards were agreed for foreign investment and energy transport. Both these areas, as well as the issue of transit of energy products, will be developed on a more binding basis in future (Energy Charter Secretariat, 2000).

Environmental aspects of energy policy are framed in general terms and set out as recommendations on energy efficiency, external costs, clean technologies and cooperation on environmental standards, etc. More detailed provisions are included in an additional environmental protocol, the Energy Charter Protocol on Energy Efficiency and Related Environmental Aspects (PEEREA), although this has no binding legal force.

GATT AND WTO RULES
For a long time, the members of the World Trade Organization were restrained about the inclusion of energy carriers in the GATT rules, although in principle, they fell within the scope of these provisions. This restraint was due to the specific role of the energy sector in national security of supply, the OPEC countries' non-membership of the WTO, and the regulation of the energy sector at national level, especially the status of the state-sector energy monopolies. With regard to the electricity industry in particular, international trade in electricity was not envisaged when GATT was established, and even today, it is generally deemed to be a service at most, not a good, implying that GATT does not apply. It was only with the conclusion of the Uruguay Round

in 1994 that the members integrated a number of energy carriers, especially coal, gas and oil and partly also electricity, more fully into the world trade regime. This can be explained, among other things, by the accession of various OPEC countries to the World Trade Organization and the progressive liberalization of the energy markets. Substantial cuts in tariffs on mineral oil, petroleum products and other energy carriers were not achieved, however, although trade facilitation schemes for petrochemical products such as plastics were agreed (, 2000).

Until now, electricity and energy-related services have barely been covered by the GATT Agreement. The specific monopoly status of the vertically highly integrated and mainly nationalized energy supply sector has not only impeded trade; it also explains states' restraint on the issue of concessions on market liberalization (WTO, 1998). However, the latest negotiating proposals indicate that market access for energy services will be eased across the board in future (WTO, 2001).

The WTO Secretariat views subsidies on energy provision and consumption as the most important barrier to liberalization of the trade in energy. The full integration of this sector into the WTO rules and the stringent application of the WTO Agreement on Subsidies and Countervailing Measures would help to cut subsidies and thus make a major contribution to climate protection at the same time. According to the WTO, the dismantling of all subsidies by 2010 – supported by appropriate flanking measures in the environmental policy field – could prevent around 6 per cent of CO_2 emissions worldwide (WTO, 2001). From an environmental policy perspective, however, it is unsatisfactory that the full integration of the energy sector into the WTO would restrict the scope for subsidies and other measures designed to promote 'green' technologies.

The energy policy relevance of the WTO Agreements extends beyond the applicability of their rules to trade in energy, petroleum products and energy-related services: As well as the impact of economic globalization on energy supply and demand (Section 2.1.6), other relevant factors are, firstly, the impact of the Agreement on Trade-Related Aspects of Intellectual Property Rights (TRIPS) and, secondly, the potential for conflict between the flexible mechanisms defined in the Kyoto Protocol and the WTO rules (Greiner et al., 2001; Box 5.3-2).

UN FRAMEWORK CONVENTION ON CLIMATE CHANGE AND THE KYOTO PROTOCOL
The United Nations Framework Convention on Climate Change (UNFCCC) defines objectives and principles, bodies and mechanisms for international climate protection policy. It entered into force in 1994 and has been ratified by 184 states around the world, including all the major industrialized and developing countries.

Climate policy entails, above all, a radical reduction in CO_2 emissions worldwide. This can only be achieved through the comprehensive restructuring of global energy systems (Chapter 4). Here, UNFCCC has a key role to play in international energy policy: It is the driving force and the most significant international forum where states can discuss the interface between environmental and energy policy and adopt major decisions.

The Kyoto Protocol to the UNFCCC was adopted in 1997 by more than 160 nations. The Protocol sets out specific reduction commitments for a defined group of greenhouse gases. It commits the industrialized countries to reducing greenhouse gas emissions by at least 5 per cent by the 2008–2012 period against the baseline year of 1990. Annex I of the Protocol stipulates a precise reduction target for each industrialized country (EU -8 per cent, USA -7 per cent, Japan -6 per cent, Australia +8 per cent, and Russia 0 per cent). The developing countries point out that the main producers of climate change (i.e. the industrialized countries) should take the lead, and have not undertaken any specific reduction commitments themselves to date.

The Protocol allows the industrialized countries some flexibility in implementing their commitments. In a process known as emissions trading, an industrialized country which exceeds its reduction target – for example, because emissions reductions can be achieved at low cost – can sell its surplus emissions permits to countries which have difficulty meeting their commitments. However, industrialized countries can also earn credits by investing directly in emissions reduction projects – such as modernizing an old power station – in other developed countries that have taken on a Kyoto target (Joint Implementation, JI). The Clean Development Mechanism (CDM) is a way for industrialized countries to earn credits by investing in emissions reduction projects in developing countries while complying with specific rules and modalities.

The provisions of the Kyoto Protocol were further elaborated at Conferences of the Parties to the Convention (COP). It was not until the 7th Conference of the Parties (COP7) in 2001, building upon the 2000 Bonn Agreement, that the way was clear for the Kyoto Protocol's rapid entry into force. The requirement for entry into force is ratification by 55 states representing at least 55 per cent of the CO_2 emissions in the baseline year of 1990. This means that entry into force is possible even without the USA and Australia, which have now pulled out of the Kyoto process. The Protocol will enter into force with the announced ratification by the Russian Federation.

The Kyoto Protocol enables states to earn credits through afforestation, reforestation and deforestation (ARD) and other forestry and agricultural measures. For the first commitment period defined in the Kyoto Protocol (2008–2012), states with reduction commitments are expected to produce CO_2 emissions amounting to 14Mt from ARD (Table 2.7-1; Schulze et al., 2002), but this will be more than offset by the eligibility of around 70Mt of CO_2 through land and forestry management. The sustainability of carbon sequestration in soils of regenerative forests requires more detailed study, however. Some studies indicate that carbon storage takes place primarily in the layer of organic matter and topsoil (Thuille et al., 2000), but it remains unclear whether storage occurs in the form of stable humus compounds in the deeper soil layers which has permanence even in a changed climate (Section 3.6).

In practice, even as an unratified text, the Kyoto Protocol has already had an impact on international energy policy. The EU has adopted a climate protection programme which contains Europe-wide provisions on energy efficiency, renewables, demand-side management and energy use in the transport sector. Many industrialized countries are producing and implementing climate protection programmes, and in some cases, this has already led to a greater focus on the role of renewables in energy policy. With the CDM, climate policy has developed a transfer instrument which effectively enables the industrialized countries' modern technologies to be made available more widely in the developing countries.

At COP7, the implementing rules for the CDM were set out in detail, resulting in the registration of the first batch of CDM projects. The rules give priority, among other things, to certain categories of small-scale CDM projects. It is difficult to predict whether the CDM will play a major role in driving forward capital transfer for clean energy from the industrialized to the developing countries – firstly because it is still unclear what proportion of the industrialized countries' reduction commitment will be fulfilled through the CDM. For example, the emission rights which Russia has not required due to its economic recession are sufficient to cover the reduction commitments of all OECD countries except the US (Jotzo and Michaelowa, 2002). In this case, the CDM's role would be negligible. Secondly, it is unclear how the CDM will be divided between sink and energy projects. The Marrakesh Accords permit sink projects amounting to 1 per cent of the investor country's 1990 emissions. Since COP8 there have been signs that sink projects may be less sustainable and therefore less attractive, despite their low costs, than hitherto assumed. Therefore, the likely level of investment in sustainable CDM energy projects is uncertain. Simple estimates suggest that it is possible for around one-third of all reduction commitments to be fulfilled via the CDM. Of this, around two-thirds could be energy projects. With a possible volume of around 100MtC per year and an estimated price of around US$1 per tonne carbon, the total investment volume from CDM could thus amount to around US$1,000–2,000 million annually (Jotzo and Michaelowa, 2002).

Will international climate policy continue to have a positive impact on energy policy? Can it evolve into a driving force for change? This depends substantially on the progress made in the climate negotiations over the coming years. Key issues on the agenda will include the adoption of more stringent, while appropriate, targets for the industrialized countries and the inclusion of the developing countries in a way which gives them scope for development while guiding them, at an early stage, towards a sustainable energy policy course.

Country/ Group of countries	Sinks (Article 3.3 - ARD) [Million t carbon]	Sinks (Article 3.4 (Forest management) [Million t carbon]	
		Country reports	Eligible sinks
EU (15)	-1	39	5
Russian Federation	-8	117	33
USA	-7	288	0
Canada, Japan, Australia, New Zealand	3	25	26
Eastern Europe, Switzerland, Liechtenstein, Monaco, Iceland	0	31	5
Ukraine	0	7	1
All Annex B states	*-13*	*507*	*70*

Table 2.7-1
Sink potential of individual (groups of) countries through afforestation (Article 3.3 of the Kyoto Protocol) and forest management (Article 3.4). Negative values denote a carbon source; positive values denote a sink.
Source: Schulze et al., 2002

2.7.2.3
Operational and coordinating activities of the international organizations

UN AGENCIES
At operational level, various UN agencies, including UNEP, UNESCO and UNDP, are involved in practical energy projects, e.g. to promote renewables and increase energy efficiency.

UNEP's activities aim to increase the use of renewables and improve the efficiency of existing energy systems. At political level, UNEP aims to incorporate environmental aspects into energy policy and improve analysis and planning in the energy sector.

With its Energy and Atmosphere Programme, UNDP is pursuing an integrated development strategy which is designed to take account of the many social, economic and ecological aspects of energy policy. The Programme has been tasked with promoting sustainable energy policies and carrying forward the implementation of UNDP's energy programmes.

The World Solar Programme 1996–2005, which is organized under the auspices of UNESCO, aims to promote the adoption and wider utilization of renewable energy sources, particularly in rural areas which currently have no access to electricity, through coordinated efforts at national, regional and international level. UNESCO also funds individual projects, supports developing countries in accessing sources of funding, and advises states on the development of legal frameworks which promote the use of renewables and promote wider distribution of the relevant technologies. Through the GREET (Global Renewable Energy Education and Training) and IREICS (International Renewable Energy Information and Communication System) programmes, UNESCO will focus more strongly on information and training in future (UNESCO, 2001; UN Ad Hoc Inter-Agency Task Force on Energy, 2001).

The other international agencies which play an active role in sustainable energy policy include the UN Regional Economic Commissions, the UN Committee on Energy and Natural Resources for Development, the United Nations Industrial Development Organization (UNIDO), the World Health Organization (WHO), the World Meteorological Organization (WMO), the United Nations Population Fund (UNFPA), the Food and Agriculture Organization (FAO) and the International Atomic Energy Agency (IAEA).

This multitude of United Nations agencies with a global energy policy responsibility requires better communication and coordination. For this reason, the Ad Hoc Inter-Agency Task Force on Energy was set up in 1997, with an additional remit to prepare case studies and an overview of the activities of the various UN agencies in the field of sustainable energy supply.

The United Nations Department of Economic and Social Affairs (UNDESA) coordinates energy policy with other United Nations policy areas. UNDESA also supports the work of the UN Commission on Sustainable Development (CSD) and promotes the implementation of sustainable energy policies in developing countries, e.g. through a technical cooperation programme funded by UNDP, GEF and the World Bank, among others.

The IAEA has a specific role to play in that it deals exclusively with atomic energy. The structure and statute of this autonomous UN organization, which was founded in 1957, are geared towards promoting and monitoring the civilian use of nuclear power. The political lobbying carried out by this well-established organization (annual budget around US$300 million; approx. 2,200 staff) has traditionally aimed to boost the use of nuclear power.

Knowledge transfer is supported, among others, by the Climate Technology Initiative (CTI) which provides information about climate protection technologies, particularly to transition countries in Eastern Europe, newly industrializing countries in Asia, and the developing countries in Africa, and supports capacity-building. The CTI was launched by 23 IEA/OECD countries and the EU Commission at the 1st Conference of the Parties to the UNFCCC in Berlin in 1995.

EUROPEAN UNION
The European Union has no separate Community competence for energy policy. Until now, the internal market rules, competition law, subsidy controls and environmental legislation have provided the legal framework for the EU's policies on energy and the energy sector. However, the EU's energy ministers meet regularly at the EU Energy Council, coordinate national energy policies, and deliberate the Community measures proposed by the Commission. Such measures aim to achieve the energy policy objectives defined by the EU: Competition, security of supply, and protection of the environment. The primary issue for the Energy Council is the liberalization of the electricity and gas markets in the EU with the aim of establishing an internal market in grid-based energies. Other key issues include measures to improve energy efficiency and promote renewables.

Initiatives by the Commission to encourage EU-wide harmonization of energy policies have so far collapsed in the face of opposition by the Member States. In order to avoid competitive disadvantages for their energy industries in a fiercely competitive market, however, the Member States will probably have to fall into line soon.

Energy policy already plays a major role in the EU's external relations: Through its funding programmes, the loans and credits provided by the European Investment Bank and the European Bank for Reconstruction and Development, and political treaties, the EU helps to improve the safety of nuclear installations in Central and Eastern Europe, for example, and supports the development of an environmentally compatible energy industry in the Mediterranean seaboard states.

2.7.3
Financing structures

GLOBAL ENVIRONMENT FACILITY (GEF)
The Global Environment Facility (GEF) is a financial instrument which is designed to address six critical threats to the global environment: Climate change, biodiversity loss, ozone depletion, degradation of international waters, persistent organic pollutants (POPs), and land degradation. The GEF is the designated financial mechanism for the UNFCCC and the Biodiversity Convention. From 1996, projects launched under the Desertification Convention could only be funded by GEF if there were synergies with climate protection and biodiversity conservation, but since 2003, desertification projects are now directly eligible for funding. GEF is jointly implemented by UNDP, UNEP and the World Bank. In 1994, after a three-year pilot phase, the donor governments, mainly from the industrialized countries, initially provided US$2,000 million for the fund; the same amount was made available again for Phase 2 (1998–2002), rising to US$2,920 million for Phase 3 (2003–2006). GEF is the most important funding mechanism for projects aimed at improving energy efficiency and promoting renewables in developing countries. To the end of 2000, a total of nearly US$570 million had flowed into 48 renewable energy projects in 47 states from GEF funds, with a further US$2,500 million in co-financing from other institutions. GEF also supports numerous energy efficiency projects. This funding is targeted primarily towards pilot projects which demonstrate the opportunities associated with the utilization of renewable energies or energy-efficiency technologies. In project implementation, particular emphasis is placed on training and support for local personnel as well as institution-building in order to promote wider use of the new technologies. For this reason, GEF also targets its funding towards the integration of the private sector in project development and implementation (GEF, 2000; UN Ad Hoc Inter-Agency Task Force on Energy, 2001).

KYOTO FUNDS
As the funding available from GEF for the establishment of a sustainable global energy supply is far too low, the 'Kyoto Funds' were launched during the negotiations on the Kyoto Protocol. The Special Climate Change Fund and the Least Developed Countries Fund will finance developing countries' activities relating to climate change in the areas of adaptation, promote North-South technology transfer, and assist developing countries whose economies are highly dependent on income generated from fossil fuels in diversifying their economies.

The Adaptation Fund established under the Kyoto Protocol is intended to support developing countries suffering from the impact of climate change. It will be funded from the share of proceeds from the Clean Development Mechanism. Provided that the CDM functions in practice, it could become an important financing mechanism.

One should be under no illusions, however, about the level of funding provided to these Funds from the public purse or their effectiveness in restructuring energy systems. All the Funds will focus primarily on adaptation measures rather than emissions reduction and technology transfer. At COP6 in Bonn in 2001, the EU and a number of other states pledged to provide a total annual contribution of US$410 million to these Funds by 2005 or provide bilateral support for the purposes stated. For the Adaptation Fund and the Special Climate Change Fund, no negotiations on funding had taken place by the start of 2003. Only for the Least Developed Countries Fund have there been initial indications from the developed countries that it will run to several tens of million euro per year (BMZ, personal communication).

WORLD BANK
Alongside GEF, the World Bank is the most important provider of energy policy financing, especially for developing countries. Since the World Bank's establishment, the energy sector has been a key area for lending, with up to 25 per cent of the World Bank's total lending occurring in this sector for many years. However, energy lending for electricity, fossil energy carriers and mining dropped by almost half from 1998 to 2001, combined with a fall in the number of projects from around 160 to 110. The World Bank cites a number of reason for this, including the replacement of World Bank financing by private investment, the introduction of lending by regional development banks, and some countries' aversion to risky energy sector reform (World Bank, 1993, 2001b).

In response to this development, the World Bank sought to exert greater influence over the economic viability of energy supply companies, especially in the

1980s. Overall, it was unable to halt the further decline in energy lending during this period, although current lending to the energy sector at the start of the 1990s still amounted to around US$40,000 million, i.e. 15 per cent of all lending. The World Bank's financing of the energy sector further decreased during the 1990s, especially after the Asian crisis in 1997, standing at around US$12,000 million in 2001, i.e. approximately 6 per cent of lending (Table 2.7-2).

The World Bank has set itself the following global energy policy goals for the next ten years:

- Increasing the share of households with access to electricity from 65 per cent to 75 per cent.
- Increasing the share of large cities with acceptable air quality from 15 per cent to 30 per cent.
- Increasing the share of developing countries where industrial consumers have a choice of suppliers from 15 per cent to 40 per cent, and increasing the share of developing countries where the power industry stops being a burden on the budget from 34 per cent to 50 per cent (World Bank, 2001b).

Although the more detailed definition of these targets by the World Bank through specific indicators (Table 2.7-3) is welcome, it remains unclear, especially in view of the current balance sheet, precisely which instruments will be deployed to achieve these objectives. The decline in demand for energy lending from the World Bank indicates that there are still substantial political concerns about the restructuring or, indeed, the privatization of the energy sector in most developing countries. The success achieved by the World Bank will therefore continue to depend on the energy policies pursued by the developing countries.

Table 2.7-2
Lendings by the International Bank for Reconstruction and Development (IBRD) and the International Development Association (IDA), which are part of the World Bank Group, for power plants and oil/gas development during the trading years 1990–2001. Share of total IBRD and IDA lending: power plants 9.2 per cent; oil/gas development 2.1 per cent. Source: World Bank, 2002b

Region	Power plants [US$ 1,000 million]	Oil-/gas development [US$ 1,000 million]
Sub-Saharan Africa	1.92	0.68
East Asia/Pacific	10.29	0.93
Europe/Central Asia	3.69	1.94
Latin America/Carribean	2.01	0.55
Middle East	0.71	0.28
Southern Asia	5.67	1.04
Total for sector	*24.29*	*5.42*

DEVELOPMENT COOPERATION AND PRIVATE DIRECT INVESTMENT

The target adopted by the international community to spend 0.7 per cent of GDP on official development assistance (ODA) has not been reached by Germany (0.27 per cent), the EU (<0.3 per cent), or the USA (0.1 per cent). In 2000, just 7.9 per cent of total ODA, i.e. US$3,760 million, flowed into infrastructure projects in the energy sector (OECD, 2001). In contrast to the decrease in ODA, private investors stepped up their commitment to the energy sector in the 1990s. More than 600 electricity projects with private financing and a total investment value of US$160,000 million were implemented in more than 70 developing and newly industrializing countries in the 1990s (Fig. 2.7-2). Private investment in energy projects peaked (temporarily) in 1997, but then slumped as a result of the Asian crisis and negative developments in Latin America.

EXPORT CREDIT AND INVESTMENT INSURANCE AGENCIES

The OECD countries' Export Credit and Investment Insurance Agencies (ECAs) play a key role in trade relations between industrialized and developing countries, especially in investment. The agenda of these institutions, which are generally publicly funded, is to boost domestic industry in foreign markets. On payment of a relatively low fee, ECAs provide insurance for exports to high-risk countries and foreign direct investment through state-guaranteed loans, guarantees, financing and credits. The ECAs are particularly active in the energy sector of developing and transition countries. Here, fossil-fueled power generation and major dam projects are the main focus of their work. Electricity generation from solar and wind energy and biomass has rarely been promoted to date. Nonetheless, export promotion can also have a positive effect on the utilization of fossil energy carriers, for example if old production plants are replaced with new and more efficient technologies or if energy efficiency is increased.

Foreign investment in the construction of power plants in the developing and transition countries totalled US$115,600 million during 1996 and 2001. Of this figure, US$50,000 million was supported by export credit agencies. Of US$97,800 million invested in oil and gas development, the ECAs supported a total of US$60,600 million.

On the recipient side, a strong geographical concentration can be observed. The leading destinations for the ECAs' financing – and, indeed, for private investment – are a handful of countries which will play a key role in the development of global greenhouse gas emissions in future: China, Indonesia, India, Mexico, Brazil, the Philippines and Turkey.

Table 2.7-3
Change in the World Bank policies on the energy sector (selection). The International Finance Corporation (IFC) is part of the World Bank Group and the largest source of lending worldwide for projects in the developing countries' private sector. Source: World Bank, 2001b

Former priorities	More recent priorities	New priorities
CHANGES IN THE INTERNATIONAL FINANCE CORPORATION'S PRIORITIES		
Financing major power plants which sell to a state monopoly at guaranteed prices	Shift towards reforms in individual sectors, environmental protection and access to energy services	Supporting reforms in the energy sector and promoting competition
		Improving environmental protection in energy supply
		Supplying people with no access to energy services
CHANGES IN THE PRIORITIES IN OIL AND GAS DEVELOPMENT		
1970s: Supporting public investment		Greater integration of activities in:
		• Environmental protection (clean fuels, gas)
1980s: Reforms in individual sectors, liberalization, improving parameters for and actively promoting private investment		• Social embedding (best practice, partnerships)
		• Further reforms (privatization, more competition)
1990s: Reforms in the transition countries' oil and gas sector		• Governance (tax administration, transparency)
		• Financing of selected private sectors
CHANGES IN THE PRIORITIES IN MINING DEVELOPMENT		
To 1990: Investment assistance and technical support for lending to develop the mining industry	1991–2000: Promoting better parameters for private investment in mining	Sustainable mining
	Privatization, restructuring and closure of mines (Poland, Romania, Russia, Ukraine)	Regional/local development through private investment in the mining industry

Table 2.7-4 shows the financing provided by the ECAs from the USA, Japan and Germany as a percentage of investment in the developing and transition countries' energy sector. In terms of their commitment, the German agencies lag well behind those of the USA and Japan.

The relevance of trade promotion measures to climate protection has been largely ignored until now. One exception is a report released by the Institute for Policy Studies (IPS), which reveals that from 1992 to 1998, US credit agencies underwrote financing for fossil energy projects around the world which, over

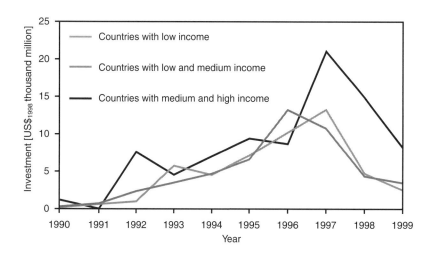

Figure 2.7-2
Total investment in energy projects with private financing in developing and newly industrializing countries. Countries with low income: per capita GDP 1998: < US$760. Countries with low and medium income: per capita GDP 1998: US$761–3,030. Countries with medium and high income: per capita GDP 1998: US$3,031–9,360. Source: Izaguirre, 2000

Table 2.7-4
Financing provided by the ECAs from the USA, Japan and Germany for the developing and transition countries' energy sector (power plant construction, oil and gas development: 1996–2001). *KfW* Kreditanstalt für Wiederaufbau, *JBIC* Japanese Bank for International Cooperation, *NEXI* Nippon Export and Investment Insurance, *Exim-Bank* Export-Import Bank, *OPIC* Overseas Private Investment Corporation.
Source: Maurer, 2002

Country	Export credit agency	Power plant construction		Oil and gas development	
		Guarantees/ insurance [US$1,000 million]	Co-financing [US$1,000 million]	Guarantees/ insurance [US$1,000 million]	Co-financing [US$1,000 million]
Germany	Hermes	1.52	-	0.55	-
	KfW	-	2.20	-	1.00
Japan	JBIC	1.36	2.50	0.34	3.83
	NEXI	1.20	0.10	0.20	0.30
USA	Exim-Bank	3.72	0.50	3.36	0.55
	OPIC	1.40	1.20	0.44	0.20
Total		*9.20*	*6.50*	*4.89*	*5.88*

their lifetimes, will release an additional 29,300 million tonnes of CO_2 into the atmosphere (IPS, 1999).

2.7.4
Fragmented approaches to global energy policy

The United Nations has a key role to play in the expansion of the energy supply in developing countries. However, energy policy is currently not a priority at UN level. The brief overview of the activities of the various UN units and agencies (Section 2.7.2) clearly shows that the UN has no identifiable or coherent strategy for the energy sector. The attempt to coordinate the various programmes and projects through a Task Force, comprising representatives of the various units, is unlikely to be successful, firstly because the various agencies, programmes and actors generally pursue their own particular interests, and secondly because the Task Force's mandate is inadequate. Mechanisms which could help convert the findings obtained within the CSD framework into political criteria for the UN's work are still in their infancy. The absence of a clear policy programme setting out the cornerstones of a UN strategy on the expansion of sustainable energy in developing countries also contributes to the fragmentation of the various UN units' energy-related activities. The failure to achieve effective policy coordination between the UN's various individual strategies, donor countries and key funding agencies further exacerbates the situation.

The overview of the existing legal and institutional bases of international energy policy illustrates the current priorities and shortcomings in global policy-making (Fig. 2.7-1): Whereas international trade law is increasingly focussing on the progressive liberalization of national energy markets and facilitating trade in energy goods and services, the environmental and development dimensions of a sustainable energy policy do not match up to what is required. The UNFCCC and the Kyoto Protocol are to be welcomed as a successful starting point for the reduction of greenhouse gas emissions. For the second commitment period from 2012, the aim must be to achieve a clear increase in reduction commitments and to progressively include, in an appropriate manner, some of the developing countries in the process as well. In particular, the institutional fragmentation, lack of an overarching political strategy, and inadequate financial resources impede the more effective use of the available instruments.

2.8
Interim summary: The starting point for global energy policy

- Energy is a key prerequisite for social development and poverty reduction. Population growth and economic and technological developments have resulted in a substantial rise in worldwide energy consumption in recent centuries and decades, which has also increased environmental pollution.
- The availability of high-quality forms of energy is unevenly distributed around the globe. Around one-third of the world's population, primarily in the developing countries, has no access to electricity. These people, who are often exposed to major health risks as a result of their reliance on fuel-

wood or dung for cooking and heating, face major obstacles to their development.

- Energy systems in the industrialized countries are geared to three main characteristics: Security of supply, affordability, and environmental compatibility. In the past, fossil and nuclear energy carriers were often subsidized by the state, and the electricity supply was subject to comprehensive state control. More recently, there has been a move away from subsidies on fossil and nuclear energy carriers, a liberalization of the grid-based energy markets, and greater government support for renewable energies.

- Transition countries in eastern and south-eastern Europe are noted for their heavy reliance on fossil energy carriers. Despite their major potential, renewable energies – such as biomass, wind and solar power – play only a minor role. The weakness of the economies during the last two decades has resulted in a drop in energy production. Urgently needed investment to maintain and modernize the energy sector was not carried out, resulting in very low efficiency of production plants. Major causes of this inefficiency are the subsidies on energy use and the close interdependencies between political and economic interests in the energy sector.

- Energy systems in developing and newly industrializing countries have no uniform pattern. There are substantial disparities between continents, countries, regions, urban and rural areas, and landscape types. In general, demand for commercial energy rises along with urbanization, and an increase in energy demand can generally be observed as per capita income rises. However, the increase in energy use is restricted by the limited availability of financial resources and the sluggish expansion of the energy supply. The energy policies adopted by most newly industrializing and developing countries do not make provision for efficiency strategies or investment in renewables. Instead, they rely on fossil energy carriers to cover most of the demand.

- Global energy policy is influenced substantially by increasing economic and technological interdependencies. This facilitates the transfer of energy efficiency standards and technologies, for example. On the other hand, the intensive transportation of goods and persons around the globe and the transfer of Western lifestyles to less industrialized regions of the world lead to a rise in global demand for energy.

- Within the UN agencies and the World Bank, a globally coordinated energy policy is still in its infancy. Institutional fragmentation and inadequate financing mechanisms make it more difficult to achieve an energy reform aimed at sustainability.

- Many developing countries are still in the process of establishing a viable commercial energy system. As they can draw on a more extensive portfolio of technology options than the industrialized countries at the same stage of development, there is a chance for the parameters for a sustainable energy system to be incorporated into this new structure (Goldemberg, 1996; Murphy, 2001). For example, investment in renewable energy systems enables distributed electricity supply to be brought on stream quickly and cost-effectively. However, this requires the rigorous rechannelling of private direct investment, government loans and credit guarantees in order to overcome, in developing countries, too, the path dependency on fossil fuels.

Technologies and their sustainable potential

3.1
Introduction

The following discussion of a transformation of the global energy system identifies technology options and presents promising solutions, some of which are little known. A great array of technologies is available at all levels, from primary energy generation to energy services. Their deployment and mix is evaluated here from a sustainability perspective. This analysis starts with an examination of the available energy sources and carriers. The realizable and sustainable potential (see definitions in Box 3.1-1) for fossil, nuclear and renewable energy is presented, conversion technologies are described, and environmental and social impacts evaluated. Moreover, the current and future costs of the technologies are discussed. While an increasing shortage of supplies will make conventional forms of energy more expensive in the long term, an opposite trend of continuous cost reductions is foreseeable for renewables. Any evaluation should therefore not be limited to the present. The analysis delivers the boundary conditions defining the realm within which the Council's exemplary path for a sustainable transformation of energy systems can be elaborated (cf. Chapter 4).

In addition to utilizing available energy sources sustainably, it is essential to deploy technologies with maximum efficiency for all conversion processes, from primary to useful energy and at the end user. To deploy fluctuating renewable energy sources, technologies for compensating such fluctuations or for energy storage are key. The discussion of this issue is organized around three aspects: first, distributed combined heat and power, second, energy distribution/transport and storage and, third, demand-side energy efficiency. Further sections of Chapter 3 examine the options for decarbonizing the energy system through secure and long-term carbon storage (sequestration), and the prospects for sustainable energy solutions in the transport sector.

3.2
Energy carriers

3.2.1
Fossil fuels

3.2.1.1
Potential

Today, petroleum, coal and natural gas dominate the supply of heat, electrical energy and fuels in energy systems worldwide. These fossil energy carriers cover 90 per cent of primary energy consumption worldwide (petroleum 40 per cent, coal 27 per cent, natural gas 23 per cent; BGR, 1998). In terms of their economic and technological recoverability, a distinction is made between reserves, resources and additional occurrences (BGR, 1998; Nakicenovic et al., 1998):

- *Reserves* are defined as known occurrences that have been recorded very accurately and can today be recovered technologically and economically at any time.
- *Resources* are defined as occurrences that have been identified or are believed to exist, and which are not yet recoverable with today's technology and under current economic conditions, but which are seen as potentially recoverable.
- *Additional occurrences* are those that cannot be classified as reserves or as resources. The existence of such occurrences is suggested by geological studies, but their size and the technological and economic conditions for their recoverability are still very uncertain.

The sum of reserves and resources is known as total resource. Resources can become reserves if, for example, fuel prices rise or recovery costs fall due to technological progress. Similarly, through technological advances at least some of the additional occurrences can move into the resources category in the long term, thereby increasing the total resource. The distinction between resources and additional occur-

Box 3.1-1

Types of potential

The following terms are usually used in the discussion of the potential of different energy carriers: theoretical potential, technological potential and economic potential. The Council felt that it was necessary to introduce the additional terms of conversion potential and sustainable potential. This report is based on the following definitions:

THEORETICAL POTENTIAL
The theoretical potential identifies the physical upper limit of the energy available from a certain source. For solar energy, for example, this would be the total solar radiation falling on a particular surface. This potential does therefore not take account of any restrictions on utilization, nor is the efficiency of the conversion technologies considered.

CONVERSION POTENTIAL
The conversion potential is defined specifically for each technology and is derived from the theoretical potential and the annual efficiency of the respective conversion technology. The conversion potential is therefore not a strictly defined value, since the efficiency of a particular technology depends on technological progress.

TECHNOLOGICAL POTENTIAL
The technological potential is derived from the conversion potential, taking account of additional restrictions regarding the area that is realistically available for energy generation. The criteria for the selection of areas are not dealt with consistently in the literature. Technological, structural and ecological restrictions, as well as legislative requirements, are accounted for to a greater or lesser degree. Like the conversion potential, the technological potential of the different energy sources is therefore not a strictly defined value, but depends on numerous boundary conditions and assumptions.

ECONOMIC POTENTIAL
This potential identifies the proportion of the technological potential that can be utilized economically, based on economic boundary conditions (at a certain time). For biomass, for example, those quantities are included that can be exploited economically in competition with other products and land uses. The economic boundary conditions can be influenced significantly, particularly through political measures.

SUSTAINABLE POTENTIAL
This potential of an energy source covers all aspects of sustainability, which usually requires careful consideration and evaluation of different ecological and socio-economic aspects. The differentiation of the sustainable potential is blurred, since ecological aspects may already have been considered for the technological or economic potential, depending on the author. The Council proposes an exemplary transformation path for the global energy system (Chapter 4) based on the sustainable potential, the activation of which is regarded as realistic within certain timescales.

Section 3.2 examines the potential of the energy carriers and energy sources from a global point of view. In particular, global maps of the conversion potential for the following technologies were calculated for this report:
- photovoltaic modules (without optical concentration),
- solar thermal power plants,
- solar collectors for heat generation,
- wind energy converters.

Based on these maps, the regional distribution of the energy that can be recovered and utilized in principle can be estimated. The calculations were carried out based on a global grid with a resolution of 0.5 ° longitude and latitude. The resulting potential is specified as average power density per surface area or per tilted module/converter area, so that the unit of measurement is always 'output per area'.

rences is much less clear than that between reserves and resources (UNDP et al., 2000).

Between 1860 and 1998, the cumulative global energy consumption was 13,500EJ (UNDP et al., 2000). The energy stored in today's reserves is of at least the same order of magnitude (Table 3.2-1). If the deposits that are classified as resources are also taken into account, the stored quantity of energy increases 20-fold. Optimistic figures are usually substantiated based on the assumption of large non-conventional oil and gas occurrences. The difference between conventional and non-conventional occurrences is that the latter occur in significantly lower concentrations, require unusual or very high technological effort for recovery and sophisticated conversion techniques, or have significant environmental implications. Examples are oil shale, tar sands and heavy crude, as well as gas in coal seams, tight sand formations and gas hydrates (Nakicenovic et al., 1998). With high oil prices, some of these non-conventional deposits (e.g. tar sands) can already be

recovered economically today and can therefore already be classified as reserves (Table 3.2-1).

Coal deposits make up the lion's share of the fossil reserves and would suffice to cover the expected energy consumption far beyond 2100. With current technology, i.e. without the application of coal hydrogenation, coal is insignificant as an energy source for the rapidly growing transport sector; growth predictions for coal are therefore lower than those for oil. According to most scenarios (Chapter 4), the proportion of coal in the fossil energy mix is likely to fall by 2100.

At present rates of consumption, the gas reserves would be able to meet demand for another 60 years or so, but if the resources are also considered, the range extends to 170–200 years (IEA, 2002c). However, since gas consumption shows the strongest growth rates of all fossil energy sources, with a doubling of consumption expected between 2000 and 2030 (IEA, 2002c), gas reserves are likely to run out faster. A potentially very large source for methane is

Table 3.2-1

Reserves, resources and additional occurrences of fossil energy carriers according to different authors. *c* conventional (petroleum with a certain density, free natural gas, petroleum gas, *nc* non-conventional (heavy fuel oil, very heavy oils, tar sands and oil shale, gas in coal seams, aquifer gas, natural gas in tight formations, gas hydrates). The presence of additional occurrences is assumed based on geological conditions, but their potential for economic recovery is currently very uncertain. In comparison: In 1998, the global primary energy demand was 402EJ (UNDP et al., 2000).
Sources: see Table

Energy carrier	Brown, 2002	IEA, 2002c	IPCC, 2001a		Nakicenovic et al., 1998		UNDP et al., 2000		BGR, 1998	
			[EJ]							
GAS										
Reserves	5,600	6,200	c	5,400	c	5,900	c	5,500	c	5,300
			nc	8,000	nc	8,000	nc	9,400	nc	100
Resources	9,400	11,100	c	11,700	c	11,700	c	11,100	c	7,800
			nc	10,800	nc	10,800	nc	23,800	nc[a]	111,900
Additional occurrences				796,000		799,700		930,000		
OIL										
Reserves	5,800	5,700	c	5,900	c	6,300	c	6,000	c	6,700
			nc	6,600	nc	8,100	nc	5,100	nc	5,900
Resources	10,200	13,400	c	7,500	c	6,100	c	6,100	c	3,300
			nc	15,500	nc	13,900	nc	15,200	nc	25,200
Additional occurrences				61,000		79,500		45,000		
COAL										
Reserves	23,600	22,500		42,000		25,400		20,700		16,300
Resources	26,000	165,000		100,000		117,000		179,000		179,000
Additonal occurrences				121,000		125,600				
Total resource (reserves + resources)	*180,600*	*223,900*		*212,200*		*213,200*		*281,900*		*361,500*
Total occurrence				*1,204,200*		*1,218,000*		*1,256,000*		

[a] including gas hydrates

represented by the gas hydrate deposits in the sea bed and in permafrost soils, which are usually classified as additional occurrences. If they could be exploited, the fossil resource base would multiply. However, currently the usability of the methane hydrate deposits is viewed rather cautiously. There are neither secure scientific statements about the size of the deposits nor attractive technical proposals for their recovery (UNDP et al., 2000), and further basic research is required in this area.

3.2.1.2
Technology/Conversion

Currently, large central power stations dominate the power supply in the electricity grids of industrialized countries. For example, in Germany during 1999, such power stations (>100MW) provided around 80 per cent of the total installed electrical capacity of approximately 119,000MW, of which coal-fired power stations provided approximately 52,000MW and gas-fired power stations around 16,000MW. Conventional steam plants are currently the predominant technology used for converting coal, while combined cycle plants tend to be used for natural gas (Hassmann, personal communication, 2000). For base and intermediate loads, both power plant types are operated without heat extraction, which means there is significant potential through increased efficiency. In this operating mode, CO_2 emissions are proportional to the electrical efficiency. On average, the efficiency of German fossil fired power stations is only approximately 39 per cent, while electrical efficiencies of more than 45 per cent for coal-fired plants and nearly 60 per cent for combined cycle gas turbine (CCGT) plants have been reported, highlighting significant potential for reducing CO_2 emissions. In addition to 'classic' power stations, further types are developed with a view to utilizing coal in gas tur-

Table 3.2-2
Further development of modern fossil power plants.
Source: Hassmann, personal communication, 2002

Fuel	Plant	Capacity	Net efficiency	CO$_2$ emissions	Estimated costs	R&D status comments	Timescale
		[MW]	[%]	[g CO$_2$/kWh]	[€/kW]		
Coal	Steam power plant with high maximum steam conditions 350 bar/700 °C	700–900	>50	168	>700	Proposed for around 2010, material development	Around 2010
	Combined cycle plant (gas and steam turbine) with circulating fluidized bed (2nd generation)	700–800	54–55	150	650–700	Hot gas cleaning, robust gas turbine	After 2010
	Combined cycle plant with pressurized coal gasification (IGCC)	400–500	>45	184	approx. 1,000	Low availability	Pilot plants have already been realized
	Combined cycle plant with pressurized pulverized coal firing	400	54–55	150	>750	Hot gas cleaning, robust gas turbine	After 2015
Natural gas	Advanced combined cycle gas and steam turbine power plants (CCGT)	400	>60	80	approx. 500	Further development required for: compressor, gas turbine (little), NO$_X$ burner, material development for catalytic combustion, hot parts (combustion chamber lining, transition elements, blades, blade cooling), compressor	From 2005

bines, with associated increases in efficiency and reductions in emissions (Table 3.2-2).

3.2.1.3
Environmental and social impacts

ANTHROPOGENIC CLIMATE CHANGE
In 1990, global CO$_2$ emissions from the utilization of petroleum, coal and natural gas were approximately 6.0GtC (IPCC, 2001a), with the combustion of petroleum having the greatest share at 44 per cent. In addition, there were 1.8GtC$_{eq}$ of methane and 2.5GtC$_{eq}$ of nitrous oxide emissions, mainly from agriculture (IPCC, 2000a). During the last 100 years, the average air temperature near the surface and in the lower atmosphere increased by 0.6 ± 0.2°C. Without climate policy measures, a global increase in temperature between 1.4 and 5.8°C is expected by 2100. This increase leads to increased atmospheric humidity and also often to higher precipitation, changes in atmospheric and oceanic circulation, melting of sea ice and snow, and a rise in sea levels (IPCC, 2001a).

The predicted shift in climatic regions and more frequent extreme weather events such as flooding and droughts are expected to have negative ecological and social impacts (IPCC, 2001b). In sensitive ecosystems, the damage can already be identified (Section 4.3.1.2). The risk of irreversible damage of ecosystems increases with increasing temperature and increasing rate of warming (IPCC, 2001b).

Weather-related damage has been on the increase for 25 years, with large annual fluctuations. The mounting economic costs arising in the wake of flooding and droughts can be measured in many regions (IPCC, 2001b; Münchner Rückversicherung, 2001; Swiss Re, 2001). According to estimates by the International Red Cross, during the past 26 years such events have affected approximately 2,500 million people (IFRC-RCS, 2001). Most of the damage occurred during the 1990s, which were the warmest decade since weather data started to be recorded (Milly et al., 2002; Münchner Rückversicherung, 2001). In 2000 and 2001, a significantly higher number of people were affected by droughts globally (176 and 86 million compared with 20 million in 1998), fewer were affected by flooding (62 and 34 million compared with 290 million in 1998), and a similar number were affected by severe storms (15 and 29 compared with 26 million in 1998) (CRED, 2003). It is likely that there is a relationship between the increasing incidence of severe weather and observed climate change.

Due to the geographic position and inadequate adaptability of developing countries, climate change is threatening to cause particularly severe damage (IPCC, 2001b). In some cases, whole states could be obliterated by rising sea levels. Up to now, these countries have hardly developed any prevention and emergency response structures.

AIR POLLUTION
Emissions of benzene, soot and other particles from industrial combustion processes, power stations and traffic have numerous ecotoxic effects. Furthermore, nitrogen oxides, hydrocarbons and carbon monoxide change the oxidation capacity of the atmosphere, which can not only lead to local production of low-level ozone, but can change the self-purifying capacity of the atmosphere overall. Combustion processes cause the emission of large quantities of nitrogen and sulphur oxides, which together with ammonia from intensive animal husbandry are a key factor in the transformation of bio-geochemical cycles by humans. These precursor substances for acids are chemically converted in the atmosphere and enter the soil through 'acidic rain'. While in industrialized countries the problem has been successfully reduced through desulphurization and denitrification systems, hardly any such measures have so far been implemented in developing and newly industrializing countries.

POLLUTION CAUSED BY THE RECOVERY AND TRANSPORT OF FOSSIL FUELS
The production of fossil energy carriers affects the soil: On the one hand, large volumes of soil are moved, particularly in open-cast coal and ore mining, which changes the morphology of the soil and leads to subsidence effects in the land surface. On the other hand, there are significant effects on hydrological processes such as drainage, the sediment load of rivers and the ground water table, with possible effects on soils and ecosystems. In nearly all industrialized countries, the intermediate storage of the soil during open-cast mining is now a legal requirement.

Oil leakages cause particularly severe ecological damage during recovery and transport. In western Siberia, oil leakages are estimated to have amounted to 2.8 million tonnes of petroleum per annum between 1980 and 1990, causing the destruction of 55,000km² of the permafrost ecosystem (Stüwe,

1993). Between 1967 and 2002, 22 large tanker accidents (loss of oil >10,000 t) were recorded, in which more than 2.4 million tonnes of petroleum were spilled into the sea (Greenpeace, 2002), causing significant ecological damage. In addition, approximately 520,000 tonnes of oil are drained into the sea each year through the cleaning of tankers and illegal pumping out of machine oil. This is twice the amount entering the sea naturally through the sea bed. A further 57,000 tonnes per year enter the sea through offshore oil installations (Feldmann and Gradwohl, 1996).

EFFECTS ON HUMAN HEALTH
The utilization of fossil fuels and wood for energy generation is one of the main sources of air pollution, for example through NO_X, SO_2, CO, polyaromatic hydrocarbons or formaldehyde. These substances have health effects for a large number of people. Dusts in the form of soil particles, mineral ashes or other small particles also have a detrimental effect on health. Further, more than 1,100 million people are subjected to concentrations of aerosol particles above the levels specified in WHO guidelines (UNDP et al., 2000), with exposure being particularly serious for the population of large cities, mainly in the rapidly growing mega-cities of Asia, Africa and Latin America.

Air pollution can trigger a series of acute and chronic diseases such as respiratory diseases (asthma, lung irritation and lung cancer) and diseases of the cardiovascular system. According to the World Health Organization (WHO, 2000, 2002b)
• there is a direct relationship between mortality rates and the daily exposure to aerosol particles. Every year, 0.8 million people die as a consequence of urban air pollution;
• the life expectancy within a population that is subjected to high concentrations of particles in the air is reduced significantly;
• in strongly polluted regions, air pollution is thought to be responsible for 30–40 per cent of asthma cases and 20–30 per cent of all respiratory diseases.

3.2.1.4
Evaluation

The consequences of the recovery, transportation and, above all, utilization of fossil energy carriers today affect every human being on the planet. Many become ill directly as a result of the inhalation of pollutants in the air, and everyone is affected by climate change e.g. through increasingly frequent extreme weather events. Ecosystems also suffer severe damage, from oil disasters after tanker accidents to the acidification of inland waters.

The most serious consequences arise from the climate change that would result from the undiminished and continued utilization of fossil fuels (Section 4.3.1.2). The existing reserves of fossil energy carriers therefore cannot be utilized fully, since carbon dioxide emissions will have to be limited due to their effect on climate change, and because the capacity to store carbon dioxide is very limited for both technological and economic reasons (Section 3.6). In order to achieve and maintain stabilization of the atmospheric carbon dioxide concentration according to Article 2 of the UN Framework Convention on Climate Change, anthropogenic emissions need to be reduced in the long term, i.e. over several centuries, to such low levels that persistent natural sinks are able to absorb them. These levels are estimated to be very small (less than 0.2GtC per year, compared with 6.3GtC per year of emissions from fossil fuels and cement, averaged over the 1990s; IPCC, 2001a). The Council therefore concludes that in the long term we will have to move away from fossil energy sources. However, it is unrealistic to try and achieve such a transformation by 2100, and for the long-term stabilization of the CO_2 concentration a longer timescale is in fact adequate.

In the short term, better environmental standards and environmentally more benign recovery, transportation and utilization technologies are important elements of environmental protection. What is required is a timely switch to less damaging bridging technologies (e.g. the replacement of coal and petroleum through natural gas), investment in increased efficiency during energy conversion and utilization, and in the long term the replacement of fossil energy carriers by renewables (Chapter 4).

3.2.2
Nuclear energy

3.2.2.1
Potential

NUCLEAR FISSION
Energy is released when heavy atomic nuclei are split. Some of the heavy chemical elements offer the opportunity to trigger a controlled chain reaction that enables the release of very large amounts of energy from small quantities of fissile material. In nuclear power plants, the energy thus generated is initially converted to heat and then into electricity. Modern nuclear technology is based on uranium as the fissile material, with the radioactive isotope U-

235 being used as fuel. Natural uranium contains only about 0.7 per cent of U-235. In nuclear reactors, the uranium is bombarded with neutrons. This leads to the formation of plutonium, which is also fissile. The binding energy of the atomic nuclei that is released in a common nuclear reactor therefore originates from the fission of uranium and plutonium. For a comprehensive estimation of the potential offered by nuclear energy, thorium should also be considered in addition to uranium and plutonium, since the spontaneous fission of thorium can also trigger a chain reaction, although this would require a different type of reactor from those currently in use.

Currently, around 2.5 PWh of electrical energy are generated annually worldwide in 440 nuclear power plants (mainly light water reactors, LWR) with an installed electrical capacity of 354GW at an average utilization rate (based on continuous operation at rated capacity) of approximately 80 per cent, corresponding to 17 per cent of global electricity generation. For generating 1TWh of electricity, approximately 22t of (natural) uranium is required (UNDP et al., 2000), resulting in an annual demand of around 55,000t of natural uranium. In its present form, the utilization of nuclear fission energy is therefore limited by the Earth's natural uranium deposits. Worldwide, 3.2 million tonnes of uranium reserves (price per kilogram below US$130; UNDP et al., 2000) have been identified, which would last 60 years at current consumption rates. If the presumed resources are also considered, global deposits increase to approximately 20 million tonnes or approximately 360 years at current consumption. This would correspond to an electricity production of approximately 3,200EJ. These figures could be increased significantly, if one were to succeed in utilizing the uranium that is dissolved in sea water for energy generation (uranium concentration 3mg per tonne, approx. 4.5 million tonnes in total). However, the large-scale applicability of associated extraction techniques has not yet been proven. Without considering unquantified deposits in China and in the CIS, thorium reserves are estimated to be around 4.5 million tonnes.

Plutonium is practically non-existent naturally, but is generated in nuclear reactors using uranium. After reprocessing into mixed oxide (MOX) fuel rods it can be reused as reactor fuel, thereby substituting around one-third of the natural uranium that would otherwise have to be used. Furthermore, plutonium from nuclear weapons can also be returned to the civilian fuel cycle, which could also extend the available resource.

The figures discussed above could be increased significantly if breeder technologies were used. However, to date these have not been mastered properly anywhere in the world. The fundamental process of breeder technology is the formation of fissile plutonium from the stable uranium isotope U-238. In this way, 50–100 times more energy can be generated from 1kg of natural uranium. Apart from unresolved technological problems, the particular risk of proliferation associated with the generation of such large quantities of plutonium should be stressed in this context (Section 3.2.2.3).

NUCLEAR FUSION

The fusion of light atomic nuclei releases energy. While the 'fusion reactor' of the Sun uses ordinary hydrogen, in associated technological processes on Earth the hydrogen isotopes of deuterium and tritium are merged to form helium. For a hypothetical future utilization of nuclear fusion to meet the total current global electricity demand of 15 PWh per year, approximately 600t of deuterium and 900t tritium as fusion fuel and around 2,000t of lithium for breeding of tritium would be required. If the current electricity demand was met by nuclear fusion, the deuterium reserves in the sea water (33g per tonne) would last several million million years, the lithium reserves in the Earth's crust (170g per cubic metre of rock) would last several thousand years or several million years if the lithium dissolved in sea water could be utilized. The theoretical potential of nuclear fusion is therefore almost unlimited. However, since appropriate power plants will not be available before the second half of the 21st century at the earliest – if at all – and furthermore are likely to create a serious risk potential of their own (see below), the Council's view is that it is currently not justifiable to base future energy strategies on nuclear fusion, even in part.

3.2.2.2
Technology/Conversion

NUCLEAR FISSION

Of the nuclear power plant capacity installed worldwide, 88 per cent is made up of light water reactors, with three different types of reactor being particularly noteworthy: pressurized water reactor (PWR), boiling water reactor (BWR) and the Russian graphite-moderated boiling water pressure-tube reactor (RBMK) (Table 3.2-3). These types today achieve electrical efficiencies of 30–35 per cent (heat to electricity). The following improvements in cost-effectiveness and safety have been proposed, amongst others:

- For water-cooled reactors, the main issue is the introduction of so-called passive safety systems.
- In gas-cooled reactors, the safety features of ceramics-coated fuel particles could be utilized, and higher efficiencies and process heat utilization would be realized at significantly higher operating temperatures.

Until a few years ago, some countries pursued the development of breeder reactors, although these efforts have now largely been abandoned for safety and cost reasons.

NUCLEAR FUSION

Triggering of the energy-releasing fusion reactions between the two hydrogen isotopes deuterium and tritium requires the temperature and density of the fuel to exceed certain values (Hamacher and Bradshaw, 2001). Worldwide, two concepts are being pursued to achieve this: magnetic inclusion of the fuel or inertia fusion. In the first case, strong magnetic fields enclose the fuel in the form of hot plasma and keep it away from the walls. In inertia fusion, small fuel pellets are made to implode through bombardment with particles or through electromagnetic waves.

In the EU, the most advanced project is the JET joint experiment. As early as 1997, the project succeeded in generating significant fusion power of

Table 3.2-3
Current and possible further development of nuclear fission technologies. 'Fast' in this context refers to high-energy 'fast' neutrons. 'Sub-critical' reactors require an external neutron supply for maintaining a chain reaction. *LWR* light water reactor. *HTR* high temperature reactor. Source: Kröger, personal communication, 2002

	2000	2020	2050
Main technology	LWR	LWR, HTR	Also fast critical and sub-critical plants
Efficiency [%]	30–35	40–45	60 (with nuclear combined cycle)
High-level and long-lived medium-level radioactive waste [mg/kWh]	9–11	2.4	0.5 (with comprehensive separation and transmutation of actinides)
Production costs [€-Cent/kWh]	3–5	< 4	no information available
Capacity range [MW$_e$]	1,000–1,500	150–1,500	(150–1,500)

12MW over approximately 1 second, with 65 per cent of the power required for heating the plasma being recovered through fusion (Keilhacker et al., 1999). The aim of the next major step of global fusion research (IAEA, 1998, 2001), the ITER international experimental reactor, is to demonstrate that significantly more fusion power can be generated than is required for heating the plasma. Once the ITER experiment has been evaluated, construction of a first prototype fusion power plant could commence in approximately 25 years, with commercial power plants expected to be operational in approximately 50 years (Bosch and Bradshaw, 2001). According to current studies, the electrical output of such fusion power plants is expected to be around 1–2GW. The efficiency of power generation in water-cooled power plants is likely to be around 33 per cent, in helium-cooled plants between 38 and 44 per cent. Since no pilot plants are in existence as yet, any cost estimates for fusion technology are inherently very uncertain.

3.2.2.3
Environmental and social impacts

SAFETY AND SOCIO-POLITICAL ACCEPTANCE
In its 1998 annual report, the Council classified nuclear technology between the categories of 'normal' and 'not acceptable' in terms of global environmental risk (WBGU, 2000). With over 30 new nuclear power plants being commissioned every year, the number of new plants peaked in 1984 and 1985. Even prior to the accident at Chernobyl in 1986, several countries regarded nuclear power to be unacceptable for fundamental reasons. In Austria, a referendum in 1978 prevented the country from getting involved in the technology. After the accident at Chernobyl, people became increasingly sceptical vis-à-vis nuclear power. As a result, the number of reactors coming on stream every year fell continuously, and some countries decided to abandon nuclear energy technology altogether (e.g. Germany, Belgium, Sweden).

COST-EFFECTIVENESS
In liberalized markets, private investors determine the generation side of the electricity sector. For them, nuclear energy is becoming increasingly unattractive for several reasons:
• The ratio of capital and operating costs is poorer for nuclear power than for other conventional energy sources (delayed return). Analyses show that, apart from few exceptions, in OECD countries electricity from nuclear power plants is more expensive than from coal or gas-fired power

plants, due to the high capital costs (IEA, 1998; COM, 2000).
• The high absolute investment costs require a large number of contract parties, leading to complex investment and management structures. The safety regulations require long approval timescales for the industry.
• If the operators of nuclear power plants had to take out insurance against all possible risks similar to what is required from the operators of fossil plants, the financial burden for the plant operator could be extremely high.

DEVELOPING COUNTRIES
To date, nuclear power has hardly been utilized in developing countries for the following reasons:
• The often decentralized supply structure is at odds with the centralized supply system which characterizes nuclear power due to the Gigawatt size of nuclear power plants.
• Construction, maintenance and operation of nuclear systems require strict safety specifications, good management and supervision. The World Bank and the European Commission found that developing countries usually have significant problems in meeting these prerequisites (World Bank, 1991; COM, personal communication, 2002).
• With typically tight national budgets, nuclear power plants are difficult to finance (COM, personal communication, 2002).

PERMANENT DISPOSAL
The safety of potential sites for the permanent storage of nuclear waste is difficult to determine today. One of the main problems with the permanent disposal of nuclear waste is the extremely long period over which safe enclosure must be ensured. Plutonium-239, for example, has a half-life of 24,000 years, although of course a halving of the radiation alone does not mean that the material no longer needs to be stored safely. After 10 half-lives, the radioactivity will only have fallen to 0.1 per cent, which still is a very dangerous level. The radioactive elements with atomic weights above that of uranium (transuranic elements) have to be stored safely for approximately 1 million years. The Council feels that one-off storage over such long periods cannot guarantee the protection of the biosphere.

In principle, the waste could be converted into more short-lived radioactive isotopes through bombardment with particles from accelerators, which would significantly shorten the periods over which nuclear waste would have to be stored safely. Since the technological application of this process may release more energy than that required by the parti-

cle accelerators, one could imagine an almost residual heat-free fission power plant without the risk of a
chain reaction. However, such a technology has not
yet been demonstrated successfully at pilot scale.

REPROCESSING
Currently, there are three large commercial plants
for reprocessing spent nuclear fuel: La Hague
(France), Windscale-Sellafield (Great Britain) and
Chelyabinsk-Ozersk (Russia). These plants process
approximately 25 per cent of the spent fuel rods
worldwide. During reprocessing, the quantity of radiation released has exceeded the permissible limits on
several occasions (EU Parliament, 2001). Today the
associated technology cannot be regarded as safely
controllable.

PROLIFERATION AND TERRORISM
The construction of nuclear weapons requires only
relatively little know-how (UNDP et al., 2000). The
main problem for the production of such weapons is
the availability of weapons-grade plutonium or
highly enriched uranium. Both are also created during civilian utilization of nuclear energy, for example
as part of reprocessing. Nuclear research is a further
potential source of plutonium. The G8 countries currently hold approximately 430t of plutonium,
another 800t are present in spent fuel rods (ISIS,
2000). Since only 1/4 of the spent fuel rods are
reprocessed, the quantity of plutonium increases by
approximately 10t every year. On the other hand, for
the construction of a nuclear bomb only approximately 6kg of plutonium are required (Froggart,
2002). The G8 countries regularly discuss the problem of the disposal of weapons-grade material, without a solution having yet been found or the financing
of a solution having been agreed.

The Non-Proliferation Treaty was signed in 1968
with the aim of preventing and controlling the circulation of military nuclear technologies and fissile
material. So far, 182 countries have ratified the treaty,
although India, Pakistan and Israel, who have
nuclear weapons, have not. In January 2003, North
Korea declared its intention to withdraw from the
treaty. The International Atomic Energy Agency carries out relevant inspections without actually being
able to meet its inspection remit, according to statements from within the organization (IAEA, 2001).
Since 1993, more than 550 cases worldwide have
been entered in the IAEA database on the black
market for nuclear material, of which 16 cases related
to plutonium or enriched uranium. The number of
unrecorded cases is unknown, and comprehensive
recording of stolen fissile material appears impossible (UNDP et al., 2000). Existing safety and registra-

tion levels vary, and there is no binding international
standard.

After 11 September 2001, potential terrorist
attacks on nuclear power plants became a policy
issue, although such attacks had already been threatened and/or carried out as early as 1972 in Argentina,
Russia, Lithuania, France, South Africa and South
Korea (Bunn, 2002; WISE, 2001). Studies and tests
show that nuclear power plants are highly vulnerable
to commercial aircraft, for example, but also to internal sabotage or attacks (Bunn, 2002).

OUTLOOK ON THE SPECIFIC RISKS OF NUCLEAR
FUSION
Since no pilot plants or actual fusion power plants
exist as yet, it is very difficult to predict the risks associated with this technology. Studies on the environmental effects of fusion currently focus on the possible risks and effects of the radioactive inventory of
power plants, i.e. tritium as radioactive fuel and the
radioisotopes in the reaction chamber wall that are
generated through nuclear reactions between the
wall materials and the neutrons released during
fusion (Cook et al., 2001; Raeder et al., 1995).

Due to the small energy inventory, the effects of
any incidents are likely to remain restricted to the
plant interior. The quantity and toxicity of the
radioactive substances that are created in a fusion
power plant strongly depend of the chosen composition of the materials. The properties of wastes from
nuclear fusion and fission differ significantly. This is
shown clearly by the decay characteristics of
radiotoxicity – a measure for the biological risk from
the substances. The radiotoxicity of the majority of
the waste generated during nuclear fission remains
almost constant over many centuries. In contrast, the
radiotoxicity of fusion wastes decreases by three or
four orders of magnitude during the first 100 years.
However, this technology would also require safe
permanent disposal of large quantities of radioactive
waste over several hundred years.

3.2.2.4
Evaluation

While the theoretical potential of nuclear energy is
large, due to the unacceptable risks of utilization the
Council's recommendation is to phase out existing
nuclear power plants at the end of their current operating licenses and not to build any new plants.

The risk potential of fusion power plants also
appears to be significant. However, since such power
plants will not be available before the second half of
the 21st century at the earliest (if at all), the recom-

mendation of the Council is to not consider fusion power plants as part of any energy transition strategy.

The Council therefore assumes the sustainable potential of nuclear energy to be zero. However, due to the existing path dependencies, any realistic global phase-out scenario is unlikely to reach this value much before 2050. It is assumed that the nuclear power plants currently being constructed in Asia and central and eastern Europe will come on stream. The maximum contribution of nuclear energy towards global electricity supplies between 2010 and 2020 could be 12EJ per year (Table 4.4-1).

3.2.3
Hydropower

3.2.3.1
Global potential

Today, around 45,000 large hydroelectric dams are in operation worldwide, of which approximately 300 are 'mega dams' (ICOLD, 1998). With nearly all large dams, electricity generation is an important objective in addition to flood protection, water storage, irrigation agriculture and improvement of navigable waterways (WCD, 2000). Small hydroelectric plants require higher investment costs per installed capacity, which is why 97 per cent of hydropower is supplied by large hydroelectric plants with more than 10MW capacity (UNDP et al., 2000). The Earth's theoretical hydropower potential is estimated to be approximately 150EJ per year, of which approximately 50EJ per year could be classified as technological potential and approximately 30EJ per year as economic potential (Horlacher, 2002; Table 3.2-4). Other authors have reported similar figures (UNDP

et al., 2000; IPCC, 2001c). Approximately one-third of the global economic hydropower potential has so far been utilized, with significant differences in the degree of utilization between different countries and regions. Significant hydropower potential remains untapped in Africa, Asia and South America, while in North America and central Europe (including Germany) the potential has largely been utilized. According to some forecasts, the installed capacity could be more than doubled within 50 years to more than 1,400GW worldwide (Horlacher, 2002). Mega-projects with capacities of more than 10GW will be the exception. The majority of new projects are likely to have capacities between 0.1 and 1GW.

3.2.3.2
Technology

The technology used in hydroelectric plants is mature and has a reputation for being extremely reliable. Depending on the circumstances, a watercourse is either dammed up to achieve the head required for utilizing the potential energy contained in the water, or large quantities of water are fed directly through turbines with small gradient (run-of-river plants). Hydroelectric plants require very high investments for construction, but benefit from long service life (≥100 years), low operating costs, low maintenance effort and very high efficiency. Water as the operational resource is renewable and free of charge. Power plants with impounding reservoirs benefit from quick operational readiness (e.g. 1GW in approximately 5–10 min) and can therefore be used to generate peak electricity and to compensate extreme load changes within electricity grids (Section 3.4.3). Pumped storage systems are able to store large quantities of energy with very little losses.

Table 3.2-4
Hydropower potential by continent. See Box 3.1-1 for a definition of the different types of potential. Source: Horlacher, 2002

Region	Theoretical potential [EJ/a]	Technological potential [EJ/a]	Economic potential [EJ/a]	Already utilized potential [EJ/a]	Currently installed capacity [GW]	Under construction or planned [GW]
Africa	14.0	6.8	4.0	0.3	20.6	76.8
Asia	69.8	24.5	13.0	2.9	241	223
Australia	2.2	1.0	0.4	0.2	13.3	0.9
Europe	11.6	3.7	2.8	2.1	176	10
North and Central America	22.7	6.0	3.6	2.5	158	16
South America	22.3	9.7	5.8	1.9	111.5	50.2
World	*143*	*51.7*	*29.5*	*9.9*	*720*	*377*

3.2.3.3
Environmental and social impacts

An important motivation for large hydro projects is the increasing demand for electricity and irrigation agriculture, often coupled with the desire for effective flood control and the expansion of navigable waterways. Many large dams have indeed met these expectations and brought significant socio-economic benefits and made important contributions to development, although in many cases insufficient consideration is given to the ecological and social disadvantages (WCD, 2000).

EFFECTS ON ECOSYSTEMS
Large dams often trigger complex side effects on landscape and ecosystems (WCD, 2000; McCully, 1996; Pearce, 1992). First of all, a reservoir causes the direct loss of land and its ecosystems. In addition, the blocking of a river section and its conversion into a reservoir has far-reaching hydrological and ecological consequences. The storage or diversion of water by the dam drastically changes the quantity, quality and dynamics of the drainage and of the sedimentation regime.

Reservoirs act as sediment traps, so that worldwide 0.5–1 per cent of the capacity of impounding reservoirs is lost due to siltation every year (Mahmood, 1987). Downstream of the dam, the reduced sediment quantity is leading to changes in sediment dynamics, which not only has a negative influence on the ecology of the river bed itself, but can also cause significant damage at the mouth of the river due to coastal erosion (e.g. Nile: Stanley and Warne, 1993; Indus: Snedacker, 1984). Other important factors (nutrients, temperature and water chemistry) are also changed, so that negative ecological effects are experienced a long way downstream. Overall, dams contribute significantly to the worldwide threat to the biodiversity of freshwater fauna and flora (McAllister et al., 2000).

GREENHOUSE GAS EMISSIONS
The simplistic assumption that hydropower is climate-friendly cannot be justified for all projects, because the breakdown of biomass in the impounding reservoir results in the release of the greenhouse gases carbon dioxide and methane into the atmosphere (WCD, 2000). Damming often leads to the replacement of natural forests that can form a greenhouse gas sink by a reservoir representing on the one hand an emission source, but which on the other ahnd is also able to store carbon through sediment formation (Raphals, 2001). These opposite effects strongly depend on the climate, the topography and the greenhouse gas balances of the flooded ecosystems and the emerging reservoir. For example, shallow, tropical reservoirs can lead to higher emissions than fossil power plants with identical output (Fearnside, 1995, 1997; IPCC, 2001b).Whereas, for deep reservoirs in higher geographical latitudes, the fossil option is likely to have a significantly greater effect on the climate. This is also true in cases where large quantities of greenhouse gases were emitted from ecosystems before they were flooded (Svensson, 1999). For assessing the effect of hydroelectric plants on the climate, the long-term greenhouse gas balances before and after flooding have to be compared for each individual case, taking due account of secondary effects (e.g. forest clearance triggered by resettlement, changes in carbon flows upstream and downstream of the dam; WCD, 2000). The preparation of comprehensive greenhouse gas balances for hydroelectric projects remains an important research task for the future.

TECHNOLOGICAL RISKS
Dams can fail, leading to the sudden release of large quantities of water, potentially claiming many victims and causing severe damage. The worst dam disaster to date occurred in August 1975 during a typhoon in Henan, China, where 62 dams were destroyed The failure of the Banqiao dam alone released 500 million m^3 of water, villages and small towns were obliterated, and more than 200,000 people lost their lives (McCully, 1996). Of all dams built before 1950, 2.2 per cent have failed, while for dams that were built later the figure is less than 0.5 per cent, i.e. significantly lower (WCD, 2000). Since the design of dams is usually based on previous long-term average climatic and hydrological conditions, global climate change can bring additional safety risks through changes in extreme precipitation events. Further risks are the possibility of deliberate destruction of dams during military conflicts or through terrorism.

EFFECTS ON HUMAN HEALTH
Reservoirs and associated irrigation projects cover large areas of land with stagnant water. In the tropics, this leads to an increased risk of waterborne infectious diseases (Nash, 1993). The construction of dams in the tropics often leads to significantly increased bilharziosis infection rates. The construction of the Akosombo dam in Ghana, for example, led to an increase of this rate in children from below 10 per cent to 90 per cent (1966–69; McCully, 1996). Malaria, encephalitis, Rift Valley fever, filarioses, poisoning through toxic blue-green algae and through mercury leaching from the flooded ground are further examples of life-threatening, directly associated health risks (WCD, 2000; McCully, 1996). Furthermore, the

indirect consequences of the poor water quality of the stagnant water (diarrhoea) and malnutrition as a result of the destruction of social structures, of flooding of fertile soils and of the resettlement of the local population also need to be considered in the assessment (Lerer and Scudder, 1999). In advantageous cases, impounding reservoirs can improve the water supply and enable irrigation agriculture and fishing, with associated positive effects on the security of food supplies.

SOCIAL IMPACTS

Hydropower provides approximately 19 per cent of global electricity and is therefore currently by far the largest renewable energy source for electricity production. In 24 countries, hydropower contributes more than 90 per cent of the total electricity supply. While the estimated cost and timescale for the construction of large hydroelectric plants is often exceeded, the scheduled electrical output and economic profitability are usually achieved (WCD, 2000).

However, large dams inevitably also create losers, mainly among the population that is forced to relocate, often involving significant violations of human rights. During the 20th century, 30–80 million people were adversely affected by the construction of large dams, and the trend at the start of the 21st century is similar: The Three Gorges Dam in China will expel more than 1.1 million people, the Pa-Mong Dam (Laos and Thailand) 500,000 people (WCD, 2000; UNDP et al., 2000). Often, the affected sections of the population neither receive appropriate compensation for the financial losses they suffer, nor are they offered appropriate agricultural land in compensation for any land they may lose, particularly if they live some way downstream of the project. It is impossible to put a monetary value on the loss of cultural and religious values and of social cohesion and identity. In particular, this applies to the indigenous communities, whose culture and lifestyle is rooted in tradition and is very closely linked to the location and its natural ecosystems (McCully, 1996). An analysis of case studies shows that participation of the affected people has hardly played a role in previous dam projects, compensation payments were usually inadequate, and the above-mentioned social effects were regularly ignored in the plans of the dam builders (WCD, 2000).

SUSTAINABILITY OF HYDROPOWER

Forced relocation, lack of participation, unfair distribution of economic benefits and the negative ecological consequences of dams create potential for social conflicts (Bächler et al., 1996). As a consequence, the political resistance against dams has increased over recent decades (UNDP et al., 2000). This has also had an influence on lenders and international institutions. First a slow rethink occurred, and then an open discussion process. The World Bank, for example, which played a significant role in the financing of many large dams in developing countries, retrospectively reassessed the projects it had funded. Today, environmental and social compatibility of new projects has a much higher status within the multilateral financing institutions.

For all hydraulic engineering projects, compliance with internationally agreed (e.g. World Bank, OECD) guidelines for sustainability should ensure their ecologically and socially compatible implementation. Under these guidelines, hydroelectric projects are to be avoided if alternative energy options can be developed that are more sustainable and are not significantly more expensive in the long-term. These international guidelines are not necessarily in harmony with often less demanding national legislation, which is frequently applied to the detriment of the affected population and of nature conservation. Half of all large dams, for example, were built without consideration of the environmental consequences for the downstream ecosystems (Dixon et al., 1989).

On an international level, the highlights of the sustainability discussion to date are the analyses and recommendations of the World Commission on Dams (WCD, 2000). Within this difficult environment, the Commission was able to develop a basis for the evaluation of large dam projects, by bringing together representatives with different interests on an international level within an open-ended and consensual process (WCD, 2000). The result is impressive: despite the fact that some countries (e.g. China, India, Turkey) and players (e.g. International Commission on Large Dams – ICOLD, International Hydropower Association – IHA, International Commission on Irrigation and Drainage – ICID; Varma et al., 2000) were not happy with all results, the Commission's report and the recommendations contained therein were received positively overall. The problem is often not a lack of awareness of the problem or of guidelines for sustainability, but the lack of a basic institutional framework. To date, sustainability guidelines could therefore only rarely be implemented coherently in practice. The following preconditions must be met, if an increasing proportion of the economically attractive projects is to be designed and implemented in a sustainable manner over the coming decades:

• *Ensuring nature conservation:* A global system of protected areas for the purpose of preserving the natural heritage (Section 4.4.1.3; WBGU, 2001a) should ensure that a certain proportion of the different types of river ecosystems (including their catchment area) remains free of intervention, i.e. above all free-flowing. Previous experience shows that precautionary protection of ecologically valuable regions has to be implemented quickly, particularly in the catchment areas of possible future hydroelectric projects, otherwise – regardless of any guidelines – 'facts' can be created in advance, e.g. in the form of logging.

- *Creating a scientific basis:* There is often a lack of ecological, social and other location- and case-specific basic data for sustainability analyses and for comparisons with alternative options. Significant investments in a better scientific database over the coming 5–15 years are therefore a central prerequisite for a more sustainable expansion of hydropower (Section 6.3.1). This database should be developed by independent regional research centres on the basis of the catchment areas (e.g. INPA in the Amazon region or ICIMOD in the Himalayas; von Bieberstein Koch-Weser, 2002), in a manner independent of individual projects.
- *Ensuring participation of the affected population:* Many negative effects could be contained with preventive and compensation measures and through detailed preliminary work and participation of the affected population. While during previous consultations the concerns, demands or protests of those affected tended to be made public, project management or government agencies often failed to take them into account properly.
- *Rectifying local institutional deficiencies:* Stronger mutual trust and better acceptance can be achieved through an efficient mediation and jurisdiction system. Environmental impact assessments (EIAs) should not be carried out retrospectively to justify projects, but should be evaluated prior to any decision in favour of a particular project option. The relevant government authorities in developing countries should be able to check and examine EIAs at a high technological level and with adequate knowledge of the locality. A need for significant investment in capacity-building thus remains. For cross-border catchment areas, cross-country regional institutions for the development of hydroelectric installations should be created. These could assist in the analysis of alternative locations within the region that also consider indirect and cumulative effects (e.g. for a series of projects on the same river).

3.2.3.4
Evaluation

Not all dam construction projects should be seen as negative (WCD, 2000). The implementation of the

recommendations presented in Section 3.2.3.3 can make additional sustainable hydropower potential accessible, although this would require a high degree of long-term and international cooperation and close integration of development policy, export financing and energy planning (e.g. World Bank, regional banks and export credit agencies).

It is difficult to make general statements about the globally available and sustainable hydropower potential, since it depends on many factors and on the development of the above-mentioned scientific and institutional framework. While large technological hydropower potential (Section 3.2.3.1; Table 3.2-4; Horlacher, 2002) remains untapped, its realization over the coming decades will only be justifiable in exceptional cases and with due consideration of sustainability criteria (von Bieberstein Koch-Weser, 2002).

Overall, the Council estimates the usable potential to be lower than other sources (e.g. UNDP et al., 2000), since the lack of a framework for the application of internationally recognized sustainability criteria seriously limits the scope. Most of the economically more attractive and less controversial projects have already been realized in the past. In North America and central Europe (including Germany) for example, the additional sustainable potential is very small. Furthermore, hydroelectric projects will be significantly more difficult to implement in future due to the rightly increased requirements for environmental and social compatibility. Many of the remaining project options are located in poorly accessible tropical forest or mountain regions, where the complexity of the ecosystems (South America, South-East Asia, Africa), the vulnerability of the indigenous population (Colombia, Brazil, Laos, Vietnam) or geological risks (Himalayas) present serious challenges. Others project options are located in densely populated regions, where large relocation programmes would be required (India, China, southern Brazil).

If the required framework (investments in research, institutions, capacity building, etc.) can be created during the next 10–20 years, and a circumspect approach is taken, approximately one-third of the hydroelectric potential utilized today could additionally be made accessible in a sustainable fashion, i.e. 12EJ per year overall by 2030. This figure could be increased to approximately 15EJ per year by 2100, but only if the above-mentioned preconditions are met (Table 4.4-1).

3.2.4
Bioenergy

3.2.4.1
The potential of modern bioenergy

According to the World Bank, traditional biomass utilization currently contributes 7.2 per cent of the global primary energy use. In the developing countries, 35 per cent of the energy is generated from biomass, and in some African countries even up to 90 per cent. Around 2,400 million people exclusively depend on traditional biomass utilization for their energy supply (IEA, 2002c). This includes the use of firewood, charcoal and dung for cooking and heating in private households. The fact that traditional biomass is mainly used in the poorest countries, with associated disadvantages for the environment and for health, is the reason for its poor reputation as an outdated energy source. Despite this, biomass offers significant potential for future energy generation, which could be utilized more efficiently and largely free from negative effects on health.

TYPES OF BIOMASS THAT CAN BE USED FOR ENERGY
GENERATION AND ASSOCIATED TECHNOLOGIES
'Modern' biomass that can be utilized for energy generation comprises the following categories:
- Agricultural waste products (e.g. straw, dung, rice husks), as far as they can be utilized without depriving farmland of nutrients;
- Timber residue and small branches, etc., insofar as they do not have to remain in the forest for ecological reasons, or are used for different purposes, e.g. for economic reasons;
- Industrial residue and recycled wood (also with economic restrictions);
- Annual or perennial energy crops that are specially cultivated for the purpose of energy generation.

The options for energy extraction from biomass depend on the range of the bioenergy carriers used. In addition to the combustion of biomass for generating heat and/or electricity, various other technologies are currently at an experimental stage or en route to commercialization. The further development of power generation through gasification of solid biomass, and possible coupling with the hydrogen economy are also being examined.

TECHNOLOGICAL AND ECONOMIC POTENTIAL OF
BIOMASS UTILIZATION IN GERMANY
Table 3.2-5 summarizes the technological and economic potential of biomass for energy generation in Germany. The detailed assessment for Germany illustrates the principle that was used to assess the potential on a European and global level.

AREA DISTRIBUTION
Germany has a total surface area of 35.7 million hectares, comprising farmland (53.5 per cent), woodland (29.4 per cent), settlements (12.3 per cent) and other areas (4.7 per cent) (Statistisches Bundesamt, 2002). Nature reserves (without Wadden Sea areas) make up 2.6 per cent of the area, national parks and biosphere reserves a further 6.4 per cent (BfN, 2002). Since there is some overlap between nature reserves and the core zones of the biosphere reserves and national parks, the European Environment Agency estimates protected terrestrial habitats to have a 8.3 per cent share of the total area (Moss et al., 1996). A recent amendment to the Federal Nature Conservation Act sets the target of creating a habitat network comprising at least 10 per cent of the land area.

Setaside farmland totalling 2 million hectares (Kaltschmitt et al., 2002) could be used for the cultivation of energy crops, as nature conservation areas or for afforestation for the purpose of carbon storage under the Kyoto Protocol. The different types of land use would result in different technological and economic potential for bioenergy generation or for saving carbon dioxide emissions.

POTENTIAL FROM FORESTRY
In Germany, less than 17 million tonnes of the annual wood increment (40.3 million tonnes dry matter – UN-ECE and FAO, 2000) is utilized as merchantable timber. The quantities theoretically available for utilization in energy generation are therefore 9.6 million tonnes of residual forest timber, 7 million tonnes of small branches etc. and 6.6 million tonnes of unused increment (Table 3.2-5). For forestry and economic reasons, the sustainable and economic potential for residual timber is only approximately 10 million tonnes per year. In addition, there is approximately 8.2 million tonnes of industrial timber and recycled wood. Retrieval of the 0.2 million tonnes of timber that accumulates during landscape management does not appear to be profitable. From an ecological point of view, the utilization of unused increment for energy generation cannot be justified. Of a total of 31.7 million tonnes per year, only approximately 18 million tonnes of dry matter are therefore available as economically usable potential. This situation is unlikely to change significantly until 2030, since an increasing demand for material extraction (paper, packaging, etc.) is to be expected. The energy potential of wood-based biomass is therefore reduced to approximately 340PJ per year, equivalent to approximately 6.8 million tonnes of carbon. If the

Table 3.2-5

Technological and economic bioenergy potential in Germany. The carbon equivalent figures show the quantity of greenhouse gas emissions (carbon dioxide, nitrous oxide, methane, all expressed in C) that are avoided compared with the utilization of fossil energy carriers (based on the current fuel mix in Germany). Total area of Germany: 35.7 million hectare. The economic potential refers to the year 2001. *DW* dry weight.
Sources: Kaltschmitt et al., 1997; Hanegraaf et al., 1998; Freibauer, 2002; Kaltschmitt et al., 2002

	Area	Utilizable quantity		Biogas	Calorific value	Energy potential	Carbon equivalent
	$[10^6$ ha]	$[10^6$ t$_{DW}$/a]	[t/ha/a]	$[10^6$ m^3/a]	[MJ/kg]	[PJ/a]	$[10^6$ t/a]
TECHNOLOGICAL POTENTIAL							
Forest	10.5						
Timber residue		9.6	0.9		18.6	179	3.6
Small branches etc.		7.0	0.7		18.6	130	2.6
Increment		6.6	0.6		18.6	123	2.5
Industrial timber		3.1			18.6	58	1.2
Recycled wood		4.3–6.0			18.6	80–112	1.6–2.4
Landscape management	0.4	0.2	0.5		18.6	4	0.1
Total timber	*10.9*	*30.8–32.5*				*573–605*	*11.5–12.1*
Agriculture							
Permanent grassland	5.1						
Meadows	4.1	0.9–1.4	0.2–0.3	750–1,100	17.7	16–24	0.3–0.5
Others	1.0						
Arable land	11.8						
Grain, corn, rape	8.1	7.6	0.9		17.0	130	2.5
Other arable	3.7	0.8–1.5	0.1–5.0				
Other agric. areas	2.2						
Total agriculture	*19.1*	*9.3–10.5*		*750–1,100*		*146–154*	*2.8–3.0*
Settlements and others							
Settlements	4.4						
Others	1.7						
Landscape management		0.4–0.9		280–560	14.2	6–12	0.2
Excrement		15.5		4,500	6.2	97	1.8
Municipal waste		1.5		580	8.3	13	0.2
Industrial waste		0.5–1.0		300–375	12.5	6–12	0.2
Digester/landfill gas		2.0		2,450–3,050	18.8	35–41	0.7
Total from waste		*20–21*		*8,110–9,065*		*156–174*	*3.1*
Setaside land[a]	2						
Short-rotation forests	2	18	9.0		18.5	333	>6.4
Energy grasses	2	24	12.0		17.6	422	7.5–8.1
Grain crops, whole plant	2	20	10.0		17.0	340	3.3–6.5
Rape oil	2				37.3	102	1.8
Total energy crops	*2*	*18–24*	*9.0–12.0*			*102–422*	*1.8–8.1*
Overall total		*77–88*		*8,860–10,165*		*977–1,355*	*18.9–25.8*
ECONOMIC POTENTIAL							
Forest		*18.2*				*339*	*6.8*
Timber residue		6.5			18.6	121	2.4
Small branches etc.		3.5			18.6	65	1.3
Industrial timber		3.1			18.6	58	1.2
Recycled timber		5.1			18.6	95	1.9
Agriculture and wastes		*29–35*				*315*	*6.0*
Grassland		1.1			17.6	20	0.4
Straw		7.6			17.0	130	2.5
Biogas		20.5		8,588		165	3.1
Energy crops		*18–24*			*17.0–18.6*	*102–422*	*1.8–8.1*
Overall total		*65–77*				*756–1,076*	*14.6–20.9*

[a] alternative utilization

nutrient supply of forests is considered, the utilization of dead wood and small branches is not sustainable in the long term. The ecologically sustainable potential is therefore approximately 20 per cent less than the economic potential. Overall, a maximum of approximately 15 million tonnes of carbon equivalent could be saved in Germany per year.

POTENTIAL FROM AGRICULTURE

A sustainable energy potential of 315PJ per year (corresponding to 6 million tonnes carbon per year) can be expected from agriculture, with 10 per cent of the mowings from permanent grassland, 20 per cent of the straw, excrement and various wastes being suitable for energy generation (e.g. for the production of biogas; Kaltschmitt et al., 2002). Depending on the energy carrier, this would result in an additional energy potential of 100–420PJ per year (1.8–8 million tonnes carbon), if setaside areas were to be utilized for energy crops. The large range is due to different growth rates and different effort required for the cultivation of energy crops. However, any assessment of this potential should consider the fact that the current setaside practice is to be replaced by a long-term ecological setaside scheme that no longer offers the option of financial support for the cultivation of energy crops (EU Commission, 2002).

Subsidies, operational flexibility and other factors result in farmers preferring the cultivation of annual plants that require the application of pesticides and fertilizers. However, for bioenergy purposes perennial plants are preferable, since they offer higher energy yields with lower quantities of fertilizers and pesticides and less intense soil cultivation (Börjesson et al., 1997). For annual species, the ecologically sustainable potential is approximately 30 per cent lower than the economic potential.

In Germany, bioenergy could compensate a maximum of approximately 11 per cent of the energy-related carbon dioxide emissions (based on the year 2000) and cover 7–9 per cent of the energy demand (technological potential; Table 3.2-6).

POTENTIAL FOR BIOMASS UTILIZATION AND
CARBON STORAGE IN THE EU

Even for a region with good statistical documentation, any estimate of the technological potential of biomass utilization in the European Union (EU-15) will have a high degree of uncertainty. Estimates of the potential range from 4,300 to 10,100PJ per year, with a median of 5,700PJ per year and a mean value of 6,100±1,900PJ per year. The following observations relate to Kaltschmitt et al., 2002 (with additions), who came up with a value of 5,200PJ per year, which is below the mean value of the estimates found in the literature (Hall and House, 1995; EU Commission, 1998; AEBIOM, 1999; Grassi, 1999; Ministry of Trade and Industry, 1999; FNR, 2000; fesa, 2002), but is relatively close to the median value.

The technological and economic potential of bioenergy and of carbon storage through modified management techniques within the EU is shown in Table 3.2-7. This potential amounts to 5,224PJ per year, which could cover 8.6 per cent of the energy used in 2000 (60,926PJ; Eurostat, 2002). The total carbon savings potential through the utilization of bioenergy carriers and the creation of sinks through modified management techniques is approximately 160 million tonnes of carbon equivalent (14 per cent of the energy-related emissions in 1990). The eco-

Table 3.2-6
Summary of the technological and economic potential for the utilization of biomass for energy generation and carbon storage in Germany. The energy use figures for Germany in 1990 and 2000 are shown for comparison.
Sources: Kaltschmitt et al., 1997; Hanegraaf et al., 1998; Freibauer, 2002; Kaltschmitt et al., 2002

	Energy balance		Carbon balance		Contribution total emissions	
	[PJ/a]		[10^6 t C_{eq}/a]		[% C_{eq}]	
Base year 1990	17,402		330		100	
Base year 2000	14,278		270		82	
	technological	economic	technological	economic	technological	economic
Bioenergy	*977–1,355*	*756–1,076*	*19.2–26.3*	*14.6–20.9*	*5.8–8.0*	*4.4–6.3*
Forestry	573–605	339	11.5–12.1	6.8	3.5–3.6	2.1
Agriculture	302–328	315	5.9–6.1	6.0	1.8–1.9	1.8
Energy crops	102–422	102–422	1.8–8.1	1.8–8.1	0.6–2.5	0.6–2.5
Carbon storage			*14.8*	*11*	*4.5*	*2.8–2.9*
Forestry			8.5	8.5	2.6	2.6
Afforestation			1.2	0.0	0.4	0.0
Agriculture			5.1	0.5–1	1.5	0.2–0.3
Total			*34–41*	*26–32*	*10–13*	*7.2–9.1*

Table 3.2-7
Technological biomass
potential for energy
generation according to
material groups in the EU.
In the base year of 1990,
1,157 million tonnes of
carbon were released from
fossil fuels. The carbon
equivalent figures show the
greenhouse gas emissions
(CO_2, N_2O, CH_4), that were
avoided compared with the
utilization of fossil energy
carriers. The ARD balance
indicates the sink potential
from afforestation, reforestation and deforestation.
Sources: Freibauer et al.,
2002; Kaltschmitt et al.,
2002; Schulze et al., 2002

	Calorific value [MJ/kg]	Quantity [10^6 t$_{DW}$/a]	Energy potential [PJ/a]	Carbon equivalent [10^6 t C$_{eq}$/a]	Area [10^6 ha]
ENERGY POTENTIAL					
Forestry		*171.6*	*3,192*	*63.8*	*113*
Timber residue/firewood	18.6	44.5	828	16.6	
Small branches etc.	18.6	25.0	465	9.3	
Industrial wood residue	18.6	67.0	1,246	24.9	
Old timber	18.6	26.8	498	10.0	
Landscape management	18.6	8.3	154	3.1	
Agriculture		*63.8*	*1,098*	*20.3*	*74*
Straw	17.2	53.2	915	16.9	36
By-products/wastes	17.0	10.6	183	3.4	38
Energy crops	*17.7*	*52.8*	*935*	*17.8*	*7.4*
Total (technical)		*288*	*5,225*	*101.9*	
Total (economic)			*3,134*	*61.1*	
SINK POTENTIAL					
ARD balance				*1.4*	*7.4*
Forestry management				*39.4*	*108*
Agricultural management				*16.4–19.1*	*74*
Total				*57.2–59.9*	
Total savings potential				*119.6*	

nomic potential is significantly less than this value, since sink capacity and bioenergy partly compete for the same areas. Ecological restrictions, such as the increased occurrence of nitrous oxide emissions with reduced ploughing, further limit the potential (Freibauer et al., 2002). If 60 per cent of the techno-logical potential was utilized in a sustainable fashion, bioenergy would only contribute 3,134PJ per year or 5.1 per cent of the primary energy use (figures for 2000). The economic sink potential would be 119 million tonnes of carbon or 10.3 per cent of the emissions.

The utilization of biomass for energy generation replaces approximately the same quantity of fossil carbon as the carbon that could potentially be stored through cultivation measures. In this calculation, the contribution that can be attributed to storage was limited in the 2001 Bonn Agreements, as part of the Kyoto follow-up process. Forest increment in the EU is equivalent to 164 million tonnes of carbon per year, 103 million tonnes of which is felled (UN-ECE and FAO, 2000), and the remaining 60 million tonnes carbon per year are stored in the biomass as volume growth. For operational reasons, this quantity cannot be harvested for the purpose of energy generation, but forms a carbon sink. The actual effect of forestry is therefore about 30 per cent higher than estimated (60 million vs. 39.4 million tonnes carbon; Table 3.2-7).

Taking account of storage through management is very significant for the Earth's carbon balance, since it is the only mechanism through which the carbon

deposits in the soil can be protected. With 1,500–2,000Gt, an exceedingly large quantity of carbon is stored in the soils of the terrestrial ecosystems, corresponding to approximately 300 years of emissions from the utilization of fossil fuels at current consumption rates. Since this carbon can partly be released through cultivation measures, the protection of these deposits has to be a primary target for the preservation of the basis of human existence and of climatic conditions (Section 3.6).

GLOBAL BIOENERGY POTENTIAL
The Council estimates the global sustainable bioenergy potential to be around 100EJ per year (Table 3.2-8), of which 40 per cent originate from forestry residues and by-products, 17 per cent from agricultural waste and approximately 7 per cent from the combustion of dung. Energy crops play an important role with a further 36 per cent.

In order to determine the potential of energy crops, the maximum area of arable land must be known. Areas to be used for food production for a growing world population and protected areas for the preservation of biodiversity and ecosystem functions have to be taken into account. Deserts (19 per cent) and mountain areas with a slope of more than 30 per cent (11 per cent) also have to be ruled out (FAO Land and Plant Nutrition Management Service, 2002).

The area available for agriculture as a proportion of the global land area is around 12.5 per cent, of which 26.5 per cent is used as permanent pasture.

Table 3.2-8
Global technological potential of biogenic solid fuels. Since the potential was estimated cautiously, the specified values can be regarded as sustainable.
Sources: FAO, 2002; Kaltschmitt et al., 2002

	Area	No. of animals	Utilizable yield		Total quantity	Technological potential
	$[10^6$ ha]	$[10^6$ animals]	$[t_{dry\ weight}/ha/a]$	$[t/animal/a]$	$[10^6\ t_{dry\ weight}/a]$	[PJ/a]
Forestry	4,173		0.5		2,237	41,600
Agriculture	1,505		0.7		994	17,200
Energy crops	322		6.6		2,113	37,400
Dung		1,599		0.8	1,220	7,600
Total						*103,800*

Woodland covers approximately 30 per cent of the land area. If about 20 per cent is designated as protected forests and natural grassland, a maximum of 10 per cent remains for the cultivation of biomass. In order to assess the sustainability of this type of utilization, the potential for the individual continents has to be considered separately (Table 3.2-9). The table shows that in Asia the existing bioenergy potential is already being over-utilized.

The calculated technological potential for energy crops requires an area of 322 million hectares, i.e. approximately 2.5 per cent of the land surface, if moderate bioenergy crop yields of approximately 6–7t dry weight per hectare per year are to be achieved in industrialized countries and in Latin America. In Africa, due to poor soils and traditional farming structures, yields of this order of magnitude are often only achievable using fast growing trees such as eucalyptus, with yields of 0.5–30t (on average 8.5 t) per hectare per year depending on precipitation, while grain yields are less than 2t per hectare per year (Marrison and Larson, 1996; FAO, 2002).

In addition to energy crops, agricultural and forestry waste products contribute significantly to the global bioenergy potential. Here ,too, sustainable utilization is a prerequisite for the calculation of the potential. FAO data on land use, agricultural production and timber production (FAO, 2002) were used for the extrapolation.

APPRAISAL
Compared with other estimates of the global bioenergy potential, the values assumed by the Council are rather low. One reason is the fact that competing land use demands were considered. IPCC (2001c) estimated the global bioenergy potential for 2050 to be 396EJ per year, with very high proportions of the total land area assigned to the cultivation of energy crops (16 per cent in Africa, 32 per cent in Latin America). In Latin America, 30 per cent of the area is currently used as permanent pasture, and in view of increasing meat consumption this figure is unlikely to fall. According to IPCC, the required arable farmland area will even increase to 15 per cent, while 9 per cent of the continent consists of arid areas. In order to achieve the IPCC values, the natural forest area of South America would have to be reduced from currently 46 per cent to approximately 16 per cent,

Table 3.2-9
Geographic distribution of the technological energy potential of biogenic solid fuels.
Source: Kaltschmitt et al., 2002

	Europe	Former USSR	Asia	Africa	Middle East	North America	Latin America	Total
				Total energy potential [PJ/a]				
Wood	4,000	5,400	7,700	5,400	400	12,800	5,900	41,600
Stalks	1,600	700	9,900	900	200	2,200	1,700	17,200
Energy crops	2,600	3,600	1,100	13,900	–	4,100	12,100	37,400
Dung	700	300	2,700	1,200	100	800	1,800	7,600
Total	*8,900*	*10,000*	*21,400*	*21,400*	*700*	*19,900*	*21,500*	*103,800*
Currently utilized	2,000	500	23,200	8,300	–	3,100	2,600	39,700
				$[10^6$ ha]				
Area for energy crops	22	32	10	124	0	36	108	332

which from a nature conservation point of view can definitely not be considered sustainable (WBGU, 2001a). A similar argumentation applies to the situation in Africa.

In terms of yields, IPCC assumes a figure of around 15t per hectare per year. This approximately corresponds to the value of 19.5t per hectare per year achieved for sugar cane (Kheshgi et al., 2000). For *Miscanthus*, 2–44t per hectare per year are assumed depending on region and soil conditions (Lewandowski et al., 2000), for switchgrass (*Panicum virgatum*) 4–34.6t per hectare per year (Paine et al., 1996; Sanderson et al., 1996). The yields of woody energy crops in moderate and northern latitudes are assumed to be 7–10t per hectare per year for poplars (Hanegraaf et al., 1998; Kheshgi et al., 2000), for willows 4.7–12t per hectare per year (Tahvanainen and Rytkönen, 1999; Goor et al., 2000). Only few authors assume yields of more than 15t per hectare per year for willows (Boman and Turnbull, 1997). For many regions, the yields assumed by IPCC are therefore too high. This applies in particular to traditional farming in Africa. The Council therefore assumed average dry matter yields of only 6–7t per hectare per year.

With a comparable land distribution, other authors have also calculated a bioenergy potential of 350–450EJ per year (Fischer and Schrattenholzer, 2001). However, their figure for the potential of agricultural residues (35EJ per year) is twice the value assumed by the Council, since twice the value was assumed for the recovery per unit area. While a yield of 1.2t per hectare per year of agricultural residues is realistic in temperate latitudes, this value is not realistic in the tropics, where the soil structure has to be stabilized by large quantities of carbon being fed back into the soil. While the Council assumed approximately 0.5t per hectare per year of recoverable forestry residues, taking account of ecological and economic restrictions (no utilization in primeval forests or in regions with poor infrastructure), Fischer and Schrattenholzer (2001) assumed 1.4t per hectare per year of forestry biomass that can be utilized for energy generation. Although this figure may be justified for the commercial forests of temperate regions, for tropical and boreal regions the value appears too high. For energy crops, the average yields of 4.7t per hectare per year assumed by Fischer and Schrattenholzer are moderate, although the inclusion of the global grassland area in the utilization for energy generation would violate ecological principles (WBGU, 2001a).

3.2.4.2
Environmental and social impacts of traditional biomass utilization in developing countries

EFFECTS ON THE NATURAL ENVIRONMENT
Biomass is becoming scarce, particularly in arid regions such as the Sahel or in the steppes of Asia (BMZ, 1999), since more material is taken here than grows back. In Asia, 1,700PJ per year is generated from non-sustainable timber utilization, corresponding to approximately 20 per cent of the energy generated from biomass. In Africa the proportion generated in a non-sustainable fashion is 30 per cent, in Latin America 10 per cent (Kaltschmitt et al., 1999). Non-sustainable biomass utilization destroys forests, degrades soils, reduces biodiversity and damages water resources.

EFFECTS ON HUMAN HEALTH
According to UN estimates, 1.6 million people worldwide die of the consequences of indoor air pollution every year (WHO, 2002b). Approximately half the world population is subjected to the damaging effects of traditional biomass utilization, the majority of them women and children (Bruce et al., 2000; Table 3.2-10). Particular risks are associated with the incomplete combustion of wood or dung in traditional, technologically inadequate stoves, with emissions of soot, suspended matter and carbon monoxide significantly exceeding safe values (UNDP et al., 2000). Health risks are associated with small particles (particles with a diameter of less than 2.5 µm are particularly dangerous), SO_x, NO_x, O_3 and polycyclic aromatic hydrocarbons. The susceptibility to acute respiratory tract infections is significantly higher for children who are subjected to fumes and exhaust gases from the combustion of biomass than for children in households with modern fuel utilization (Behera et al., 1998; Smith et al., 2000). Mothers and children also suffer an increased risk of chronic obstructive pulmonary diseases, lung cancer, tuberculosis, asthma or ischaemic heart diseases (Smith et al., 2000; Smith, 2000; Box 3.2-1).

3.2.4.3
Evaluation

The Council estimates the global potential of modern bioenergy to be approximately 100EJ per year, made up of 20 per cent from the utilization of agricultural waste products and approximately 40 per cent each from forestry waste products and energy crops. Such an expansion could only be achieved over several decades (Table 4.4-1). The long-term potential of traditional biomass utilization is approximately 5EJ per

Fuel cycle stage	Potential health effects
Production	
Dung as fuel	Infections
Charcoal	Carbon monoxide poisoning, burns, traumas
Collection of fuel	Reduced child care
	Less time for food preparation
	Poorer family nutrition
Combustion	
Smoke (acute effects)	Inflammation of the conjunctiva
	Irritation/inflammation of the upper respiratory tract
	Acute respiratory tract infections
	Positive effect: deterrent for insects, spiders, etc.
Smoke (chronic effects)	Lung cancer
	Chronic obstructive pulmonary diseases, chronic bronchitis
	Tuberculosis
	Cor pulmonale
Toxic gases	Poisoning
(e. g. carbon monoxide)	Foetus: low birth weight, damage
Heat	Burns (acute effect)
	Cataract (chronic effect)

Table 3.2-10
Health threats during different stages of the biomass fuel cycle. *Cor pulmonale:* right ventricular failure due to chronic pulmonary disease.
Source: WHO, 2002a, c

year. The Council's estimate places stronger emphasis on the sustainability of biomass utilization than comparable studies, e.g. by avoiding the conversion of natural ecosystems to cultivate energy crops, and ensuring an adequate return of nutrients to woodland and farmland soils. The figures are therefore significantly lower than other current potential estimates, such as those presented by IPCC (2001c) or Fischer and Schrattenholzer (2001).

3.2.5
Wind energy

3.2.5.1
Potential

For calculating the onshore and offshore wind energy potential, it is assumed that advanced multi-megawatt wind energy converters are used (Fig. 3.2-2). The

Box 3.2-1

Biomass stoves cause disease: The example of India

Three-quarters of households in India (approximately 650 million people) depend on biomass, which meets 85–90 per cent of the energy demand. Due to inadequate combustion technology, this energy source represents a significant health threat from emissions, mainly for women and children. In India, the health risks due to indoor air pollution are significantly greater than the risks due to outdoor pollution, even in large cities (Fig. 3.2-1 and Section 4.3.2.7). According to estimates, approximately 500,000 women and children under 5 die early due to the utilization of solid fuels in households. This is equivalent to 5–6 per cent of the national disease figures and therefore exceeds the much more frequently mentioned risks of smoking or malaria.

In India, government and private programmes support the introduction of improved stoves. It is estimated that 7.6 million efficient stoves were used in households in 1992.

Sources: Terivision, 2002; Smith, 2000; UNDP et al., 2000; Murray and Lopez, 1996.

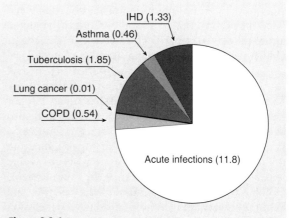

Figure 3.2-1
Estimated distribution of annual health impact, expressed in DALYs (Disability Adjusted Life Years), attributable to indoor air pollution caused by cooking in India. *COPD* chronic obstructive pulmonary diseases; *IHD* ischaemic heart diseases (e.g. heart attack).
Source: Smith, 2000

conversion potential was calculated based on wind speeds interpolated from meteorological data for the appropriate hub height. Average values over a 14-year period were used (1979–1992). However, in addition to the conversion potential, further limitations have to be considered for calculating the global technological potential (Box 3.1-1). Urban areas, forest areas, wetlands, nature reserves, glaciers and sand dunes, for example, were excluded from the calculation. Agriculture, on the other hand, was not regarded as competition for wind energy in terms of land use. However, the installation of wind power plants may be prohibited by the prevailing wind conditions due to the topology (e.g. ravines, basins) or the slope of the terrain (problem with foundations). Furthermore, certain minimum distances to settlements, for example, must be adhered ,too. Sea depths of more than 40m are currently not considered to be suitable for offshore installations. The average annual ice coverage of the sea and a regionally varying minimum distance from the coast (0–12 nautical miles) were also taken into account. For both offshore and onshore applications, local exclusion criteria (smaller nature reserves, infrastructure surfaces, military areas, etc.) were accounted for through correction factors derived from the respective popula-

tion density. Taking account of these, the global technological potential for onshore and offshore installations was calculated as 1,000EJ per year. The Council considers 10–15 per cent of this technological potential to be realizable in a sustainable fashion, and proposes a figure of approximately 140EJ per year as the contribution from wind energy for a sustainable energy supply that is achievable in the long term.

3.2.5.2
Technology/Conversion

Wind power plants convert the kinetic energy contained in moving air into mechanical rotation energy and subsequently into electricity. Theoretically, a maximum of just under 60 per cent of the power contained in the wind can be extracted (Dwinnell, 1949).

The global market for wind energy plants is currently structured into two fundamentally different areas of application: While in Asia tens of thousands of very small, decentralized systems are used as battery charging stations, large, grid-connected turbines are more significant globally from a quantitative and economic point of view.

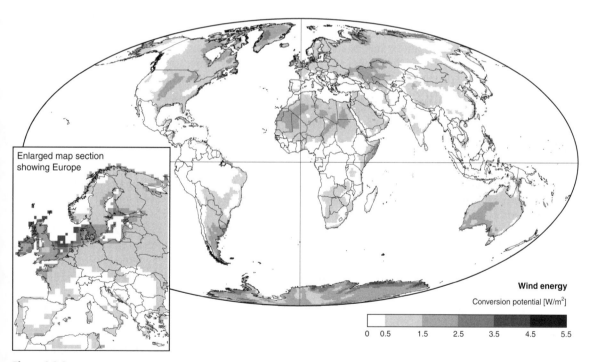

Figure 3.2-2
Global distribution of the conversion potential of onshore and offshore wind energy (in the latter case up to a depth of 40 m). The conversion potential was derived from the theoretical potential, based on the predicted annual efficiency of a multi-megawatt wind energy converter in 2050 (see Box 3.1-1). Economic and land use restrictions were not considered. The resolution of the calculation was 0.5*0.5°, approximately corresponding to 50*50km.
Source: Kronshage and Trieb, 2002

In contrast to traditional windmills that operate according to the resistance principle, modern rotors utilize the lift principle (similar to aeroplane wings). In addition to the rotor, large systems essentially consist of a generator and a tower. Different system designs have been proposed over the years, although horizontal three-blade models mounted on a tower made of steel tube have now become generally accepted. Usually, a mechanical gearbox is installed between the rotor axis and the generator for adjusting the rotor speed to the required generator speed, although more recently gearless generator technology has become established on the market. The rated power of grid-connected systems has increased from typically 30kW to up to 3MW over the last 30 years, with 5MW versions being envisaged for offshore applications.

Due to fluctuating wind speeds, the average annual output of wind turbines is only between 20 and 25 per cent of the rated power (more than 30 per cent for offshore installations). In general, the power contained in the wind is proportional to the cube of the wind speed. Modern systems start producing energy at a wind speed of approximately 3m/s. Governing sets in at wind speeds around 25m/s in order to avoid damage to the systems. Since the average wind speed at rotor height is an important parameter for the yield from wind power plants, it has significant influence on the costs of electricity production. Taking account of operation and maintenance costs, generation costs are currently between €-cents 5.5 and 13 per kilowatt hour at appropriate locations in Germany (BMU, 2002b).

Good onshore locations may become scarce with increasing utilization of wind energy, so that offshore wind parks are now being proposed, with the first such installations already in operation. While in the North Sea the number of full-load equivalent operating hours can increase to up to 4000 hours per year, the installation costs also approximately double. The main motivation for the installation of offshore applications is therefore not a possible price advantage, but the opening up of suitable new locations. For Germany, the installed capacity that could be realized largely conflict-free is estimated to be up to 25GW.

During a 20-year service life, a wind power plant can generate approximately 80 times the amount of energy that is currently required for its manufacture, utilization and disposal, depending on the location (Bundesverband Windenergie, 2001). The systems can thus recover the energy required for their construction in approximately three months. Like for many renewables, no greenhouse gas emissions arise during system operating, but due to the energy required for manufacture and disposal. Average CO_2 emissions per kilowatt hour of wind-generated electricity depend on the electricity mix in the country of manufacture and on the system. The more advanced the transformation process towards a sustainable energy supply system, the lower the specific CO_2 values. No explicit values are therefore provided.

3.2.5.3
Environmental and social impacts

The following issues occasionally lead to reservations vis-à-vis wind energy due to possible environmental and social impacts:

- *Land use:* On the one hand, wind energy is currently one of the most economic forms of renewable energy generation, on the other hand it is also characterized by comparatively low energy densities. Large areas of land are therefore required to generate significant quantities of energy, typically 0.06–0.08km^2 per megawatt (EUREC Agency, 2002). However, there is no reason why land on which wind power plants have been installed should not continue to be used for agricultural purposes, so that the actual land use is only about 1 per cent of the above figure (e.g. for foundations, access roads) and is thus very small.
- *Noise pollution:* Wind power plants generate mechanical and aerodynamic noise. Both components have already been reduced successfully through modern technology (acoustically optimized rotor profile, direct drive generators, moderate speeds). Provided adequate distances to settlements are maintained, noise emissions from modern wind power plants are therefore no longer a problem.
- *Visual intrusion:* Occasionally, wind power plants are regarded as visually intrusive. While this subjective effect is difficult to quantify, it is nevertheless one of the main obstacles for the expansion of wind energy. Shadows and reflections are also regarded as visual interference. However, these effects can largely be avoided through careful selection of the location and through appropriate technology (e.g. matt finish).
- *Nature conservation and offshore systems:* The environmental effects of offshore wind energy utilization are currently the subject of intensive ecological research (BMU, 2002c). Among the factors being examined are the location and size of wind parks, noise emissions and the effects of energy transmission on birds, marine mammals and fish. Competing utilization of the sea by, for example, the fishing industry, the military, the steel industry and shipping also have to be considered.

3.2.5.4
Evaluation

Electricity from wind energy can already be supplied cost-effectively under current political boundary conditions. The environmental relevance of the technology (resource use, greenhouse gas emissions, material recycling) is regarded as positive. The Council therefore advocates further speedy expansion of this renewable energy source. Only a certain proportion of the calculated global technological potential can be regarded as sustainable. The Council recommends a figure of approximately 140EJ per year as the contribution from wind energy achievable in the long term for a sustainable energy supply.

3.2.6
Solar energy

3.2.6.1
Potential

The solar energy potential of four different technologies was calculated:
- Centralized solar thermal power plants with optical concentration (Fig. 3.2-3),
- Centralized photovoltaic power plants without optical concentration (Fig. 3.2-4),
- Decentralized photovoltaic modules without optical concentration (Fig. 3.2-5),
- Thermal solar collectors (Fig. 3.2-6).

Without specifying a particular technology, photovoltaic systems were assumed to have an annual system efficiency of 25 per cent by 2050. For solar thermal systems, the potential for combined heat and power was not considered. For solar collectors for heat generation, an annual efficiency of 40 per cent is regarded as achievable by 2050. The technological potential was calculated based on this conversion potential, taking account of land and surface use restrictions (see Box 3.1-1 for a definition of the different types of potential). Each of the conversion technologies considered yields values that correspond to multiples of all future projections of human energy use for the respective sectors. Against this background, the global technological potential can be regarded as practically unlimited. The annual flow of solar energy is distributed relatively uniformly over the Earth's more densely populated regions. This advantageous characteristic is reflected in the maps showing the potential of technologies without optical concentration. However, considerable seasonal fluctuations occur in higher latitudes, which have to be compensated by other technologies.

3.2.6.2
Technology/Conversion

PHOTOVOLTAICS

Photovoltaic cells ('solar cells') convert light directly into electric energy. Solar cells currently consist of several layers of different semiconductor materials. They are electrically connected in series and encapsulated within modules. In this way, technologically manageable electrical voltages are generated. The encapsulation protects the semiconductor elements from environmental influences and guarantees long service life. Photovoltaic modules can be connected to module arrays and connected to consumer loads or to the grid via suitable electronics. A storage element (e.g. rechargeable battery) is usually integrated in systems that are not connected to the grid.

Since the development of the first solar cell in 1954, the dominant raw material for solar cell production has been crystalline silicon. Due to the required high purity of the material, solar cells at current cell thicknesses are rather expensive. Numerous alternative technologies are being developed with the aim of significant reductions in material use and costs (Luther et al., 2003; Table 3.2-11).

For long-term global strategies, in addition to efficiency and price, the availability of raw materials should also be considered in the assessment of solar cell technologies. Silicon as the second most frequent element of the Earth's crust is non-critical, while other elements such as indium and tellurium, which are used in some thin film technologies, could become scarce at high production rates.

A further consideration is the energy payback period of the systems. In central Europe, modern grid-connected systems recover the energy required for their manufacture within approximately three years. This period is short compared with the confirmed technological lifetime of the systems of more than 20 years (Pehnt et al., 2003). It should be noted that today's systems are not optimized in terms of energy payback period.

Like for wind energy, any greenhouse gas emissions that may occur are not generated during the operation of photovoltaic systems, but during manufacture and disposal. The average CO_2 emissions per kilowatt hour of solar electricity therefore depend on the electricity mix within the country of manufacture of the system.

Since solar cells are connected to modules, which in turn can be combined to systems of any size, photovoltaics offers a wide range of possible applications. Remote micro-systems typically supply some tens of watts, large, grid-connected power plants can be designed for the megawatt range. While photovoltaic systems in remote rural areas are today usu-

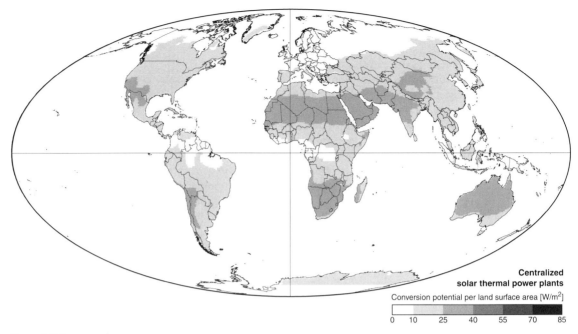

Figure 3.2-3
Global distribution of area-specific conversion potential for energy conversion via solar thermal power plants with linear optical concentration. The specified power densities relate to the surface area used. Like for photovoltaic power plants, the annual system efficiencies, defined according to the current status, depend on the latitude, since the relative active collector area decreases towards the poles due to shading. Combined heat and power was not considered. The resolution of the calculation was 0.5*0.5°, approximately corresponding to 50*50km. See Box 3.1-1 for a definition of the different types of potential.
Source: Kronshage and Trieb, 2002

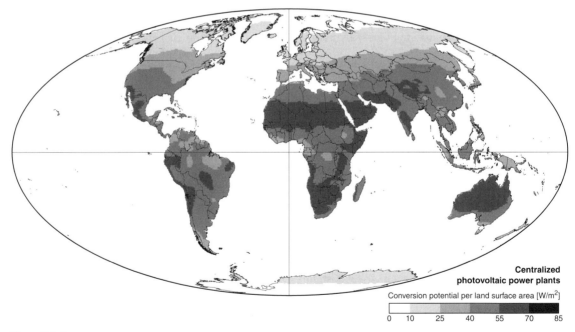

Figure 3.2-4
Global distribution of area-specific conversion potential for energy conversion via centralized photovoltaic power plants without optical concentration. The specified power densities relate to the surface area used. Like for solar thermal power plants, the annual system efficiencies, defined according to the status in 2050, depend on the latitude, since the relative active module area decreases towards the poles due to shading. The resolution of the calculation was 0.5*0.5°, approximately corresponding to 50*50km. See Box 3.1-1 for a definition of the different types of potential.
Sources: Efficiency estimation: WBGU; technical implementation of the map: Kronshage and Trieb, 2002

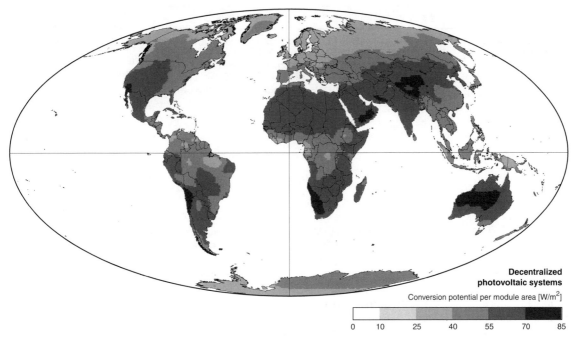

Figure 3.2-5
Global distribution of the area-specific conversion potential for decentralized solar-electric energy conversion via photovoltaic modules without optical concentration. The specified power densities relate to the tilted module area, not to the horizontal ground surface. The maps were based on the year 2050, with assumed annual system efficiencies of 25 per cent. The resolution of the calculation was 0.5*0.5°, approximately corresponding to 50*50km. See Box 3.1-1 for a definition of the different types of potential.
Sources: Efficiency estimation: WBGU; technical implementation of the map: Kronshage and Trieb, 2002

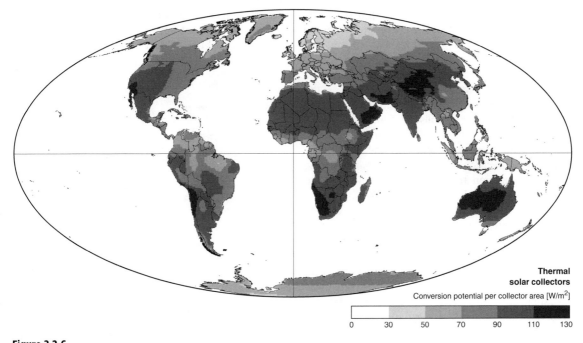

Figure 3.2-6
Global distribution of the area-specific conversion potential for decentralized energy conversion via thermal solar collectors. The specified power densities relate to the tilted collector area, not to the horizontal ground surface. The maps were based on the year 2050, with assumed annual system efficiencies of 40 per cent. The resolution of the calculation was 0.5*0.5°, approximately corresponding to 50*50km. See Box 3.1-1 for a definition of the different types of potential.
Source: Kronshage and Trieb, 2002

	2000	2020	2050
Main, market-dominating technologies	Crystalline silicon solar cells	Crystalline Si and thin film solar cells. Tandem solar cells for PV power plants with optical concentration. Organic and dye solar cells	Thin film solar cells (incl. Si), tandem solar cells, organic and dye solar cells, new concepts
Modul efficiency [%]	14–15	Si wafer module: 18–20 Thin film: 15 Organic, etc.: 10 Tandem: 40	no info available
Costs [€/kWh]	~0.6 (Location with 1000 full load hours)	~0.14 (Location with 1300 full load hours)	~0.06 (Location with 1300 full load hours)
Capacity range	W–MW	W–MW	W–MW

Table 3.2-11
Future development of photovoltaics. Due to the high modularity of this technology, applications meeting a variety of demands can be realized. It is therefore expected that the technological variety of photovoltaic electricity generation will persist. The energy costs are essentially inversely proportional to the annual insolation (Fig. 3.2-3). Source: WBGU

ally already more cost-effective than the alternative of expanding of the central grid, in situations where extensive grids already exist, such systems are far from competitive compared with large conventional power plants. As part of the exemplary Council scenario (Chapter 4), installation prices for the modules and the associated system components of €1 per watt are regarded as realistic for grid-connected systems in 2020. This is only approximately one-third of today's investment costs. For 2020, this would result in global average electricity costs of approximately €-cent 12 per kilowatt hour. The costs depend on the geographical latitude – in tropical arid regions the costs are only approximately half compared with Europe. In 2020, crystalline silicon is expected to be mass-produced and reach an annual capacity in the order of 10GW. By then, mass production should also have been achieved for various thin film technologies (Lux-Steiner and Willeke, 2001). In future, different solar cell technologies will continue to be used in parallel, with the specific choice being determined by the costs, the application area, the regional field of application and the technological availability.

In the medium and long term, further technologies that are currently being developed in the laboratory may contribute to cost reduction and tapping of new areas of application:
- Photovoltaic power plants with optical concentration and capacities from several 100kW to MW combine cost-effective concentrators (e.g. Fresnel lenses) with highly efficient solar cells (Fig. 3.2-7).
- Solar cells based on dyes or organic compounds, for example.

Basic research is presently also exploring visionary photovoltaic concepts, the potential of which is difficult to estimate at this stage (Chapter 6). Currently

achievable efficiencies of typical technologies are summarized in Table 3.2-12.

SOLAR THERMAL POWER GENERATION
In solar thermal power plants, direct sunlight is concentrated onto an absorber via optical elements. The absorbed radiation energy heats a heat transfer medium. This heat energy can subsequently be used for driving largely conventional engines such as steam turbines or Stirling motors. Solar thermal power plants are therefore closely associated with classic power plant technology, with solar energy replacing fossil fuels. All solar thermal systems built

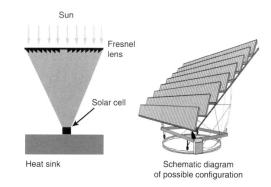

Figure 3.2-7
Schematic diagram of future solar power plants based on photovoltaic systems with optical concentration. The solar radiation is concentrated onto a very small solar cell area via cost-effective lenses (left). The concentration factor could reach a value of 1,000. The module systems have to track the sun (right). It is expected that any additional costs for the optical components and the mechanical tracking system will be compensated through savings and efficiency increases for the semiconductor solar cells. Power plants using this technology could already be used in the near future for supplying peak electricity (caused by, for example, electric cooling systems).
Source: WBGU

Table 3.2-12
Efficiencies of solar cells in the laboratory and in flat modules. The figures in brackets relate to values from pilot production or initial commercial production. *W* wafer technology, *T* thin film technology. Sources: Green et al., 2002; UNDP et al., 2000; Hein et al., 2001; Hebling et al., 1997

Cell technology	W/T	Efficiency in the laboratory [%]	Efficiency of the module [%]
Monocrystalline silicon	W	24.7	13–15
Multicrystalline silicon	W	19.8	12–14
Amorphous silicon (incl. Si-Ge-tandem)	T	13.5	6–9
Copper indium/gallium diselenid	T	18.9	(8–11)
Cadmium telluride	T	16.5	(7–10)
III-V concentrator cells (incl. tandem and triple)	W and T	33.5	(25)
Crystalline thin film silicon cell	T	19.2	
Organic and dye solar cells	T	2–11	

to date are based on strong concentration of sunlight, so that their application only makes sense in regions with a high ratio of direct solar radiation. Three different technologies have already been realized:

- *Parabolic trough power plants:* Solar radiation is focussed onto tubular light absorbers, usually containing special oil as heat transfer medium, in linear reflectors with parabolic shape that track the sun around one axis. The oil is heated to approximately 350–400°C and subsequently generates steam in a heat exchanger for a largely conventional steam turbine. Such systems can be designed for relatively large capacities, currently between 30 and 80MW. In California, such power plants have already been supplying electricity for more than 10 years, with a total installed capacity

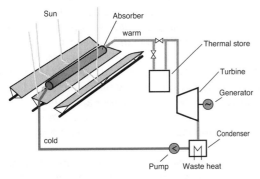

Figure 3.2-8
Schematic diagram of a future solar thermal trough power plant. In this example, the solar array is arranged as a Fresnel concentrator, with the required parabolic shape being realized as a segmented reflector array made up of smaller flat glass mirrors. In contrast to systems commonly used today, it is likely that in future water will be used as the heat transfer medium and evaporated directly in the absorber tubes. Such systems are expected to achieve lower costs and to avoid problems with thermal oils. The thermal store shown in the diagram is intended to enable operation of the power plant even after sunset.
Source: WBGU

of approximately 350MW. Further significant cost reductions could be achieved through direct vaporization of water within the absorber tubes (Fig. 3.2-8).

- *Solar power towers:* A large array of movable mirrors focuses the sunlight onto a receiver installed on a tower, where the heat transfer medium (water, salt, air) is heated to 500–1,000°C. Due to the high temperatures, the energy can, in principle, be coupled directly into a gas turbine or a modern combined cycle plant. Capacities of around 200MW have been proposed for solar power towers, which is approximately 10 times the capacity of current pilot plants.
- *Parabolic dish power plants:* This system uses parabolic mirrors to track the sun. A heat transfer medium at the focus of the mirror can be heated to 600–1,200°C. Such systems are usually rather small (some 10kW of nominal capacity). They therefore lend themselves for decentralized applications. Engines are used to convert the heat energy into mechanical energy and subsequently into electrical energy. The technology is currently at an experimental stage.

In general, the aim is to design systems with high operating temperatures, due to the higher efficiencies that can thus be achieved during the conversion of heat into electricity in thermodynamic machines (Table 3.2-13).

Medium-term heat storage (hours, days) can significantly extend the scope for solar thermal power plant technology. Using melted salts as a storage medium, solar thermal systems have already been realized that generate electricity around the clock without conventional auxiliary heat supply.

The close relationship between solar thermal systems and conventional power plants enables the integration of fossil heating and solar thermal technology in so-called hybrid power plants. For regions with high incident solar radiation, the extension of exist-

	2000	2020	2050
Main technologies	Parabolic troughs	Parabolic troughs and solar power tower	Parabolic troughs and solar power tower
Electrical efficiency [%]	14	20–25	25–30
Costs [€/kWh]	0.14–0.2	~0.07–0.14	~0.06
Features		Integrated heat store for several hours	Integrated heat store for several hours
Capacity range	Several tens of MW	Several tens up to 100MW	Several tens up to 100MW

Table 3.2-13
Efficiencies, costs, capacity range and special features of solar thermal power plants in pure solar operation. The term 'parabolic troughs' also includes systems based on Fresnel collectors (Fig. 3.2-8). The efficiency figures relate to the electrical annual system efficiencies per aperture area and per direct vertical insolation on the converter surface ('direct normal incidence', DNI)
Source: WBGU

ing fossil power plants with a solar thermal component is seen as an important short and medium term market for renewables.

Another possibility is the combination of solar thermal power plants with thermal biomass utilization. This would enable continuous operation of multi-megawatt power plants exclusively based on renewables.

Solar thermal power plants can also be operated in combined heat and power mode. An example is electricity generation and the simultaneous extraction of potable water through sea water desalination (e.g. via thermal distillation processes). Solar efficiencies of up to 85 per cent are possible in combined heat and power mode.

In the long term, the concepts of parabolic troughs or solar power towers appear more promising than smaller parabolic dish systems, since compared with photovoltaics the latter have the disadvantages of two mechanical systems that are subject to wear and tear (optical alignment of the parabolic dish, thermodynamic machine). In sunny regions, parabolic troughs and solar power towers are currently by far the most cost-effective option for generating electricity using solar energy in large power plants. Due to their different features, it is expected that both technology options will be used widely, depending on the underlying conditions (size of power plant, infrastructure, local electricity demand, options for the transmission of electricity over long distances, incident solar radiation, etc.).

SOLAR HEAT
Solar collectors convert solar radiation into heat. Typical applications are swimming pools, domestic or industrial hot water, space heating, or process heat, with the required temperature increasing in the above order. The higher the required temperature level, the more complex the collector, since more effort is required to avoid heat losses. A distinction is made between the following basic types of thermal collector:

- *Collectors with unglazed absorber:* In this simplest form of solar collector, a heat transfer medium (e.g. water) flows through black plastic mats without cover. High heat losses occur through convection and conduction, so that only moderate temperatures can be achieved. Due to their low costs, such collectors have become established for swimming pool heating.

- *Flat plate collectors:* In this collector type, heat losses are usually reduced by two measures. On the side facing the sun, the absorber is covered with a glass pane, and at the back it is insulated. A further option is the use of optically selective absorbers with strong absorption characteristics in the visible and near infra-red spectral range, but little emission in the thermal infra-red range. Compared with unglazed absorbers, significantly higher temperatures can thus be achieved. Flat plate collectors are currently predominantly used for heating domestic or industrial water. In Germany, typically approximately 60 per cent of the related annual heat demand can be covered by the solar contribution. While demand can usually be met fully in the summer, in the winter the water tends to be pre-warmed by the collector and subsequently heated by conventional means. Thermal collectors are also increasingly used in so-called combination systems to support space heating.

- *Evacuated tube collectors:* With this type of collector, vacuum insulation almost completely eliminates heat losses through conduction and convection from the absorbers, which are located within glass tubes. Such systems can achieve good efficiencies and high temperatures, even in winter. This type of collector is particularly suitable for higher geographical latitudes, winter conditions or process heat applications. Due to the hermetic encapsulation of the absorber, they are also beginning to be widely used to heat domestic or industrial water in countries with less developed technological infrastructure.

The heat energy collected in a solar collector system is usually transferred to a heat store, so that differences in heat demand and supply can be compensated over several days. Seasonal storage systems that enable heat collected in the summer months to be utilized during the winter have also been tested successfully in pilot plants. Since heat losses increase with the ratio of surface to volume, seasonal heat stores (as far as they are based on the utilization of sensible heat) should be comparatively large, which necessitates their integration into a local district heat network.

In principle, the modularity of solar collectors enables them to be used in any output range that may be required. For a typical location in Germany, the costs for heat generated in this way are currently €-cents 3–7 per megajoule (BMU, 2002b), for sunnier locations costs are correspondingly lower. The materials used in thermal solar collectors are usually environmentally compatible and can moreover be almost completely recycled.

COOLING WITH SOLAR ENERGY
Cooling and air-conditioning of buildings are ideal applications for solar energy, since demand largely coincides with the availability of solar energy. Two techniques are available: On the one hand, solar electricity can be used to operate a conventional compression-type refrigerating machine. On the other hand, solar heat can be used to drive thermodynamic cooling processes via sorption or adsorption techniques. Hybrid cooling systems utilizing solar and conventional energy can save more than half the primary energy required for the operation of a conventional system.

As space cooling is associated with significant energy consumption for newly industrializing and developing countries in lower geographical latitudes, solar cooling technologies are an interesting alternative: Furthermore, solar cooling technologies are well suited for the usually decentralized energy structures in these countries. In future, the significance of cooling and air-conditioning technologies utilizing solar heat may increase considerably for many areas of application and world regions.

3.2.6.3
Environmental and social impacts

Possible negative environmental effects of photovoltaic energy technology result from the manufacturing process and the materials used in the end product. While in principle the environmental impact of the manufacturing process can be reduced to low levels by implementing appropriate measures and

technologies, environmental risks cannot be ruled out completely for some thin film technologies. In principle, toxic substances could be released during accidents (e.g. fires) or in the event of damage to the solar modules. Safe recycling is an important factor for the application of these technologies. No such risks exist if silicon is used for solar cells. Solar thermal power plants are benign, as long as environment-friendly heat transfer media are used. The same is true for solar collectors. Moreover, the potential maps in Chapter 4 show that supplying Europe with solar energy through energy imports from neighbouring world regions with high incident solar radiation is easier to realize than an autonomous energy supply system (Fig. 4.4-5). The creation of an associated infrastructure in the Maghreb countries and in the Middle East, including transmission cables to Europe, would be significant from a strategic policy angle as well as from a development angle.

3.2.6.4
Evaluation

Solar energy can be used as a source for solar electricity, hot water, space heating and cooling. Suitable technologies are available for all areas of application, although in some cases they still have to go through cost-reduction processes, and further improvements are required through research and development efforts. A significant expansion rate should be the aim for the medium term, so that reasonable cost-efficient solar technologies are available, once the expansion of other renewable forms of energy reaches the limits of their sustainable potential. In contrast to all other forms of renewables, based on all future projections of human energy consumption both the technological and the sustainable potential of solar energy is practically unlimited.

3.2.7
Geothermal energy

3.2.7.1
Potential

The energy source for near-surface applications of the ground heat, e.g. in heat pumps, is the sun. In contrast, systems that utilize the heat contained in lower strata tap into the Earth's heat sources, i.e. the thermal energy dating back to the time our planet was formed, but mainly the energy from the decay of radioactive elements. The very high temperatures in the Earth's interior (probably around 5,000°C) cause a continuous heat flow of approximately 0.1W per square metre towards the surface through the Earth's crust. In regions with geothermal anomalies, e.g. in volcanic regions, high temperatures occur close to the surface of the Earth. Below 100°C, this geothermal heat is only suitable for heating purposes, above 100°C it may also be used for power generation.

It is difficult to make general statements about the potential: While in principle near-surface utilization of geothermal heat is possible anywhere, so-called hydrothermal utilization is bound to the occurrence of hot water. The utilization of the heat contained in dry, deep rock is still under development.

According to estimates, within the next 10–20 years economic reserves will reach the level of the current global primary energy use (UNDP et al., 2000). However, in order to avoid global utilization of geothermal heat becoming unsustainable, the amount of geothermal heat being extracted should be limited to the Earth's natural heat flow. The potential available regionally is often unknown. Due to open questions regarding the technological implementation and various sustainability aspects of utilization (e.g. unresolved disposal of large quantities of waste heat due to lower conversion efficiencies in low-temperature processes), the Council has assumed a realistic and sustainable potential of 30EJ per year by 2100.

3.2.7.2
Technology/Conversion

DEEP, HOT ROCK OR HOT SEDIMENTS
In the so-called hot dry rock technique, cold water is pressed through deep boreholes, and heated water is extracted. For the heat exchange with water, dry hot rock strata with cracks and fissures are required. With this technique, comparatively high temperatures of approximately 100–180°C can be achieved. At other locations, water is pressed through the pores of hot sedimentary rock, although this technique is limited to temperatures of around 100°C.

In both cases, the extracted heat can be fed into local or district heat networks, or used as process heat in industry. The higher the temperature, the more efficient the electricity generation. While at temperatures above 150°C, steam can be used directly for electricity generation in appropriately adapted conventional steam turbines, so-called binary systems will usually have to be used for lower temperatures. In these systems, the heat energy contained in the water is transferred to another liquid within a heat exchanger. Depending on the temperature of the geothermal heat, such systems only reach electrical

efficiencies of 10–16 per cent, at temperatures of about 80°C even less (BMU, 2002b). Significant quantities of waste heat are thus generated, which can be utilized locally or has to be disposed of. Current cost estimates for electricity generation vary between €-cent 7–15 per kilowatt hour. Due to the constancy of the geothermal heat flow, geothermal power plants with capacities in the megawatt range are particularly suitable for base load operation.

HYDROTHERMAL SYSTEMS
In contrast to the hot dry rock application, at other locations steam or hot water may already be present in the ground, which can be extracted through boreholes and used directly for heating or electricity production. The water should be returned to the same depth via a second borehole in order to maintain the water cycle, and to avoid the contamination of surface waters through the high mineral content of the water extracted from the ground. At temperatures between 40–120°C, geothermal heat from thermal aquifers has hitherto only been utilized for space and water heating.

NEAR-SURFACE GEOTHERMAL HEAT
The key technology for utilizing near-surface geothermal heat is the heat pump. It transforms heat from a lower to a higher temperature level, consuming additional energy in the process. Usually, geothermal heat in the 5–10°C temperature range is extracted from the ground through heat exchangers placed at a depth of 1–2m.

All heat pumps require high-quality energy for their operation, which must be taken into account of for their evaluation. The so-called coefficient of performance (COP), i.e. the ratio of useful energy (utilized heat energy) and energy used (e.g. electricity, gas), is used for this purpose. Since in conventional power plants only approximately one-third of the primary energy is converted to electricity, a heat pump running on electricity at today's energy mix should have an annual COP of significantly more than 3.6. With all other heat pumps, for which the energy balance does not include waste heat losses at power plants, a COP of 1.1 is adequate. Current prototypes and small versions of thermally driven heat pumps (natural gas) achieve annual COPs of 1.3. Electrically driven compression heat pumps achieve COPs of more than 3.6.

In combination with heat pumps, the ground can be used as a heat store, if the same systems are used for cooling/air conditioning in the summer. In this case, the waste heat generated during the cooling process is stored in the ground, thereby increasing the temperature for heating operation during the winter.

3.2.7.3
Environmental and social impacts

Hot water extracted through deep boreholes has to be re-injected, since it not only contains minerals, but also hydrogen sulphide, ammonia, nitrogen, heavy metals and carbon dioxide. Suitable technology already exists. For electricity generation with thermodynamic machines based on volatile media, the organic medium used should be non-toxic and not have significant greenhouse potential. Due to the low flow temperatures and low conversion efficiencies, large quantities of waste heat are generated locally during geothermal electricity generation, which have to be disposed of.

For electric heat pumps, the total energy balance has to be considered carefully. Low-temperature heat sources also have to be selected carefully: Strong cooling of ground water should be avoided, but the extraction of heat from rivers subjected to waste heat loads may in fact be ecologically sensible.

3.2.7.4
Evaluation

Geothermal heat has large technological potential. In contrast to solar and wind energy, it is continuously available. The Council nevertheless estimates the sustainable potential by 2100 only very cautiously at 30EJ per year.

Geothermal heat from large depths at high temperature levels can be used for electricity generation, local and district heat networks, or a combination of both. On the other hand, only thermal applications are suitable for lower temperature levels. The Council recommends the further development of relevant technologies and the promotion of their more widespread use. Care needs to be taken that heat pumps using near-surface heat have adequate COPs.

3.2.8
Other renewables

In addition to the technologies for utilizing renewable energy sources described so far, for which large-scale application within a sustainable energy system can be assumed, first attempts are being made to convert renewable energy sources that are at an early stage of development.

The marine energy sources of tidal and wave energy fall into this category. Under certain geomorphological conditions, where a flow of water is restricted and thus accelerated, the velocity of the tidal water flows may be sufficient for energy gener-

ation purposes. Due to the significantly higher density of water compared with air, much lower flow velocities are required than for wind energy. Tidal flows with speeds as low as approximately 1m per second appear to be suitable for utilization. In estuaries and river mouths, the tidal amplitude can be more than 2m and can be used for the operation of turbines in tidal power plants. Wave energy is created through the interactions of the surface of the sea and the wind. The energy density increases with the distance from the coast. The technological challenge is therefore to develop systems for locations at significant distance from coastlines. A variety of concepts have already been proposed, some of which are being tested.

In many processes of the chemical, petrochemical or associated industries, fossil energy carriers are not only used as raw materials for products, but are also utilized to some extent for energy generation. In order to decouple energy generation and the utilization of fossil energy carriers for non-energy purposes, the sun can take over the energy function directly in many cases. For high-temperature processes, for example, technological concepts can be used that are similar to those used in solar thermal power plants (Section 3.2.6.2).

Sunlight can also be used in photochemical and catalytic applications, which are currently dominated by artificial light sources. In principle, the photochemical synthesis of liquid energy carriers should be possible. Other promising approaches for the future utilization of solar energy are membrane systems employing processes similar to photosynthesis, or photochemical and biological production of hydrogen.

The research section of this report (Section. 6.3.1) addresses energy conversion concepts that the Council considers to be capable of developing into marketable technologies over the next 10–20 years. Overall, the Council cautiously estimates the potential of these emerging technologies at 30EJ per year by 2100.

3.3
Cogeneration

3.3.1
Technology and efficiency potential

In combined heat and power (cogeneration) plants, fuels are fired not just to generate electricity but also to utilize heat that would otherwise go to waste. This heat can then be used for e.g. heating purposes. In this way, such plants attain a high degree of utilization of the fuel employed, possibly up to 80 to 90 per cent with well designed plants. Cogeneration is therefore an important technology for conserving primary energy.

Cogeneration plants may be employed wherever, alongside electricity generation, there is a demand for low-grade heat (up to around 120°C) or process heat (up to around 200°C). There is a wide selection of cogeneration technologies available in a power range from $1kW_e$ to several hundred MW_e (Table 3.3-1). The electricity utilization factor today attainable during cogeneration operation ranges from 15 per cent for smaller steam turbines to 45 per cent in high efficiency engine-generator sets and, in future, this could climb to 60–65 per cent, for example in combination power plants employing fuel cells when this technology matures. The electricity-to-heat ratio varies accordingly between 0.20 and 1.50 and, over the long term, may even attain 2.50. High ratios favour the application of this technology, since the trend is for heat demand to reduce relative to electricity consumption at typical sites. Additionally, such plants will fare economically better, as usually revenues from electricity are higher than from heat.

As fuel, fossil energy sources, like coal, oil or gas, may be fired but fuels from renewable sources and, in future, even hydrogen may be employed. Steam turbines and Stirling engines may also fire solid fuels, like coal and wood, but all other technologies require liquid or gaseous fuels, in some cases with high purity requirements. For fuel cells, in addition hydrogen must be won from fuels containing this element. The actual energy transformation component is therefore only a part of the complete system that must always be considered as a whole in terms of efficiencies and costs.

As a rule, cogeneration plants are more efficient in their use of primary energy than if electricity and useful heat were to be furnished separately. The energy savings that can be realistically attained and the resulting cut in CO_2 emissions depend very greatly on the size and type of the cogeneration plant, its design parameters, reference system and fuels fired. Typical primary energy savings with cogeneration plants are 15–30 per cent, corresponding to a CO_2 abatement of up to 50 per cent, when comparing natural gas-fired cogeneration with separate generation of electricity and heat from coal. When the avoided additional heat generation is credited, typical specific CO_2 emissions of cogeneration plants firing natural gas are 0.19 to 0.25kg per kilowatt hour$_{electrical}$ (Nitsch, 2002). The higher the overall degree of utilization and electricity-to-heat ratio of a cogeneration plant, the greater its energy and ecological advantages; consequently, over the longer term, relevant technologies – combined cycle, fuel cells and high efficiency engines – are more advantageous.

Table 3.3-1
Overview of technical data of systems with full cogeneration: figures in brackets are the maximum electricity-to-heat ratios that will be attainable in future. *cogen* cogeneration, *PAFC* phosphoric acid fuel cell, *PEMFC* proton exchanger membrane fuel cell, *MCFC* molten carbonate fuel cell, *SOFC* ceramic solid oxide fuel cell.
Sources: Nitsch, 2002, and WBGU

Technologies	Capacity [MW]	Electrical efficiency [%]	Electricity-to-heat ratio	Technology status and potential
Steam turbine cogen plant	1–150	15–35	0.20–0.50	Mature
Gas turbine cogen plant	0.5–100	25–35 (40)	0.30–0.60	Still development potential
Combined cycle cogen plant	20–300	40–55 (60)	0.60–1.20	Still development potential
Small scale engine cogen units	0.005–20	25–45 (50)	0.40–1.00	Still development potential, especially for small outputs
Spark-ignition	0.005–1	25–37 (45)	0.40–0.80	
Diesel	0.05–20	35–45 (50)	0.60–1.00	
Micro-gas turbines	0.02–0.5	20–30 (35)	0.30–0.50	Still significant development potential, market penetration commencing
Stirling engines	0.001–0.05	30–35 (45)	0.30–0.60	Still development potential, market penetration commencing
Fuel cells	0.001–20	30–50 (60)	0.80–1.50	Still major development potential
PAFC	0.1–0.2	35–40	0.80–1.00	
PEMFC	0.001–0.2	30–40 (50)	0.80–1.00	
MCFC	0.1–10	45–50 (55)	1.00–1.40	
SOFC	0.001–20	40–45 (60)	1.00–1.50	

3.3.2
Range of applications

In principle, by applying cogeneration it is possible to meet a very high share of the demand for space heating and hot water, amounting to some 65 per cent, and of the demand for industrial process heat, amounting to up to 80 per cent of medium-grade heat. But cogeneration technologies may also be applied for cooling, air drying, air-conditioning and seawater desalination, for example by solar-thermal power plants in countries with high insolation.

In urban areas, the possibilities for cogeneration range from supplying single- and multi-family houses, office blocks and commercial buildings through groups of buildings up to entire housing estates and industrial or commercial zones. Additional options for cogeneration result from more effective utilization of existing district heating supply networks and replacing heating plants by cogeneration plants.

In industry, the focus is on modernization of 'traditional' cogeneration, for example replacement of steam and simple cycle gas turbines by IC engine, combined cycle and, over the medium term, fuel cell plants, but also on expansion of process heat generation. Decentralized cogeneration plants, like IC engines, gas and micro-gas turbines, Stirling engines and fuel cells, are undergoing rapid technical and economic development, which means that the range of application for cogeneration is steadily growing.

At present, cogeneration in Germany supplies 165TWh heat per year, or 11 per cent of the total heat demand, and 72TWh electricity per year, or 13 per cent of total electricity generation, at an electricity-to-heat ratio of 0.43. Currently, it prevents CO_2 emissions of around 30 million tonnes per year compared to the separate generation of electricity and heat. This means that Germany is a mid-field runner among European countries. Outside of Europe, cogeneration is rather more restricted in application. But the theoretical potential of cogeneration in Germany at around 500TWh$_e$ electricity, or some 110GW$_e$ of installed capacity, is approximately seven times today's share, and thus corresponds to virtually the entire present electricity generation. For this calculation, an appreciable reduction in heat demand down to some 60 per cent of today's level and continued evolution of cogeneration technologies with a corresponding increase in the average electricity-to-heat ratio to roughly 1.0 has been assumed (Nitsch, 2002). By 2030, a relatively certain total potential of around 200TWh cogeneration electricity per year and 280TWh useful heat per year may be expected, corresponding therefore to a three-fold increase of the current level, with an average electricity-to-heat ratio of 0.7. To attain this, about 20,000MW$_e$ of additional cogeneration capacity would be required, of which some two-thirds would fall on decentralized plants (rated less than 10MW$_e$). The mobilizable CO_2 abatement potential is around 80 million tonnes CO_2 per year, and corresponds to roughly 7 per cent of German emissions in the reference year, 1990.

3.3.3
Economic performance

From the aspect of provision of electricity, a cogeneration plant is financially viable if the additional expenditures for extraction of heat and reduced electricity generation in comparison to conventional power plants are at the very least offset by revenues from heat sales. For today's larger turbine- and engine-based cogeneration plants, the electricity generation costs so determined are €-cent 3.5 per kilowatt hour$_{electrical}$ and for smaller plants €-cent 6 per kilowatt hour$_{electrical}$.

From a macro-economic viewpoint, application of cogeneration would be worthwhile even now, especially if external costs are factored in. Like in the past, it may be expected that capital and running costs will drop even more, particularly for smaller cogeneration units. But if comparisons are made on the basis of short-term marginal costs, that is the currently low costs of purchasing electricity from the present already partially depreciated power plant inventory, the financial performance of cogeneration plants will only be satisfactory in exceptionally favourable cases. The financial problems of existing cogeneration plants are due almost solely to the drop in electricity prices following liberalization of electricity markets. Consequently, the further market prospects for cogeneration plants depend greatly on the energy policy setting, for example the German law on cogeneration.

For innovative cogeneration systems, like those employing fuel cells, Stirling engines and micro-gas turbines, their introduction faces similar obstacles. To gain broad market acceptance they will have to attain as a minimum the cost level of cogeneration systems already deployed today, which means, for fuel cell cogeneration plants, installation costs of €1,000 to 1,200 per kilowatt hour$_{electrical}$, signifying a necessary cost reduction by one order of magnitude. When considering the learning curves of comparable decentralized technologies, this should certainly be attainable with larger market volumes.

3.3.4
Evaluation

Thanks to its energy and ecological advantages combined with its economic performance, especially if external costs are considered, cogeneration technology is an indispensable component of any efficiency-raising strategy for energy-transforming systems. Today's cogeneration technologies in particular fit very well into the emerging trend for greater inter-meshing and decentralization of energy supply.

Extending cogeneration is therefore a significant specific measure for raising the efficiency of energy systems on the supply side. Although from the macro-economic aspect this technology's performance is already positive, the conditions currently prevailing on the market are problematical. The Council therefore recommends that expansion of cogeneration be fostered to a greater degree by enhancing the energy policy setting.

3.4
Energy distribution, transport and storage

3.4.1
The basic features of electricity supply structures

The regional structures of global energy supply vary greatly since energy supply and demand depend on numerous local factors. Basically, a distinction can be made between contiguous, densely populated regions and expansive, sparsely populated regions. While the former often have electricity and gas grids over large areas, the latter usually have a distributed approach using microgrids and single-home power systems (off-grid concepts; Section 3.4.2). The quality of the energy supply does not necessarily differ in these two fundamentally different concepts. Supply strategies for large grids deserve special attention as they handle the largest share of energy supply by far (Section 3.4.3).

In electricity grids, the amount of electricity generated and consumed always has to be the same. If the amount generated exceeds current consumption, the frequency and voltage of the grid rise; if generation is lower, they drop. If no remedy is provided, equipment connected to the grid could be damaged. Hence, knowledge of the statistical and empirical shares of generation and consumption are decisive in controlling electricity grids; characteristic daily and annual trends help to this end. As grids consist of several voltage levels, the various load curves have to be taken into consideration at all levels; even if the high-

voltage level is balanced, a part of the grid at the medium-voltage level could be overloaded, for example. As fluctuating renewable energy sources such as wind power and photovoltaics become more important, the supply side is also becoming more dynamic. There are several strategies for the optimal coordination of fluctuating energy supply and demand:

- Adaptation of energy demand to supply (load management);
- Modification of structures on the generation side with the aim of adapting the generation of electricity to energy demand;
- Large-area networking of power producers and consumers to use statistical balancing effects to match generation and consumption;
- Storage of energy.

3.4.2
Supply strategies for microgrids

Rural areas in developing countries not only lack such basic goods as clean drinking water and telecommunications, but also energy services (Section 2.4). Expanding the grid in these regions is unrealistic in many cases as the low expected power consumption of users would not justify the high cost of extending the grid over long distances. Thus, electricity is preferably supplied from distributed sources, as are other energy services.

The technologies for distributed power supply fit into two categories: isolated systems for single users and microgrids for larger user groups, such as villages. Diesel, wind power, photovoltaics, and micro-hydropower are used to generate electricity as they can be tailored to the local needs perfectly and combined to reduce costs. Isolated systems for individual users usually only have to satisfy very small energy needs. Photovoltaics is best suited for this purpose as suitably sized PV-generators are available, and operation of the systems is low-maintenance. The typical technology for individual systems is called a Solar Home System. Such a system generally comprises a photovoltaic module, a battery, and a charge controller. While smaller systems are designed for direct current for, e.g., several fluorescent lamps and a radio, larger systems can provide alternating current and hence even run colour televisions and other appliances. One major obstacle towards the introduction of such systems is the relatively high initial investment, combined with a lack of microcredit. Hence, fee-for-service concepts are often used; here, an investor installs a system for the customer, who can only use the power for a fee. Overall, the socio-economic barriers are often greater than the technical ones when it comes to such individual solutions. If the costs of the technologies used today (primary battery, petroleum lamps, diesel generators) are calculated for the service life of systems using renewable power, photovoltaics, hydropower and wind power generally turn out to be more economical. Another type of individual application that is widespread are photovoltaic pump systems. These systems often create additional income by increasing agricultural production.

Under certain conditions (such as when the space between homes is small enough), it might make more sense to set up micro-systems for multi-user groups instead of using individual systems. The greater demand for electricity provides more flexibility in the selection of the power generator. In addition, a microgrid can be designed to allow for a connection to the national grid when it is extended.

3.4.3
Supply strategies within electricity grids

3.4.3.1
Fluctuating demand in electricity grids

The energy demand in expansive electricity grids like the European interconnected power grid consists of a large number of consumers with different consumption patterns at different times of the day and year. The consumption of most consumers can be predicted: electricity is needed at certain times of the day for cooking and lighting. While the exact times at which appliances are switched on and off cannot be predicted, the large number of spatially distributed power consumers level out the statistical fluctuations, leaving relatively smooth daily and annual curves.

The patterns of consumption vary between countries, regions and climatic zones. For example, in hot countries the use of air-conditioners affects the daily curve. In Germany, power is mostly needed during the day and in the evening. The base load is the lowest level of consumption in the daily pattern, which in Germany is approximately just below the halfway point for the daily peak. To cover fluctuating demand and ensure the power supply, base load power plants (such as coal-fired and run-of-river plants), which cannot increase and decrease output quickly, are used in combination with gas and pumped storage hydropower plants, which can be run up and down more quickly to meet fluctuating demand. Overall, enough power plants have to be available to cover peak demand. But since the plants do not generate power all the time, capacity provisioning is a cost fac-

tor. Hence, the gap between peak load and base load has been brought down in the past few decades by means of measures to control energy demand (load management of major industrial consumers, cheap night power, etc.).

3.4.3.2
Fluctuating supply from renewable sources

The supply of energy from renewable sources like wind and solar fluctuates greatly. The relevant time frames depend on the type of final energy (electricity or heat) and the potential to store it. As heat cannot be transported well, the daily load curve cannot be leveled out by means of a large heat grid. The options used to store heat are much more simple and local than those used to store electricity. The fluctuations in electricity can be viewed in seconds, minutes, hours, and over the yearly load curve.

FLUCTUATIONS IN SECONDS/MINUTES
In this time frame, random fluctuations occur with wind power and solar energy due to passing clouds and gusts of wind. These statistical fluctuations do not, however, correlate if the systems are in different locations: if many systems are networked, these short-term fluctuations are balanced (Fig. 3.4-1). Hence, in large-scale, bi-directional grids no problems are expected in this time frame. Even extreme fluctuations, such as emergency shutoffs during storms, are technically manageable. Power plants that can be easily regulated (gas turbines, pumped storage plants, etc.) have to compensate for the remaining variations in electricity production from wind and solar sources.

FLUCTUATIONS IN THE HOURLY RANGE
The power output of solar energy systems depends on the angle of solar radiance, which in turn depends on the time of day. Hence, the hourly fluctuations are easily predictable. The amount of time during which electricity can be generated from solar energy can be increased by connecting systems in an east/west direction (Fig. 3.4-2).

The use of solar thermal power plants for the generation of electricity can increase this time frame even longer. These power plants can deliver power for some five hours after sundown by means of thermal storage tanks (Nitsch and Staiß, 1997). It makes most sense to use such power plants in the western parts of a grid, for instance in Spain for Europe.

The wind speed does not have a salient daily load curve. On the average, there is more wind when there is a lot of cloud cover so that solar and wind correlate negatively. Hence, wind and solar energy converters

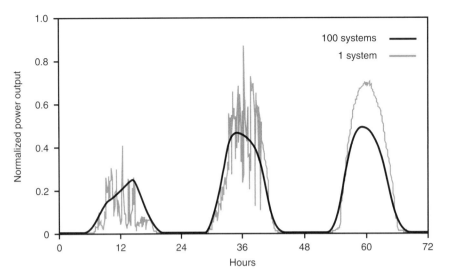

Figure 3.4-1
Levelling of fluctuations in the generation of electricity by linking a large number of photovoltaic systems. A comparison of the highly fluctuating single system and the mean of 100 systems with the same output in different locations.
Source: Wiemken et al., 2001

compensate for each other to a certain extent in the hourly range.

SEASONAL FLUCTUATIONS
The power that can be generated from solar energy varies from season to season and according to latitude. Near the equator and at high latitudes, rainy seasons and heavy clouds affect the yearly load curve additionally. In principle, a large-scale north/south grid can compensate for these effects (Fig. 3.4-3).

Seasonal fluctuations in wind power are more regional and harder to forecast than solar radiation. The supply of hydropower has a clear yearly curve (for instance, due to rainy seasons), which can partially be compensated for by storing water behind dams. As biomass can be stored readily, energy can be generated from it throughout the year.

3.4.3.3
Strategies for matching energy supply and demand

Meeting the fluctuating energy demand with the likewise fluctuating supply of energy from renewable sources also poses a considerable challenge. In the following, some tools are described to this end; all need to be adapted to special regional features.

ELECTRICITY TRANSPORT AND DISTRIBUTION
The fluctuations on the supply side (Section 3.4.3.2) recommmend having power provision technologies and regions linked in a large grid so that the energy supply can fulfil demand at any time of the day or year. As electricity can be easily transported, regions that are far from each other – and hence have differ-

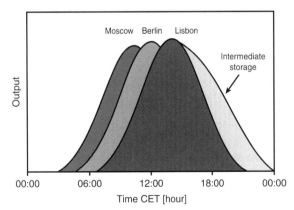

Figure 3.4-2
The supply of solar energy in Europe as a function of the time of day and location. The use of thermal intermediate storage can extend the daily operating time of solar thermal power plants.
Source: Quaschning, 2000

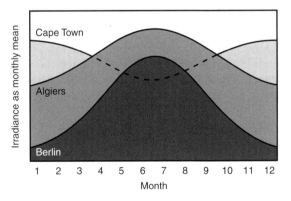

Figure 3.4-3
Annual curves of solar irradiance in the northern and southern hemisphere for Algiers, Berlin, and Cape Town.
Source: Quaschning, 2000

ent energy production times and consumption patterns – could be connected in a grid.

To help identify different regional supply strategies, a distinction needs to be made between the large-scale, global connection of regions and the finely meshed grids for the local population: one can imagine generating large amounts of power in areas that will not have any finely meshed grid for the foreseeable future. The energy could be transported from there to highly industrialized centres without any great losses. For example, the transport losses in high-voltage, direct-current lines for solar power from northern Africa to central Europe would be about 10 per cent over some 3300km. Indeed, transcontinental and intercontinental links can be envisioned all the way up to a 'global link', especially if there are breakthroughs in technology development (such as a superconductor).

CONTROL OF ELECTRICITY DEMAND, LOAD MANAGEMENT

The load curve can be adapted to the supply structure if proper incentives and technologies are used (Section 3.4.3.1). Such measures would reduce the transmission losses and the need for storage as the share of renewable sources grows. The potential is hard to estimate at present but is probably at least around 20 per cent of electricity demand in the winter and some 10 per cent in the summer, or 10 per cent of household consumption. Load-based rates are the main incentive for consumers; here, rate lamps or switches could be introduced for this purpose. This would provide the incentive for 'intelligent' appliances (refrigerators, electric cars, etc.) that only consume electricity when rates are low.

ENERGY STORAGE

In the long term, the transport and direct consumption of electricity in extended grids will probably remain less expensive than storage. Storage should therefore be kept to a minimum. Even if load management is optimized, however, fluctuations in the power grid are expected to be so great once the share of renewable energy sources has reached around 50 per cent that the daily and yearly load curves will exhibit both excess power and shortages, necessitating additional energy storage. There are a number of different technical solutions to meet the wide range of trigger speeds, output power, and storage capacity. The technologies can be roughly divided up into quite fast storage types with high output (such as capacitors, flywheels, superconductors) and slower ones with high energy content (pumped storage hydroelectric plants, compressed air tanks, electrochemical storage, etc.). To keep costs down, pumped storage hydroelectric plants are used most often to

provide large-scale grid support for mid- to long-term energy storage. Redox systems, especially those using hydrogen, are being developed for future storage systems.

SYNERGIES OF ELECTRICITY AND HEAT

In high latitudes, the demand for energy correlates with the demand for heat. If the energy supply mainly stemmed from solar energy, a shortage of electricity could be expected in winter. If wind power dominates the generation of electricity, then one would expect to have excess power in winter. The use of heat pumps can link power consumption to heat needs; greater heat demand can then be met when the supply of electricity increases.

Demand for electricity and heat can also be decoupled if the heat storage capacity increases. For example, control reserves in grids can be designed as CHP plants. The excess heat can be stored, thus increasing overall efficiency. Heat pumps can further be used to store heat from potential excess electricity.

3.4.4
Hydrogen

3.4.4.1
The basics

Hydrogen has long been known as an important, universally useable element in metallurgy and for the synthesis of chemical compounds. It was used for energy purposes in Germany in the past as an essential component of town gas. The volume of hydrogen consumed today is equivalent to around a fifth of the world's consumption of natural gas. In terms of energy, the use of hydrogen has, however, been negligible. It is important for the transformation of the energy system as basically only water and energy are required to produce it and almost no pollutants are emitted when it is used. Hydrogen technology would allow for the long-term storage of large amounts of energy, and the gas can be easily transported. In combination with renewable energy sources, hydrogen thus has the potential to become a crucial energy carrier in a future sustainable energy system.

3.4.4.2
Production

There are two basic steps to manufacture hydrogen: from organic matter (fossil resources or biomass) or by splitting water molecules using electricity (electrolysis). Hydrogen can be produced in great quantities from hydrocarbons (natural gas, oil, coal, biomass, etc.) by means of a reformation process. The heat for the high reaction temperatures (850–2,000°C) results from the partial combustion of the raw materials. Some 60 per cent of the energy in coal and up to 85 per cent of the energy content of natural gas can be stored as chemical energy in hydrogen. The energy balance sheet could be improved even further in the long run if high temperature solar heat is used. In particular, hydrogen technology would allow for biomass to be used efficiently: a synthesis gas containing hydrogen occurs during gasification and contains some 75 per cent of the chemical energy stored in the biomass; thus, if efficient energy converters are used (such as fuel cells), the overall efficiency for the generation of electricity could be as high as 40 per cent.

In addition to biomass conversion, electrolysis is the most important way to produce hydrogen from renewable energy without any by-products or toxic emissions. All renewable energy sources that can be used for the generation of electricity can thus be used to produce hydrogen. One special benefit that electrolysis offers is that even a fluctuating supply of electricity from renewable sources can be efficiently used (Section 3.4.3.2). The largest electrolysis systems currently have a connected load of $150MW_e$. While alkaline electrolysis at ambient pressure has long been used commercially and is a mature process, more advanced concepts – like high-pressure electrolysis – are still under development. Table 3.4-1 compares the most important methods of producing

hydrogen. It also contains the data that can be expected to apply when hydrogen is brought to market according to the transformation roadmap proposed by the Council (starting around 2020, Chapter 4). The efficiency of all of the processes, including construction of facilities and procurement of resources, is around 60 per cent and will increase to just below 70 per cent in future.

Facility output varies greatly across production processes. To lower costs, conversion plants for natural gas are designed for the highest output possible per facility, while transport costs for the primary energy carrier limit the size of biomass plants. Electrolysis can be used as a modular technology both for distributed power supply close to the consumer (such as at filling stations) or centrally to suit the output of the plant generating electricity.

At electricity costs of around €-cents 4 per kilowatt hour, hydrogen could cost around €-cents 7–8 per kilowatt hour in the long term (Nitsch, 2002). In contrast, hydrogen produced from natural gas is already half as expensive at €-cents 4, though CO_2 is released into the atmosphere in this process. Hydrogen produced with renewable electricity has the economic disadvantage of necessarily being more expensive than the electricity it is produced with.

3.4.4.3
Storage and distribution

Like natural gas, hydrogen can be compressed and liquefied for storage and transport in liquefied gas

	Steam reformation of natural gas		Gasification of biomass		Electrolysis (module)	
	Today	>2020	Today	>2020	Today	>2020
H₂ production [m³/h]	100,000	100,000	13,000	13,000	500	500
[MW$_{H_2}$]	300	300	40	40	1.5	1.5
Resource input [MW]	405	385	55[1]	53[1]		
Electricity required [MW]	1.5	1.5	3.0	2.8	2.1	2.0
Process efficiency [%]	74	78	73	76	73	77
Water required [m³/h]	58	58	28	28	0.4	0.4
Operating pressure [bar]	30	30	50	50	30	100
Efficiency, gaseous hydrogen at consumer [%]	64	68	60	66	63[2]	67[2]
Investment costs [€/kW$_{H_2}$]	350	350	ca. 700	ca. 500	1,000	ca. 700

Table 3.4-1
Key data for selected methods to produce hydrogen, today and in 2020.
[1] corresponds to some 12t/h of wood;
[2] without the deployment of renewable electricity, but with transport losses across 3,000km as high-voltage direct current.
Sources: WBGU with reference to Nitsch, 2002; Pehnt, 2002; Dreier and Wagner, 2000; Winter and Nitsch, 1989; BMBF, 1995; DLR and DIW, 1990

tanks and compression tanks. In addition, hydrogen can be bound e.g. to metal hydrides for depressurized storage. In the long term, hydrogen will probably be stored in very large quantities to compensate for daily and seasonal fluctuations. Here, the tried and tested technologies for the underground storage of natural gas in empty salt caverns and gas or oil storage sites are available. The storage costs of hydrogen are roughly twice as high as for natural gas due to the low energy density of hydrogen. But the effect of these costs will be minimal as these costs only make up a small portion of the overall costs. In addition to liquefaction, which consumes roughly a third of the energy content of hydrogen, high-pressure composite tanks for up to 700 bar are especially interesting for mobile applications. In addition to the specific costs for fuel, the much greater costs for the storage of hydrogen required for its mobile use in comparison to these costs for petrol and diesel will greatly step up the trend to much more efficient vehicles.

When assessing the infrastructure needed for the use of hydrogen, it must be kept in mind that its production and consumption will be closely related to the electricity supply as electrolysers can be designed for distributed power and adapted to the main points of consumption (such as combined heat and power units with local heat distribution networks, filling stations, and industrial plants). In addition, one great advantage of hydrogen as an energy carrier is that the existing infrastructure for natural gas can be used for transport and distribution. Solely hydrogen grids have also been in operation for many years. Overall, the conditions are good for the long-term, gradual transition towards hydrogen as an energy carrier for stationary applications based on the well developed natural gas infrastructure.

High-voltage, direct-current power lines are available as a tried and tested technology for the long-term, long-distance transport of energy across thousands of kilometres. Transport via pipelines will not be necessary or economically attractive until a regenerative hydrogen economy has been established on a large scale, entailing very large amounts of energy for transport. A typical pipeline, for instance one from northern Africa with a diameter of 1.6–1.8m, would supply some 23GW of H_2, which would cover some 10 per cent of Germany's current energy needs (Nitsch, 2002). The transport costs for these dimensions and a distance of 3000km would be around €-cents 1.5 per kilowatt hour of hydrogen (Winter and Nitsch, 1989). This figure takes into account energy losses of 15 per cent due to the compression and transport of the gas. Another option for long-distances is the transport of liquified hydrogen in tankers. While transport via tankers is very inexpensive and entails little loss, some 10kWh of electricity

per kilogram of H_2 is needed to liquefy hydrogen at -253°C. The overall efficiency of around 75 per cent for gaseous hydrogen then drops to around 60 per cent for liquid hydrogen. But if hydrogen is to be consumed as a fluid (for instance as fuel), large-scale liquefaction and tanker transport are nevertheless an interesting option.

3.4.4.4
The use of hydrogen

Hydrogen can be used in many of the same ways as natural gas. All of the common energy converters (flame burners for heaters, industrial and power plant boilers, turbines, and combustion engines) require only moderate adaptations to be run on hydrogen or gas mixtures rich in hydrogen.

The efficiency of hydrogen combustion in engines is comparable to that of petrol with state-of-the-art technology. As the only pollutant emitted is NO_x, emissions can be kept very low by optimizing the combustion process (Section 3.4.4.5). Both in stationary (CHP units) and mobile applications, hydrogen would largely solve the problems of the local emission of pollutants, even if only 'conventional' technologies are used, as the exhaust gas does not contain carbon monoxide, sulphur dioxide, hydrocarbons, lead compounds, or soot particles. Combustion of pure H_2/O_2 is also interesting; here, one direct by-product (i.e. without a heat exchanger) is dry steam, which can be conditioned with the addition of water. This technology is well suited for the provision of process steam in industry and to generate power for peak loads.

Furthermore, hydrogen can also be used with other technologies that are less suitable for hydrocarbons and would require that the hydrocarbons be reformed beforehand. The best known process is fuel cell technology. In addition, catalytic combustion, which occurs below 500°C and only has minimal NO_x emissions, is a noteworthy option. It allows for the construction of open heating surfaces using catalysers, for instance for space heating with 'zero emissions'.

Fuel cells are an important building block in energy systems supported by regenerative hydrogen as they can convert hydrogen into electricity and useful heat directly, efficiently and without any emissions. Fuel cells are available as pilot and demonstration systems and in small batches for a wide range of outputs from a few watts (portable systems) to units with several kW (small and mid-size combined heat and power units) and even several MW (large CHP plants). They run at temperatures of 80–800°C (Fig. 3.4-4). Combined power plants with fuel cells as auxiliary units in the 50–100MW range are also already in planning. The automotive industry is intensively developing fuel cells to make them ready for production as zero emission drive units for vehicles. The efficiency rates for commercial systems considered possible for the short term are 45 per cent (for PEMFC, probably surpassing 60 per cent in the long term) and 55–60 per cent (for MCFC, SOFC). In combination power plants, efficiency rates up to 70 per cent are considered possible (Table 3.3-1). It should be kept in mind here that these rates of efficiency are even possible for small units with only a few kW of output thanks to their modular design, thus making fuel cells well suited for efficient, distributed CHP units with a high overall degree of usage (Section 3.3). However, the proven rates of efficiency in practice still fall some 5–10 per cent short of these targets.

To become competitive in the short term, fuel cells will require further progress in technology and system design and, above all, an enhanced energy policy setting, especially for CHP units. This is essential to provide a secure basis on which developers can advance the technology and make the investments needed before products can be brought to market.

The development of energy converters is increasing the importance of efficient systems with relatively low output levels. What began over a decade ago with renewable energy technologies is now continuing with combined heat and power units, micro gas turbines, Stirling motors, and fuel cells. Power plants are also now being planned as combined-cycle (gas and steam) systems with far lower outputs of no more than 200MW. Progress in electronics and IT will allow for the combination of a growing number of small units into distributed power plants. The liberalized energy market rewards such developments as these systems can respond flexibly to market demands and investments in them are modest.

3.4.4.5
Potential environmental damage from hydrogen

The flame temperatures for the combustion of hydrogen must not be allowed to reach levels that would produce great amounts of NO_X. Solutions to this problem are, however, already available as the lack of other pollutants has allowed researchers to focus on the optimization of the combustion process to minimize NO_X emissions.

Hydrogen molecules are a natural component of the atmosphere at a concentration of around 0.5ppm near the surface. When hydrogen is used in large amounts, the atmospheric concentration may increase due to leaks, causing chemical reactions that indirectly increase the concentration of the heat-trapping gas methane. Some initial estimates revealed that the current use of hydrogen by mankind does not significantly affect the methane concentration (IPCC, 2001a). Even if H_2 fuel cells replace 50 per cent of fossil fuels and less than 3 per cent is lost due to leaks, this additional volume would not exceed the current hydrogen source (volatile organic compounds including methane) (Zittel and Altmann, 1996). To resolve the current uncertainties, these issues need to be taken up in research on atmospheric chemistry. In addition, the decomposition of hydrogen in soil needs further study.

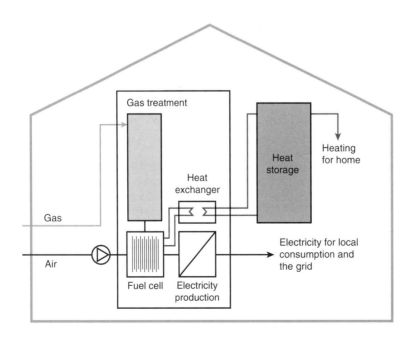

Figure 3.4-4
The principle of a home energy system based on hydrogen. Such homes can also be connected to a hydrogen grid, in which case distributed gas reformation and purification would not be necessary.
Source: WBGU

3.4.5
Electricity versus hydrogen: An assessment

The basic efficiency/cost ratios for the generation of renewable electricity and regenerative hydrogen by means of electrolysis are listed in Table 3.4-2. Gaseous hydrogen in central Europe has some 65 per cent of the energy content of solar electricity at the point of deployment. With liquid hydrogen, users still have somewhat more than 50 per cent of the original solar power.

Today, electrolysis is, beside biomass reformation, the most favourable conversion method for renewable energy from hydrogen. Regenerative hydrogen is thus provided less efficiently and at a higher cost than renewable electricity. It will hence only be important for the energy industry when it can be used in applications that make economic sense and compliment the universally useful energy source of electricity. The main argument for the introduction of hydrogen technology will be its good storage properties.

Most energy services (space heat, hot water, process heat, propelling power, light, and communication) can be provided using useful heat and electricity from renewable energy sources. Both are less expensive than regenerative hydrogen. Using hydrogen only makes sense when the direct use of electricity or heat is not possible for technical or structural reasons (such as in the transport sector or when the supply of renewable electricity exceeds demand and the excess power needs to be stored). The gains in terms of storability or applicability have to be weighed against the additional costs and conversion losses, both for individual applications (niche markets) and for the overall energy system.

The importance of hydrogen for the energy industry thus lies in the possibility of expanding the applicable limits of renewable energy sources. But first, the direct use of these energy sources will have to become more common itself, of course. The importance of hydrogen is thus directly linked to the intensity and continuity of an overall strategy for the tapping of renewable sources of energy. Various applications can be envisioned for the energy function, the storage function and the transport function of hydrogen:

- The storage of large amounts of fluctuating electricity from renewable sources once a very large share of renewable energy has been attained and conventional load management and storage no longer suffice;
- The transport of energy across long distances, even between continents;
- The requirement for zero emissions locally or in the overall electricity system (CO_2-free energy system). This will entail a suitable energy supply for users even in areas not or hardly accessible for electricity (such as transport, notably aviation, and part of industrial high-temperature heating).

Provided the share of renewable energy sources in the overall energy supply does not exceed 50 per cent, supply fluctuations can be compensated for in large interconnected grids and by combining various renewable energy sources. In addition, load management (control of user demand) and (thermal) energy storage in solar thermal power plants, for instance, could also level out fluctuations. As electricity from renewable energy sources increasingly penetrates the grid, however, large-scale technologies to store high-quality energy will probably become indispensable.

Hydrogen can perform this task. It also will benefit from the current ascent of natural gas, which is considered a form of fossil 'transition energy' towards an energy system with more renewable energy and hydrogen. This is because the investments in infrastructure, which often hamper the introduction of new energy sources, will be kept comparatively low since the natural gas infrastructure can be used very well to transport hydrogen. Hence, only distributed hydrogen grids will have to be established.

3.5
Improvements in energy efficiency

Today, in industrialized countries significant losses still occur during various energy conversion stages and during the application of useful energy:

Table 3.4-2		Degree of utilization		Costs	
Efficiency/cost ratios between renewable electricity (generation = 1.0) and regenerative hydrogen for advanced technologies. Source: Nitsch, 2002		Only production	Including long-distance transport	Only production	Including long-distance transport
	Electricity	1	0.9	1	1.5
	H_2, gaseous	0.75	0.65	1.65	1.9
	H_2, liquid	0.6	0.52	2.5	4

- approximately 25–30 per cent during the conversion from primary energy to final energy;
- on average approximately one-third during the conversion from final energy to useful energy, with high losses of approximately 80 per cent in the drive systems of road vehicles;
- approximately 30–35 per cent through unnecessarily high useful energy demand, e.g. for air conditioning of buildings and in industrial high-temperature processes (Fig. 3.5-1).

In theory, the energy demand for each energy service could be reduced by more than 80–85 per cent of current energy demand (Jochem, 1991). Within the framework of discussions on sustainable development, in Switzerland this potential became the basis for the technological vision of a '2000 watt society', which could be achieved around the middle of this century (ETH-Rat, 1998).

In addition to technological aspects of energy and material efficiency and of closed-loop materials management, the demand for energy and material services, which changes with increasing income, higher resource efficiency and the move towards a knowledge society, also needs to be made a focus of debate. The question is whether post-industrial societies need to aim for sufficiency in material goods (including mobility) in the long term. This would not lead to a risk of stagnation for the world economy, since growth in intangible goods (e.g. services) would not be restricted.

3.5.1
Efficiency improvements in industry and business

The general target of increased efficiency can be differentiated technologically as follows:

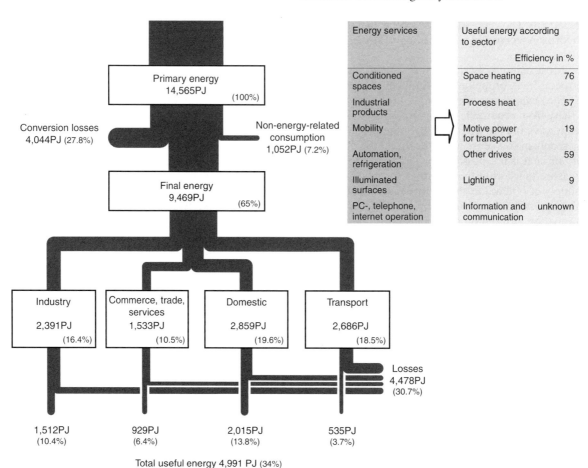

Total useful energy 4,991 PJ (34%)

Figure 3.5-1
Energy losses within the energy utilization system represented by Germany in 2001. Primary energy consumption in that year totalled 14,565PJ. After conversion losses and non-energy-related consumption, the remaining final energy consumed was 9,469PJ. The conversion into heat and mechanical energy led to significant losses totalling 4,478PJ. The box lists the efficiency of conversion from final energy into useful energy for different energy services. Ultimately, only 4,991PJ or 34 per cent of the primary energy used was converted into useful energy.
Source: IfE/TU Munich, 2003

- significantly improved efficiencies during the conversion stages of primary energy/final energy and final energy/useful energy, often using new technologies, e.g. combined heat and power or combined refrigeration and power systems, fuel cell technology, replacement of burners with heat pumps (Williams, 2000);
- significantly reduced useful energy demand per energy service through low-energy buildings, through substitution of thermal production processes with physical/chemical or biotechnological processes, through light-weight design of moving parts and vehicles, of through recovery or storage of kinetic energy (Levine et al., 1995; IPCC, 2001b);
- reduction of no-load losses, i.e. of the final energy use consumed during the lifetime of devices and systems without serving a purpose. One way of achieving this is through a reduction of idle times and no-load power by using more efficient technologies and by influencing user behaviour;
- increased recycling of energy-intensive materials and increased material efficiency through improved design or material characteristics, with significantly reduced primary material demand for each material service (Angerer, 1995);
- more intense utilization of long-lived investment and consumer goods through machine and equipment hire, car sharing and other related services (Stahel, 1997);
- the spatial arrangement of new industrial and other settlement areas according to exergy criteria (Kashiwagi, 1995), and better mixing of settlement functions in order to avoid motorized mobility.

In principle, the options for reducing the energy demand of industrial production while demand for energy services grows can be split into five categories, only two of which are subject to thermodynamic limits (Jochem, 1991).

IMPROVEMENTS IN THE EFFICIENCY OF ENERGY CONVERTERS

Energy converter systems (e.g. burners, turbines, motors, etc.) can be technologically improved, for example, through more heat-resistant materials, better controls, etc. The replacement of burners with gas turbines in medium-temperature processes, for example, or the use of heat transformers and the arrangement of new enterprise zones with cascading heat utilization open up further potential (Stucki et al., 2002; Kashiwagi, 1995).

REDUCTION OF USEFUL ENERGY DEMAND THROUGH PROCESS IMPROVEMENTS AND SUBSTITUTIONS

Process improvements and substitutions offer significant options for increasing energy efficiency. Examples are:

- substitution of metal rolling including intermediate heating furnaces through casting with near-final dimensions, and in the more distant future through spraying of shaped sheet metal components in their final form;
- substitution of thermal separating processes through membrane, adsorption or extraction techniques, as already used in the food and pharmaceutical industries;
- application of new enzymatic or biotechnological techniques for synthesis, dyeing or material separation; improvement of mechanical drying techniques or extension/combination with new concepts (e.g. ultrasonic, pulse technique);
- substitution of heat treatment techniques through techniques with higher accuracy and controllability (e.g. electrical ultra-short heating through microwaves, laser techniques);
- returning of braking energy into the grid (regenerative braking) through appropriate power electronics.

INCREASED RECYCLING AND IMPROVED MATERIAL EFFICIENCY

The production of energy-intensive materials from discarded materials often requires significantly less energy than their production from raw materials, even taking account of the energy required for the recycling processes. Relatively high recycling rates are today achieved for those materials that have already been in use for decades (e.g. in Germany: crude steel: 42 per cent, paper: 60 per cent, container glass: 81 per cent); on the other hand, the values for newer materials are much lower (e.g. plastics: 16 per cent). Through further utilization of the recycling potential, total industrial energy demand could be reduced by at least 10 per cent (Angerer, 1995). Further potential results from reducing the material demand per unit service provided by a material. This can be achieved by modifying the material characteristics and through changes in product design (e.g. thinner packaging materials, foams, flatter surface structures) (Enquete Commission, 2002).

SUBSTITUTION OF MATERIALS THROUGH LESS ENERGY-INTENSIVE MATERIALS

The substitution of materials opens up significant energy saving potential. Decisions about materials and their substitution are today predominantly made based on costs, material and utilization characteristics, as well as the image of the material and current trends. In future, low total energy demand or low total emissions should become increasingly important criteria. Materials and products that are biogenic or can be produced using biotechnological techniques (e.g. wood, flax, starch, natural greases and oils) can thus become interesting alternatives.

MORE INTENSE USE OF CONSUMER GOODS

More intense use of consumer goods can also contribute to improving material and energy efficiency. The notion of a parallel economy ('using instead of possessing') describes the idea of making goods from a pool accessible to several users. Familiar examples are the rental of construction or agricultural machines, electric tools, cleaning machines, cars (car sharing) or bicycles. In this way, a lower quantity of goods can satisfy the same social needs (Fleig, 2000).

The energy saving potential inherent in these five options requires further examination, and future technological developments are in any case difficult to forecast. Overall, the total technological energy saving potential is estimated as more than 50 per cent of today's industrial energy demand (Jochem and Turkenburg, 2003).

3.5.2
Increased efficiency and solar energy utilization in buildings

Any discussion about increases in efficiency for energy use in buildings has to take account of very different basic conditions around the globe, because economic, social and natural influences (e.g. building tradition, available materials, population density, family structures and particularly the climate) lead to different designs. Even within individual countries, the differences between poor/rich, towns/rural areas and existing buildings/new buildings mean that increases in efficiency in the building sector have to be approached from different angles.

SPACE HEATING

In high latitudes, and particularly in continental climate regions, domestic energy use is dominated by space heating, so that improved thermal insulation of buildings has to be a priority. For example, vacuum insulation systems are currently under development offering up to 10 times higher insulation with identical thickness than conventional insulating materials. They are of particular interest for building refurbishment projects. A further example of an innovative approach is transparent thermal insulation installed on the exterior walls of buildings. While sunlight can penetrate the material and is absorbed in the dark wall behind the insulation, any heat released by the wall cannot escape back through the insulation material and therefore contributes to heating the building.

Further keywords for technological efficiency improvements are efficient gas condensing boilers, avoidance of electric resistance heating and connection to local or district heating networks or cogeneration power plants. Overall, energy efficient buildings require heat supply systems that efficiently and cost-effectively meet the remaining low heat demand. Micro-heat pumps will therefore become increasingly significant in future, since they utilize the available electricity infrastructure and can be offered cost-efficiently as mass products; micro-cogeneration plants with fuel cell technology could also become interesting in future (Section 3.4.4.4).

Improved economic stimuli also have large potential. In eastern Europe, the introduction of individual bills for users of district heating systems alone reduced demand by up to 20 per cent. The conversion from manual control of district heating networks to automatic control offers similar potential.

WINDOWS

Solar and energy-efficient construction aims to create an innovative building envelope, with energy, light, sound and mass transfers adjusted according to seasons and user demands. One of the central elements are the windows, with two different requirements being in conflict with each other: While for the purposes of illumination and (particularly in domestic buildings) of heat gain, the aim is to let as much sunlight as possible enter a room via a window, the heat flow back through the window has to be minimized in order to maintain a comfortable indoor climate. In a multi-pane glazing system, for example, the glass panes facing each other may be coated in such a way that visible light can enter the room, while losses through infra-red heat radiation are largely suppressed. Through the use of selective coatings and filling with heavy inert gas, modern triple glazing can reduce heat losses to very low levels of approximately 0.5W per square metre and Kelvin (U-value: thermal transmittance of a window). In future, vacuum windows may lead to further advances.

In order to avoid overheating in the summer, optical switching functions can be implemented in windows that control their optical characteristics without significantly influencing their thermal behaviour and without using conventional shading mechanisms (e.g. slats). The trend in technological developments is towards coatings whose optical properties can be changed reversibly across a wide range (e.g. electrochromatic or gas-chromatic glazing).

HEATING OF DOMESTIC AND INDUSTRIAL WATER

Due to the relatively low temperature required for heating domestic and industrial water, solar energy is eminently suitable for this purpose (Section 3.2.6.2). If fossil or electric energy is used to complement solar energy, like for space heating, gas condensing boilers or heat pumps provide a very attractive and energy-efficient solution. Good thermal insulation of pipes and boilers and devices for water saving are simple and economic measures, but unfortunately are still not implemented as a matter of course.

SPACE COOLING

In many countries, the air inside buildings is cooled in the summer or all year round. The worldwide trend towards urbanization increases this trend. With identical temperature difference to external air, the energy consumption for cooling is higher than for heating, because the buildings being cooled often have poor thermal insulation. Proposals for efficiency improvements include:

- *Buildings:* Improved air conditioning technologies; of particular interest are automated complete solutions for commercial buildings with numerous networked individual components. Controlled ventilation in combination with heat exchangers and heat or cold stores offer significant potential. The options for active cooling of buildings using solar energy were discussed in Section 3.2.6.2. Passive cooling concepts can be used that enable a pleasant indoor climate to be achieved in mid-latitudes without the use of refrigeration machines. Examples of suitable measures are night ventilation or concrete core cooling via ground probes or heat pumps. An innovative technology is the use of micro-encapsulated phase-change materials that enable light-weight buildings to be made 'thermally heavy'. In this case, heat loads occurring during the day are stored by melting the materials at almost constant temperature, and then released to external air during the night.
- *Town planning:* More green spaces and better air flow through building agglomerations have a positive influence on the urban climate. Similar to local district heating, local district cooling networks with efficient central cooling devices could be a useful option. In analogy to CHP, combined cooling and power via appropriately modified thermodynamic heat transformers could be considered.

COOKING

Given the current energy mix in industrialized countries, electricity should not be used for cooking, since the utilization of gas is more advantageous for this purpose. In developing countries, liquid gas or kerosene cookers are preferable to traditional stoves using wood or charcoal as fuel (Section 3.2.4).

LIGHTING

Good daylight utilization in office buildings can save energy for artificial illumination and increase working comfort. Daylighting systems therefore try to reduce the difference in brightness between the areas near the windows and core areas, and to improve natural illumination so that less artificial light is required. These measures reduce the electricity demand for illumination and the associated generation of heat, which in turn does not have to be compensated through additional cooling. Where this is not possible, fluorescent lamps, which are five times more efficient, should be used instead of incandescent lamps. In developing countries, within the context of electrification the aim is to switch from kerosene lamps to fluorescent lamps.

OTHER ELECTRICAL APPLIANCES

Like all other forms of final energy, electricity should be used as efficiently as possible. While suitable appliances are often more expensive to buy, higher initial costs are usually compensated through lower consumption over the lifetime of the device. No-load losses are particularly problematical. Consumer electronics and communications devices in particular are often not isolated from the mains when operated by remote control, but remain in standby mode with reduced electricity consumption. A convenient option for avoiding unnecessary electricity consumption is the installation of an automatic switch between the device and the socket, which separates the respective device from the mains in standby mode. To increase the prevalence of efficient domestic appliances, consumers should be able to easily identify whether a particular device has low standby consumption or can be separated completely from the mains.

The electricity consumption of large domestic appliances (e.g. dishwashers, washing machines and refrigerators) on the market could be reduced significantly through obligatory EU efficiency labels. Targeted consumer information can significantly increase the efficiency of energy utilization without any technological investment. Refrigerators, for example, use significantly less energy if they are placed in a cool and well ventilated space. Mechanical spin driers use much less electricity than heated driers to extract moisture from laundry.

REMOVAL OF STRUCTURAL OBSTACLES

In the building sector, structural barriers need to be removed. Architects, for example, are usually paid according to the value of the building, and not according to its efficiency. As contact persons for clients, architects and installers play a significant role as energy advisors, for which they should be trained adequately. Another issue is the so-called landlord/tenant dilemma, with the former often not being motivated to make investments in improved insulation and heating technology, because the investment costs cannot be fully reflected in the rent, while the tenant benefits from lower energy costs. The tenant, on the other hand, is unlikely to make such investments, because the costs usually do not pay for themselves over a comparatively short tenancy period.

3.6
Carbon sequestration

Carbon dioxide can be removed from the atmosphere in three ways: through natural uptake in the biosphere, by dissolving in seawater followed by sedimentation, and by human intervention in the form of technical carbon management. Falling under the latter is capture of CO_2 before or after the combustion of fossil energy carriers, its transformation into the liquid or solid phase, and transportation to repositories for long-term sequestration in suitable geologic reservoir formations or in the deep ocean (Reichle et al., 1999; Ploetz, 2002).

3.6.1
Technical carbon management

A high rate of CO_2 capture is possible by technical means at point emission sources, like coal- and gas-fired power plants, cement plants, steelworks and oil refineries. Basically, a distinction can be made between two methods of CO_2 capture:
- flue gas scrubbing, in which CO_2 is removed from the flue gas stream by absorption or adsorption, membrane separation or distillation;

- removal prior to combustion, in which first a hydrogen-rich synthesis gas is won from coal or natural gas by coal gasification or steam reforming, followed by CO_2 removal.

Power plant efficiency is reduced by CO_2 capture and storage. The main reasons for this are the energy expended for regeneration of absorbents, membranes and solvents as well as for their manufacture and disposal, and for CO_2 transportation (Table 3.6-1).

The estimated costs for CO_2 capture including compression/liquefaction for transportation account for some three-quarters of the total costs of sequestration both in the ocean and in geologic formations (Reichle et al., 1999; Grimston et al., 2001) and are therefore the overall cost determining factor. Hendricks and Turkenburg (1997) quote, for a standard power plant, capture costs of €100 to 250 per tonne carbon and, for a combined cycle power plant with integrated coal gasification, less than €100 per tonne carbon.

Estimates of future potentials for CO_2 sequestration are currently focused on storage capacity. It is likely that these will be determined less by technical feasibility than by costs in comparison with other CO_2 abatement strategies as well as by social and political acceptance. It is estimated that, if geologic sequestration were to be applied on an industrial scale, electricity costs for the final consumer could rise by 40 to 100 per cent (Grimston et al., 2001).

Current estimates of potential appear in national research programmes as medium- and long-term objectives. Thus the American Federal Energy Technology Center (FETC) states a target for reducing CO_2 sequestration costs by a factor of 10–30 by 2015. As from 2050, around half the required emission reductions (referred to a 550 ppm stabilization scenario for CO_2) is to be attained by CO_2 sequestration. However, the US Department of Energy does not expect sequestration even to be practicable on an industrial scale before 2015 (US DOE, 1999).

Process	CO_2 retention efficiency [%]	Loss in efficiency for power generation [%]
CO_2 removal from synthesis gas following CO conversion (from coal gasification or steam reforming of natural gas)	90	7–11
Build-up of CO_2 concentration in effluent gas (usually by combustion in atmosphere of oxygen and recirculated flue gas)	~100	7–11
CO_2 removal from flue gases	no data	11–14
Carbon removal before combustion	no data	18
CO_2 retention in power plants with fuel cells	no data	6–9

Table 3.6-1
Efficiency of CO_2 retention and sacrifices in efficiency for various capture technologies.
Source: Göttlicher, 1999

Sequestration in geologic formations

The aim of CO_2 sequestration is to segregate the greenhouse gas for as long a period as possible from the atmosphere. To this end, following its capture, CO_2 must be stored where it can be isolated from contact with air. Coming into consideration as storage options are deep geologic formations, like salt caverns, deep coal seams, depleted and active gas and oil fields as well as deep (saline) aquifers. But when assessing a repository, a distinction must be made between permanent storage and applications where CO_2 is pumped in only as an additional economic measure. Thus injecting CO_2 into deep coal seams that cannot be mined serves methane recovery (EGR or enhanced gas recovery; Bachu, 2000). CO_2 also reduces oil viscosity, which is why it is used all over the world in oil wells to improve their yield (EOR or enhanced oil recovery). But the retention time of a few months up to years for CO_2 sequestered in this way is short (Bachu, 2000). Therefore, these two options have to be assessed critically with regard to their carbon balance. For estimates of storage potential worldwide, usually the theoretically available storage capacity is quoted (Table 3.6-2), and not the technical or economic potential. The figures vary greatly, in particular because just a few systematic investigations of storage capacities are available.

If large quantities of CO_2 are stored underground, there is a risk that it could be liberated if leaks occur. Because CO_2 is heavier than air, a CO_2 lake could accumulate at the ground where the gas exits, in which all life would be asphyxiated (Holloway, 1997). Therefore, the integrity of the repository is of great importance. At present, only depleted gas and oil fields may be regarded as secure repositories, and to a lesser extent also saline formations. For deep aquifers, the integrity of these repositories is still unknown, although it is assumed that this will be high. Pilot applications, like in the Sleipner Field in the North Sea appear to confirm this (Baklid et al., 1996; Torp, 2000). But there are still no integrity and monitoring guidelines or assessment criteria for storage quality requirements (Gerling and May, 2001).

Ocean sequestration

Use of the huge CO_2 repository represented by the oceans and seas with its estimated capacity for anthropogenic CO_2 much exceeding 1,000GtC (IPCC, 2001c; Herzog, 2001) involves, at today's status of knowledge, high risks with respect to longevity of storage and environmental impacts. The storage period depends on where the CO_2 is injected and on prevailing ocean currents as well as depth of injection. Simulations show that CO_2 has to be placed at greater depths to prevent its rapid release into the atmosphere. At an injection depth of 950m at favourable locations, CO_2 could be retained over long periods in the ocean, of an order of magnitude of 1,000 years (Drange et al., 2001).

Due to injection, the partial pressure of the CO_2 is increased and at the same time the seawater's pH is lowered. Although as yet the biological consequences have not been adequately investigated, they are a cause for concern. Proven are significant changes in the structures of microbiological communities, inhibition of metabolism and an appreciable sensitivity of marine organisms to lowering of pH. Upsetting the acid-base equilibrium by lowering pH could result in disintegration of calcium-containing skeletons as well as changes in metabolism that may reduce the growth and activity of organisms (Nakashiki and Oshumi, 1997; Seibel and Walsh, 2001). It has also been observed that fish find it difficult to breathe (Tamburri et al., 2000). The costs for CO_2 sequestration in the sea, including capture and transportation, are today quoted as 30 to 90 US$ per tonne CO_2 (Hendriks et al., 2001; DeLallo et al., 2000).

3.6.2
Potential for sequestration as biomass

Carbon sequestration in terrestrial ecosystems

At present, terrestrial ecosystems store some 460–650GtC in vegetation and 1,500–2,000GtC in the soil (Fig. 3.6-1; IPCC, 2000a, 2001a). 30–50 per cent of the carbon exists in an easily degradable form, which means that if no precautions are taken some 700Gt of

Table 3.6-2
Comparison of various geological sequestration options.
EOR Enhanced Oil Recovery, *EGR* Enhanced Gas Recovery.
Sources: Parson and Keith, 1998; IPCC, 2001c; Herzog, 2001

Sequestration option	Estimated capacity [GT C]	Relative costs	Repository integrity	Technical practicability
Active oil wells (EOR)	low	very low	good	high
Deep coal seams (EGR)	40–300	low	unknown	unknown
Depleted oil and gas repositories	200–500	low	good	high
Deep aquifers, caverns/salt domes	100–1,000	very high	good	high

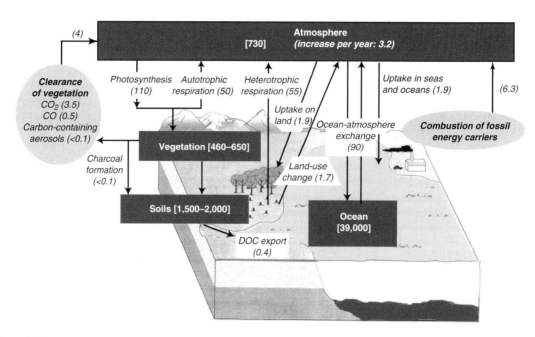

Figure 3.6-1
Global carbon inventories and flows in vegetation, soil, oceans and the atmosphere. All values in Gt carbon (inventories: square brackets) or Gt carbon per year (flows: round brackets and in italics).
Source: amended from Ciais et al., 2003

carbon could be liberated over a short time by changes in land use. On the other hand, according to simulation calculations, by 2050 these reservoirs could be augmented by some 70GtC in forests and a further 30GtC on land used for agriculture (IPCC, 2001c). Despite its substantial sink capacity, CO_2 uptake in the terrestrial biosphere can offset only to a small part anthropogenic greenhouse gas emissions, including emissions from land use, as documented by the rising concentration of CO_2 in the atmosphere. Due to the increase of biomass in the forests of North America and Europe, since the beginning of the 1990s some 10 per cent of global emissions have been absorbed (IPCC, 2001c). An analysis of carbon flows of entire continents shows that, for example, the primary forests of Siberia eliminate a large part of the CO_2 emissions of Russia and the EU due to intensified photosynthesis (Schulze, 2002).

In contradiction to the hypothesis of ecological equilibrium (Odum, 1969) and despite a relatively constant core biomass, natural ecosystems are capable of tying in large quantities of CO_2 (Schulze et al., 1999). Within the climate system, primary forests act as important CO_2 sinks, but are under increasing threat due to human intervention ranging up to their complete destruction. Their protection would have positive impacts for both climate protection and nature conservation.

The behaviour of the natural carbon reservoir in soils under the influence of temperature rise is very difficult to estimate at present, since the effects of climate on the stored amounts differ between the tropics and the boreal zones. But more serious than the possible impacts of climate change would appear to be the effects of anthropogenic changes in land use on the carbon storage capacities of soils. According to IPCC (2000a, 2001a), in the 1980s and 1990s 1.6–1.7GtC per year were liberated from terrestrial ecosystems. Even if it proves possible to greatly limit the consumption of fossil fuels, failure to protect the soil carbon inventory would completely countervail all climate protection efforts. One possibility to promote carbon storage is offered by changing the management regime of ecosystems used for agriculture and forestry to one designed for carbon sequestration. Forest measures encompass, for example, avoidance of clear-cutting combined with ecologically compatible practices. However, in agriculture there is a high degree of uncertainty concerning both suitable areas of land and the level and permanence of the attainable sequestration. Changed management of agricultural soils, for example modified ploughing techniques, might involve raising nitrous oxide emissions and increasing fertilizer application, thus resulting in liberation of carbon dioxide (Freibauer et al., 2002).

CARBON SEQUESTRATION IN MARINE ECOSYSTEMS

Single-cell algae in the world's oceans – phytoplankton – are responsible for around half of global carbon fixing by photosynthesis. In some areas of the ocean, phytoplankton growth is greatly restricted due to a lack of the micro-nutrient iron, for example in the sub-Arctic Northeast Pacific, in the Pacific near the equator and in the South Seas. Various ocean fertilization experiments have demonstrated that a local and short-duration algae bloom can be triggered by adding iron (Martin et al., 1994; Boyd et al., 2000).

Watson et al. (2000) assume when estimating the longevity of sequestration that iron will have to be added continuously to achieve permanent removal of CO_2 from the atmosphere. If the additional phytoplankton does not sink but instead remains near the surface, within a year the carbon would return to the atmosphere. Despite the large uncertainties, commercial projects are already underway (Markels and Barber, 2001).

The global potential of biological marine CO_2 sequestration is limited to ocean areas that are deficient in micro-nutrients, for example in the South Seas. Because here deep water is formed that feeds the upwelling areas in the tropics, nutrient-poor buoyant water could result in losses in primary production. Consequently, on an overall balance, fertilization with iron could be a zero-sum game. Additionally, serious consequences are to be expected for marine ecosystems as a result of fertilization with iron (Chisholm et al., 2001) in the form of a reduction in the species diversity and composition of phytoplankton communities, an increase in toxin-producing cyanobacteria as well as eutrophication and accelerated oxygen consumption in the deeper ocean layers, with the consequence of anoxic degradation processes, so possibly liberating greenhouse gases like methane or nitrous oxide.

3.6.3
Evaluation

Generally, all options for carbon management are less sustainable than measures for emissions reduction by raising efficiency and switching from fossil fuels: carbon enters, from stable fossil repositories, a cycle that leads also into the atmosphere with greater or lesser risk, where its greenhouse action comes into play. However, in many countries fossil carriers will remain as the predominant energy source over decades (Chapter 4). Therefore, end-of-pipe technologies for carbon sequestration offer an option for climate protection to hinder excessive emissions, in particular during this century. Criteria for evaluating specific carbon management options are the longevity of sequestration, its reliability, and its environmental impact.

The Council regards geological sequestration as possessing an interim potential that could be used for removing CO_2 from the atmosphere under the condition that the integrity of the repository can be guaranteed and the retention period is sufficiently great, i.e. exceeds 1,000 years. At today's status of knowledge this is the case for storage in depleted and active gas and oil fields as well as in salt caverns. The Council conservatively estimates the sustainable potential to be around 300GtC. The greater technical potential of CO_2 sequestration in saline aquifers of more than 1,000GtC is assessed by the Council as non-sustainable at present knowledge, as there is sufficient evidence neither of the longevity and integrity of storage nor of the avoidance of environmental harm. Before pursuing this sequestration approach, further research is needed.

Concerning sequestration in the ocean via both deep-sea injection and iron fertilization, on account of the environmental risks and uncertainties in the longevity of storage, in particular in the case of iron fertilization, the Council does not regard this as having any sustainable potential.

The terrestrial biosphere makes a key contribution to stabilizing atmospheric carbon dioxide concentrations. But there is hardly any scope for extending this repository, as natural ecosystems like primary forest and wetlands are limited in their area and additionally are suffering widespread destruction by human activity. Therefore, the creation of additional sinks does not represent an alternative to avoidance of fossil emissions.

3.7
Energy for transport

Due to the fact that a high proportion of energy demand is consumed for transportation, this sector plays a crucial role for the transformation of energy systems. The challenge is to enable mobility while reducing the consumption of fossil fuels. Strategies preventing traffic and shifting the modal split, and concepts improving transport efficiency provide crucial stimuli for the political debate surrounding sustainable development and climate protection. In the EU and Japan, trends towards efficiency improvements are now becoming apparent: more economical aviation engines, the marketability of ultra-efficient cars and the emerging mass production of fuel cell vehicles show that the industry already sees economic potential.

3.7.1
Technology options for road transport

The route towards environment-friendly road transport is characterized by a variety of options for efficient and renewable drive and fuel systems, although the ideal solution has yet to be found (Wancura et al., 2001). Only few of the technologies, for which broad market penetration can be expected over the next 30 years, are likely to offer relief for the environment and the climate. Some are even likely to lead to additional greenhouse gas emissions, for example the production of methanol from hard coal, or the production of hydrogen in conventional power plants (ETSU, 1998).

VEHICLES WITH FUEL CELLS
Vehicles with hybrid fuel cell drive will be able to reduce emissions of greenhouse gases and air pollutants by almost 100 per cent. The gaseous or liquid energy carrier generates electricity in an electrochemical oxidation process, with the only by-products being water and heat. The vehicle is driven by an electric motor, perhaps in combination with battery systems. Hydrogen, methanol or petrol are currently used as fuels. Hydrogen can be produced from natural gas or via the electrolysis of water, with the electricity originating from either fossil or renewable sources (Section 3.4.4). Methanol produced from natural gas and petrol have a high hydrogen content, which can be separated and utilized directly in the fuel cell.

The weak point of this technology has hitherto been the high energy consumption during the production of the fuel. In the most favourable case, a fuel cell car is estimated to reduce CO_2 emissions by about 50 per cent over its life time, compared with an average diesel or petrol car (Bates et al., 2001). Another issue is the storage of hydrogen in the car, which has not yet been solved economically. Assuming wide application of fuel cells and a high proportion of renewable energy carriers used for their production, most noxious substances can be reduced by more than 90 per cent (IABG, 2000b).

NATURAL GAS
From an environmental perspective, the utilization of natural gas for transportation is a positive bridging technology on the route from fossil fuels to renewable solutions (Section 3.2.1). Compared with petrol or diesel, natural gas can improve the greenhouse gas balance and reduce urban air pollution. Further expansion should be encouraged through a denser network of gas filling stations and vehicle conversions, particularly in towns and regions suffering

from high pollution. Bi-fuel technologies combine the utilization of two fuels (e.g. natural gas and petrol) within the same vehicle and offer an important transition option as long as area coverage of gas supply is still patchy (Halsnaes et al., 2001).

HYBRID DRIVES AND BATTERIES
Hybrid drives combine electric motors and combustion engines. They are currently being tested for cars, buses and small lorries. Hybrid drives with electric motors are expected to lead to a doubling of primary energy efficiency (Johansson and Ahman, 2002). Modern battery and storage technologies have strategic importance for the whole energy system, particularly in combination with fuel cell drives, where they play an essential role (Section 3.4; IABG, 2000b; Halsnaes et al., 2001). Before the wide market introduction of vehicles with hybrid drives and electric vehicles, further technological improvements in terms of battery capacity, charging process and charging stations are required (Fischedick et al., 2002).

IMPROVEMENTS IN EFFICIENCY FOR CONVENTIONAL VEHICLES
Today, the development of more efficient drive technologies is a standard strategy of all European car manufacturers. Improving the efficiency of combustion processes is currently the subject of intense research (e.g. through the integration of ceramic components, new ignition systems, variable valve management, improved turbo chargers; Halsnaes et al., 2001). Compared with 1995, the fuel consumption of diesel engines, and therefore the CO_2 emissions per mile, have already been reduced by 20 per cent. A reduction of the energy and environmental costs by a further 50 per cent is considered feasible (Johansson and Ahman, 2002). The advantage of optimized conventional drive systems is the option of rapid introduction on the market. Efforts are also being made in the construction and design of vehicles, with targeted modifications being introduced that lead to less energy consumption for the same mobility service. However, this efficiency potential is much lower than for drive and fuel technologies. A maximum of 6 per cent of CO_2 have been avoided through weight reductions achieved to date; reduced rolling resistance has only saved about 1 per cent (Bates et al., 2001).

DRIVE SYSTEMS WITH RENEWABLE FUELS
Renewable fuels such as biogas, biodiesel, ethanol, methanol from residual timber and hydrogen are currently being introduced on the market. They offer some reduction in environmental and energy costs, but prices are higher (e.g. for hydrogen from wind or

solar electricity), which is why subsidies for their introduction on the market are required in the medium term.

3.7.2
Improvements in efficiency through information technology and spatial planning

Information technology has already revolutionized the transport sector, so that today persons and goods can be moved much more efficiently (Golob and Regan, 2001). However, in an era of growing worldwide flows of commodities, freight transport in particular offers further potential for significant efficiency improvements. The volume of road freight traffic outside urban areas could be reduced by up to 8 per cent in the short term through the application of telematics for fleet management (Kämpf et al., 2000). The total potential for reductions through efficiency measures in road transport is estimated to be more than 60 per cent (IPCC, 2001c). However, since in Germany road transport is only responsible for approximately 5 per cent of all CO_2 emissions, a maximum of 2–3 per cent of nationwide CO_2 emissions could be saved, assuming no change in overall mileage. The best short-term result (10–15 per cent efficiency improvements) is expected from electronic road charging (ETSU, 1998).

The investments in telematics and information systems are currently motivated mainly by the desire to improve traffic flows, and less by environmental aspects. A particular side effect of telematics could prove to be problematical: improved traffic flows could lead to further increases in the volume of traffic, thus compensating any emission reduction that may have been achieved.

With the aid of information technology, the transitions between road, rail, local public transport and inland water transport are today facilitated through concepts such as multi-modality or 'combination transport'. All approaches for promoting a multi-modal infrastructure aim to increase the demand for and the attractiveness of non-polluting means of transport. The worldwide potential for efficiency improvements through the expansion of multi-modal infrastructures is very large. In this context, making rail use more attractive through political measures should have a high priority (Section 5.2.4.1). Currently, rail has a weak competitive position compared with the car. The area coverage of rail network is inadequate. Prices and transfer facilities are not very attractive. Options for rail use are limited, particularly for the populations of smaller towns.

Modern concepts for spatial, urban and transport planning offer much more than technological efficiency improvements and can achieve a net reduction in the amount of traffic per capita or per tonne of goods. New settlements are today designed in such a way that traffic is avoided right from the start, with mixed use ensuring short routes, or good local public transport connections and high population density leading to good public transport utilization. These concepts have not yet been implemented widely and offer significant potential for better energy efficiency in the transport sector worldwide.

3.7.3
Sustainability and external effects of increased transport energy demands

Road transport not only causes the familiar environmental effects associated with fossil fuels (Section 3.2.1), but has many other negative 'external effects' (accidents, land sealing, noise, etc.; UNEP, 2002). Emissions from aviation have a particularly strong effect on the climate, since at typical cruising altitudes they contribute to the greenhouse effect not only via the CO_2, but also through the formation of ozone and vapour trails. No comprehensive evaluation of different transport technologies from a sustainability perspective has been undertaken to date (Enquete Commission, 1995). In order to avoid new technologies being developed without consideration of environmental issues, the Council recommends the evaluation of options for future transport technologies in consultation with experts from the climate, energy, ecology and town planning disciplines.

While renewable fuels such as biodiesel (methyl ester), ethanol and methanol from biomass are thought to have large technological potential, their environmental balance and in particular their greenhouse gas balance needs to be examined critically. If they are produced in conventional agriculture, the emission of greenhouse gases (N_2O, CH_4) during production may even completely negate the CO_2 emission reduction effect of biogenic fuels. Their production therefore only makes sense if sustainable agricultural methods are applied, if adequate land area is available, and if a positive overall greenhouse gas balance is achieved (IPCC, 2000a). In view of the introduction on the market of biofuels and bio-oils it is sensible to promote integrated research along the chain of 'cultivation method – industrial processing – fuel utilization', in order to be able to define conditions for the sustainable production and use of such fuels.

3.7.4
Evaluation

To achieve a transition towards sustainable energy use in the transport sector, efficiency improvements for existing technologies (e.g. improved engines, fuels) and the increased utilization of renewable, more environment-friendly energy sources for drive systems should have priority. In view of the problems associated with the allocation of land (Sections 3.2.4 and 4.3.1.3) and the low natural conversion rate of only 1 per cent from solar energy to biomass, the Council's recommendation is to limit efforts associated with the technology option of 'biogenic fuels', and to reduce current levels of support. Priorities for support should be fuel cell drives, natural gas and hybrid vehicles, telematics and multi-modality. Compared with petrol or diesel, natural gas has environmental benefits, mainly for the climate and for air quality. It should therefore be supported as a bridging technology. In the long term, fuel cell engines using fuels produced from solar or wind energy open up a promising avenue towards sustainable development.

3.8
Summary and overall assessment

Chapter 3 assesses the sustainably utilizable fossil, nuclear and renewable energy resources, based on an appraisal of the specific conversion technologies and an evaluation of their environmental and social impacts.

The *fossil fuel* resource base – potential geopolitical developments aside – appears sufficient to meet a globally growing energy demand throughout and beyond the coming century. This, however, is unacceptable for climate protection reasons (Section 3.2.1). The Council views *geological storage of carbon dioxide* as providing only a limited potential of – cautiously estimated – some 300GtC if depleted oil and gas caverns are used. On the basis of the knowledge currently available, the Council does not consider the other options for CO_2 sequestration to be sustainably utilizable (Section 3.6).

In the Council's assessment, the use of *nuclear power* is not sustainable because it is associated with intolerable risks (e.g. proliferation, terrorism, and the absence of secure final repositories). Similarly, the Council currently does not see a potential for use of nuclear fusion within the context of a sustainable energy system. This technology would not be available in time to contribute to the transformation

process, and would also be associated with substantial risks (Section 3.2.2).

The use of *traditional biomass* prevailing in many developing countries is assessed by the Council to be unsustainable because, among other things, it generates substantial health hazards (Section 3.2.4.2).

The Council appraises the sustainable *hydropower* potential relatively cautiously at 15EJ annually (in 2100). This is because, in many developing countries, in particular, the preconditions are scarcely in place to meet the requirements upon their environmental and social impacts which, quite justifiably, have become stricter (Section 3.2.3).

In contrast, *new renewable energy sources* harbour a major sustainable potential for the future: solar energy, wind power, modern biomass use, geothermal energy and other sources. While those sources that can only be expanded to a limited extent (e.g. wind power, bioenergy) are often already available today at competitive prices, the sources which can be expanded practically without limit (e.g. photovoltaics, or solar thermal power generating plants) are still comparatively expensive today in a microeconomic perspective. For cost-reducing learning processes to be able to run their course in the field of solar electric energy conversion, a vigorous expansion rate would need to be ensured over the medium term. Only then will the technologies that can be expanded effectively without limit be available at sufficiently low cost at the point in time when the expansion of other types of renewable energy meets the limits of their sustainably utilizable potential.

The Council estimates the globally sustainable potential of *bioenergy* at about 100EJ per year. This is significantly lower than other recent appraisals of bioenergy's potential because the Council has given stronger weight to the limitations upon biomass usage that result from sustainability considerations (Section 3.2.4). *Wind power* can already be produced today in a manner that is both environmentally sound and, in some cases, micro-economically cost-effective. The Council estimates the sustainably utilizable potential at approx. 140EJ per year (Section 3.2.5). Against the background of all future prospects of human energy usage the sustainably utilizable *solar energy* potential is practically limited only by technological and economic restrictions upon growth in installed capacity, but not by resource availability as such (Section 3.2.6; Fig. 4.4-5). Capacity could be raised by the end of the century to more than 1,000EJ per year (Table 4.4-1). Due to technological uncertainties, the Council has estimated the sustainable potential of *geothermal energy* cautiously at 30EJ per year by 2100 (Section 3.2.7). In addition to the above sources, it is reasonable to expect that in the future currently unforeseeable technological developments will lead to the tapping of *other renewable energy sources* or novel conversion technologies (Section 3.2.8). The Council integrates this expectation within its assessment by assigning it a sustainably utilizable potential amounting to 30EJ per year in 2100 (Table 4.4-1).

The Council's assessment underscores the major potential to *improve efficiency* in conversion processes throughout the entire chain of the energy system, e.g. by expanding the use of combined heat (and cold) and power technologies (Section 3.3). Moreover, there is a major potential to improve efficiency in the use of final energy, as well as at numerous further points in industry, commerce and buildings (Section 3.5).

The increasing use of fluctuating renewable energy sources will impart growing relevance to the issues surrounding the transportation, distribution and storage of energy. Here there is major potential for technological development, for instance by shaping power demand and the ever closer networking of power production through to an interconnected worldwide system (Section 3.4). Over the long term, hydrogen can play a key role as energy carrier and storage medium. The intensified use of natural gas and the associated expansion of appropriate infrastructure will facilitate the transition to a hydrogen economy (Section 3.4.4).

In the transportation sector, the key medium-term challenges are, in the view of the Council, to improve the efficiency of existing technologies and establish new mobility concepts. The Council views fuel cell drives within the context of a hydrogen economy as a highly promising long-term option, but can lend only limited support to the large-scale use of biogenic fuels.

Current global energy systems are based essentially upon fossil fuels and nuclear power as well as, in developing countries, the use of traditional biomass. The Council's assessment of the sustainably utilizable potential of energy carriers available worldwide shows that these energy systems are in need of a transformation that must be shaped in a long-term process. This global transformation of energy systems towards sustainability will have to rely above all upon vigorous expansion of renewable energies, and upon efficiency improvements (Chapter 4). Appraisal of the sustainably utilizable potential shows that, over the long term, solar energy will need to be made the key element of global energy supply.

4.1
Approach and methodology for deriving an exemplary transformation path

The first chapters of this report discussed the initial setting (Chapter 2) and the technological and sustainable potentials of present global energy sources (Chapter 3). For several reasons, today's global energy system must be considered non-sustainable. In particular, its climate impacts jeopardize the life-support systems on which human existence depends, air pollution and the unsustainable use of biomass generate substantial health problems and some 2,000 million people still lack access to modern forms of energy.

APPROACH
Chapter 4 elaborates one of many possible scenarios for the transformation of present energy systems towards a sustainable energy future. The stress here is on 'possible'. Many developments are conceivable that would reconfigure present worldwide energy systems in a sustainable fashion. Insofar, the scenario derived in this chapter is not to be understood as prescriptive, but illustrative. It illustrates that it is both technologically and economically feasible to turn global energy systems towards sustainability.

METHODOLOGY
To derive a transformation path, the German Advisory Council on Global Change applies the principle of setting normative guard rails which it has already used previously (WBGU, 1997b; Toth et al., 1997; Petschel-Held et al., 1999; Bruckner et al., 1999). This is based on the concept of demarcating potential future development trajectories by guard rails. Guard rails thus deliver criteria that a scenario must meet if it is to be sustainable (Fig. 4.1-1). Compliance with guard rails is a necessary condition for the sustainability of a path, but not a sufficient one, as guard rails can shift, for instance through new knowledge, or can be joined by entirely new guard rails. The Council takes this approach because it is generally

difficult to provide a positive definition of sustainable futures. It is easier to demarcate the realm perceived as unacceptable. Within the sustainable realm, under the restrictions posited, there are no further requirements upon a scenario for the future. The scenario can follow any trajectory within that realm. As long as it does not collide with a guard rail, it remains sustainable.

The guard rail approach for selecting sustainable scenarios can be compared to a filter that tests a series of plausible scenarios for the future in terms of their compatibility with a set of guard rails. At the same time, using a modelling approach, simulations are computed that identify sustainable path trajectories. The guard rails themselves are formulated by the WBGU (Section 4.3). The filtering method does not lead to any single viable path, but merely sets limits to the diversity of possible futures.

The guard rails can also be used to identify the areas in which measures can be taken – either to lead the system out of the non-sustainable into the sustainable realm, or to change the direction of a trajectory which, while presently still in the sustainable realm, would collide with a guard rail if it were to continue unaltered (Fig. 4.1-1). Such measures are discussed in Chapter 5.

Figure 4.1-2 concretizes the analysis philosophy of the guard rail concept for the example of the climate system. Here the guard rail method is applied as follows:

1. First, a basic set of scenarios for the future is presented (Section 4.2).
2. The Council selects a scenario (Section 4.2.6) that is conservative in terms of the transformability of its structures towards less energy-intensive products and services. If the transformation towards sustainability can be demonstrated for such a reference scenario, then the condition of sustainability will also hold for scenarios that are less conservative with regard to these structures.
3. In the next step, the guard rails are discussed, and compliance of the selected scenario with these is tested (Section 4.3). This reveals certain problems that attach to this scenario. In particular, the sce-

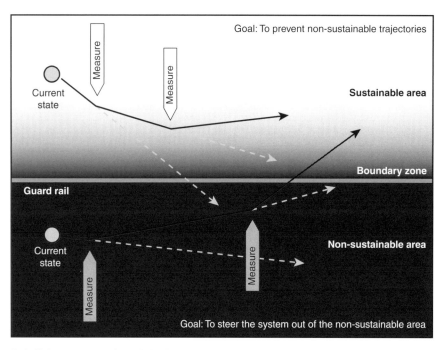

Figure 4.1-1
Connection between guard rails, measures and future system development.
The figure shows possible states of a system in terms of its sustainability, plotted over time. The current state of a system relative to the guard rail can be in the green area, the 'sustainable area' according to best available knowledge, or in the red area, the 'non-sustainable area'. If a system is in the non-sustainable area, it must be steered by appropriate measures in such a way that it moves 'through' the guard rail into the sustainable area. The guard rail is thus permeable from the non-sustainable side. If a system is in the sustainable area, there are no further requirements upon it at first. The system can develop in the free interplay of forces.

Only if the system, moving within the sustainable area, is on course for collision with a guard rail, must measures be taken to prevent it crossing the rail. The guard rail is thus impermeable from the sustainable side. As guard rails can shift due to future advances in knowledge, compliance with present guard rails is only a necessary criterion of sustainability, but not a sufficient one.
Source: WBGU

nario transgresses the climate guard rail (Section 4.3.1.2), as do all the other basic scenarios examined in the first step.

4. Subsequently, the selected scenario is modified in such a way that it complies with the guard rails. This delivers an exemplary transformation path (Section 4.4).

5. Finally, simulations are carried out using another, newly developed modelling approach, in order to underpin the exemplary path. Here cost-effective paths complying with the climate guard rail are identified, and the scope for action explored that results if various guard rails are stipulated. These additional analyses are used to discuss the properties of the exemplary transformation path (Section 4.5).

4.2
Energy scenarios for the 21st century

From the numerous energy scenarios that are available, the WBGU chose the IPCC scenarios as the basis for its analysis. These scenarios focus on the climate problem, which corresponds to the Council's priorities. The IPCC scenarios are recognized by the international scientific community and are based on

consistent assumptions about the driving forces for greenhouse gas emissions.

4.2.1
SRES scenarios as a starting point

The analysis of possible long-term developments of the energy system is based on a range of scenario groups developed by the IPCC (2000b, 2001c): The non-mitigation emissions scenarios – i.e. scenarios in which additional climate policy measures are assumed to be absent – described in the IPCC Special Report on Emissions Scenarios, SRES (IPCC, 2000b; referred to in the following as 'SRES scenarios') serve as reference scenarios for the IPCC climate change mitigation scenarios that build upon them ('post-SRES scenarios'; IPCC, 2001c). The SRES scenarios show the wide range of plausible future developments, resulting from the uncertainties attaching to the driving forces and their interactions and to the mechanisms simulated in the different models (IPCC, 2000b). Many of the scenarios assume strong environmental or social interventions, which distinguishes them from conventional business-as-usual scenarios.

A total of 40 scenarios were grouped into four families. All scenarios within a 'family' have a char-

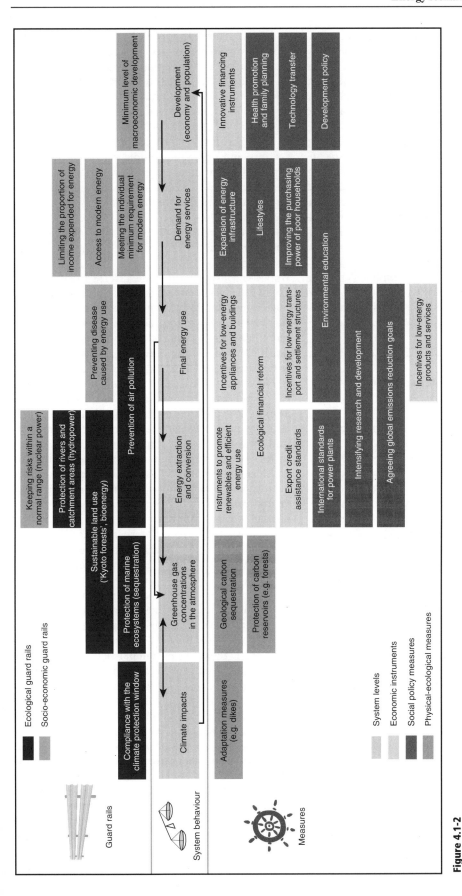

Figure 4.1-2
Application of the guard rail approach to the coupled energy-climate system and measures that follow from this for the sustainable transformation of energy systems. The uppermost row contains the ecological and socio-economic guard rails stipulated normatively by the German Advisory Council on Global Change (WBGU) for the climate-energy system (Section 4.3). The middle row represents the system behaviour of the coupled climate-energy system. The arrows represent relevant causal links. Due to population growth and the need for economic development, demand for energy services grows, which, through the agency of energy use and the resource extraction and conversion that this requires, leads to elevated greenhouse gas concentrations in the atmosphere. Through feedback loops, climate change affects global development and other points in the chain. For instance, global warming can generate additional greenhouse gas emissions, e.g. through the thawing of permafrost soils or increasing forest fires, and can influence demand for energy services (e.g. more heating or cooling). To be able to comply with the guard rails, various parameters need to be steered by specific measures (Fig. 4.1-1). For instance, technology transfer, incentive policies or international standards for power plants, as well as measures such as geological carbon sequestration, affect the various points in the chain between humankind and climate impacts.
Source: WBGU

acteristic storyline, i.e. a description of the relationships among influencing factors and their development. For simplicity, the four families can be distinguished along two dimensions: In the first dimension, a distinction is made between a world focussing on strong economic growth (A), in contrast to a world focussing strongly on sustainability (B). In the B world, environmental measures are considered, e.g. for air pollution control, but not measures specifically aimed at climate protection (e.g. CO_2 taxes). The second dimension distinguishes between a world with increasing economic convergence and social and cultural interaction among regions (globalization, 1) and a world with stronger emphasis on regional differences and local solutions (regionalization, 2).

This leads to four scenario families: A1 (high growth), B1 (global sustainability), A2 (regionalized economic development), B2 (regional sustainability).

4.2.2
Basic assumptions of the SRES scenarios

A1 WORLD: HIGH GROWTH

The A1 storyline has the following characteristics: strong market-orientation, sustained economic growth (worldwide approximately 3 per cent per year, corresponding to the growth rate over the last 100 years), strong emphasis on investment and innovation in education, technology and institutions, rapid introduction of new and more efficient technologies, increasing mobility and increasing social and cultural interaction and convergence among regions (e.g. in terms of per capita income). In the long term, the current demographic developments in industrialized nations (very low fertility rates, high degree of aging) are assumed to be transferred to the developing countries, due to the global convergence assumed in the A1 world. Following an increase of the world population to approximately 9,000 million, a reduction to approximately 7,000 million in 2100 is expected from 2050. This population development is in the lower range of existing projections, but above the lowest UN projection (IPCC, 2000b).

Energy productivity will increase by approximately 1.3 per cent, which is faster than the average over the last 100 years. However, low energy prices offer little incentive for the efficient utilization of final energy, so that very high primary energy consumption is assumed, with a strong increase in motorization and urban sprawl worldwide. In effect, the scenarios transfer the economic development in Japan or South Korea after World War II or in China over recent years to all developing countries (Roehrl

and Riahi, 2000). In terms of economic growth and global convergence of per capita income, the scenarios are therefore very optimistic.

Within the A1 scenario family, a distinction was made between four different paths, depending on the assumed technology development: the carbon-intensive path A1C, the oil- and gas-intensive path A1G, path A1T with a high proportion of non-fossil energy carriers, and finally the middle path A1B, for which similar improvement rates were assumed for all energy carriers or technologies. This differentiation highlights the influence of technological development, with the other driving forces remaining unchanged (in particular identical economic development) (Section 4.2.5).

B1 WORLD: GLOBAL SUSTAINABILITY

The B1 scenarios assume the same population development and similarly strong economic growth as the A1 scenarios. Here ,too, the development in the different regions is assumed to converge ('globalization'). Disparities in income are assumed to narrow at the same speed as in the A1 scenarios.

However, the B1 world is distinguished from the A1 world by a strong social and environmental awareness – de Vries et al. (2000) characterize it as 'affluent, just and green'. The world is also characterized by high efficiency improvements in the energy sector. A high proportion of increases in production and income are utilized for the expansion of social institutions, redistribution measures and environmental protection. Economic structures are characterized by rapid change towards a service and information society, in which materials are used sparingly. Clean and efficient technologies are introduced rapidly. Values shift towards a non-material mindset.

Despite strong economic growth, energy demand is low, i.e. only approximately one-quarter of the energy use in the A1 scenarios in 2100. On average, energy intensity decreases by approximately 2 per cent per year over the next 100 years. Compared with the historic rate of 1 per cent per year this represents a significant improvement, driven particularly by high energy prices. This world is characterized by high income transfer and high taxes. Globalization and liberalization are combined with a strong international sustainability policy. Strong support for research and development is provided. Cities develop in a compact way and with a high proportion of non-motorized traffic. Furthermore, the trend towards urbanization is dampened. Even without deliberate climate policy measures, these developments lead to low greenhouse gas emissions, because they are in themselves very effective in terms of climate protection.

A2 WORLD: REGIONALIZED ECONOMIC DEVELOPMENT

The A2 world is heterogeneous, since the regions want to preserve their national, cultural and religious identity and take different development paths (Sankovski et al., 2000). Separate economic regions develop. Economic growth and the speed of technological developments is therefore lower than in other scenario families. Technologies spread more slowly, and trade flows are smaller than in A1 scenarios. Per capita income converges less strongly than in the A1 or B1 scenarios. The scenario is based on very high population growth (15,000 million by 2100), since fertility patterns do not converge, in contrast to the A1 and B1 scenarios. Energy productivity increases by only 0.5–0.7 per cent per year, and energy demand is high, although not as high as in the A1 scenarios. The energy systems of the A2 world are very heterogeneous. The energy carrier mix in the individual regions strongly depends on resource availability.

B2 WORLD: REGIONAL SUSTAINABILITY

The B2 storyline describes a future in which local and regional solutions for sustainable development play a significant role. International institutions and structures become less important. The emphasis is on environmental protection, but only at a national and regional level. The growth in population is lower than in A2 scenarios (approximately 10,000 million by 2100). Economic growth is moderate, technological development is less pronounced than in the B1 or A1 world. Many projections match today's trends, for example in terms of population or economic growth, or the increase in energy productivity. Energy demand is lower than in the A1 and A2 scenarios, but higher than in the B1 scenarios. The current trend of declining investment in research and development is assumed to continue.

4.2.3
Emissions in the SRES scenarios

There is a strong variation in the emission of greenhouse gases and noxious substances between and within the scenario families. The fossil-intensive growth scenarios A1C and A1G show the highest CO_2 emissions, although the A2 scenarios also have very high emissions: Whilst economic growth is less strong, slower technological development leads to a smaller reduction in carbon and energy intensity. A1B and B1 scenarios show a change in trend towards lower emissions from about 2050. This can be attributed to a trend reversal in population development, and also to improvements in productivity. These trends more than compensate the economic

growth. Conversely, the A2 and B2 scenarios show continuously growing CO_2 emissions. The B1 and A1T scenarios have the lowest CO_2 emissions of all scenarios. They are both based on rapid development of non-fossil technologies, although there are strong differences in energy use.

The scenarios also differ in terms of land use and changes thereof: The trend towards a reduction in global forest areas is reversed in most scenarios, particularly in B1 and B2 scenarios. Due to the assumed lower growth in population and the reduction in population after 2050, and due to increased productivity in agriculture, methane and nitrous oxide emissions are much lower in A1 and B1 scenarios than in A2 and B2 scenarios. Sulphur emissions are generally lower than in previous projections, since it is thought that local and regional air pollution will be abated much earlier than previously assumed.

According to these model calculations, the global average ground-level temperature will increase by 1.4–5.8°C between 1990 and 2100 (IPCC, 2001a). This range is a result of uncertainties both in the climate system and in the socio-economic driving forces. Even the B1 and A1T scenarios with the lowest emissions violate the guard rail of the WBGU climate window (Section 4.3). The IPCC climate change mitigation scenarios, which are based on these SRES scenarios, are therefore introduced below.

4.2.4
IPCC climate change mitigation scenarios ('post-SRES' scenarios)

The IPCC Third Assessment Report (TAR) developed possible paths for achieving different stabilization targets for the CO_2 concentration of the atmosphere (between 450ppm and 750ppm) based on the SRES scenarios (IPCC, 2001c). The assumptions about the main driving forces (population, economic growth, demand for energy services) match those of the respective SRES scenarios. Stabilization of the CO_2 concentration by 2150 at the latest was specified as an additional condition. However, the reduction only applies to energy-related greenhouse gas emissions: CO_2 emissions from changes in land use and (non-energy-related) emissions of other greenhouse gases are identical to those in the reference scenarios.

Even for a stabilization level of 450ppm, the expected global warming during the 21st century will only remain below the WBGU guard rail (global warming less than 2°C relative to pre-industrial values; Section 4.3.1.2) for medium to low climate sensitivity values. For the long-term equilibrium, global warming above the limits specified in the WBGU climate window is expected, even for low climate sensitivity (IPCC, 2001d). If the climate window guard rail specified by the WBGU is to be complied with, stabilization levels of 450ppm or less are therefore required. However, no post-SRES stabilization scenarios with lower target levels are available, although other scenarios (e.g. Azar et al., 2001) show that stabilization levels of 350ppm can be achieved, for example through increased utilization of biomass in combination with carbon storage. By choosing a 450ppm scenario, the WBGU does not wish to give the impression that this is a safe greenhouse gas concentration level in the sense of Article 2 UNFCCC. Compliance with the WBGU climate window will rather require an analysis of integrated climate protection strategies (not only energy policy) and the development of appropriate scenarios.

Yet existing scenarios show (IPCC, 2001d) that in order to reach stabilization levels of 450ppm CO_2 or below, the increasing trend of global emissions has to be reversed very quickly, i.e. within 10 or 20 years, after which rapid reduction needs to continue over the following decades. If additionally the long investment cycles for power plants and distribution networks are considered, it becomes clear that the next 10–20 years are the crucial time window for the transformation of energy systems.

4.2.5
Technology paths in the A1 world

The A1 scenarios show the different technological paths that are conceivable under the same economic, social, political and demographic driving forces. For all paths, stabilization at 450ppm can be achieved, albeit with very different energy strategies, costs and risks.

4.2.5.1
Comparison of energy structures and climate change mitigation strategies

The A1-450 stabilization scenarios and their respective reference paths within the A1 scenario group are examined more closely below. The MESSAGE dynamic optimization model, coupled with the MACRO macro-economic model was used for quantification purposes (Messner and Schrattenholzer, 2000). MESSAGE minimizes the aggregated costs of energy production with given demand for energy services (preset by the macro-economic model) and calculates a cost-optimized energy carrier mix on this basis. Changes in demand resulting from measures for limiting CO_2 emissions (e.g. through a CO_2 tax) were not considered in the scenarios analysed in this

report. In the stabilization scenarios, primary energy use relative to the respective reference scenario does not decline. The stabilization scenarios with intensive use of fossil energy carriers even show a strong increase in primary energy demand. This can be attributed to the application of energy-intensive carbon dioxide separation technologies for carbon storage (Section 3.6.1; Table 3.6-1).

Depending on the assumptions about the technological paths in the reference scenarios, the development paths for the energy systems in the A1 scenarios differ for the same stabilization target. This highlights the path dependency associated with a preference for certain technologies in the individual reference scenarios. In the A1T path, for example, the proportion of solar energy increases due to climate policy measures, whilst the 'balanced' A1B scenario and coal-intensive A1C scenario show a strong increase in the proportion of nuclear energy (Roehrl and Riahi, 2000).

COAL AND NUCLEAR ENERGY-INTENSIVE PATH: A1C
The A1C scenarios are characterized by the utilization of coal technologies with reduced noxious substance emissions, which, without additional climate policy measures, lead to very high greenhouse gas emissions, although they are environment-friendly if the climate change problem is left aside. They are based on the assumption that conventional oil and gas reserves will decline quickly, requiring strong investment in cost-intensive new coal technologies (high temperature fuel cells, coal gasification and liquefaction). But nuclear technology is also developed further (for example uranium extraction technologies), since intensive use is made of nuclear energy, particularly in regions with low coal deposits. In the A1C reference path, coal is the main energy carrier, with a primary energy share of 47 per cent in 2100. The nuclear energy share is 18 per cent. Due to the high demand for coal, which cannot be satisfied with domestic coal in all regions, an intensive global methanol trade will develop, since methanol (generated from coal) is required, particularly for the transport sector. The main mitigation measures for achieving the 450ppm stabilization target are carbon storage and increased efficiency, but these scenarios also require strong expansion of nuclear energy.

OIL- AND GAS-INTENSIVE PATH: A1G
The main characteristic of the A1G scenarios is the utilization of non-conventional oil and gas resources, including oil shale, tar sands and methane hydrates (Section 3.2). Rapid technological progress in extraction and conversion technologies for oil and gas is assumed. Global trade in oil and gas will increase

strongly, and new gas pipelines will be built from 2010 or 2020. Due to the energy demand for extraction and gas transport, primary energy demand is particularly high. In the A1G reference path, gas is the main energy carrier in 2100 (45 per cent share of primary energy), followed by renewable energy sources (25 per cent) and oil (14 per cent). However, nuclear energy also has a high share (12 per cent). With a cumulative consumption of approximately 34,000EJ of oil and 59,000EJ of gas, even these oil- and gas-intensive scenarios use during the 21st century only a fraction of total fossil fuel occurrences (Nakicenovic and Riahi, 2001). It is assumed that a small proportion of deposits that are today classified as additional occurrences will already be recoverable during the 21st century (Table 3.2-1).

Here ,too, carbon storage and increased efficiency are the main climate change mitigation measures for reaching the stabilization target of 450ppm CO_2. The re-injection of CO_2 into oil and gas fields plays an important role. However, drastic structural changes would be required during the 22nd century, since the capacity limits for re-injection into gas fields would be reached.

MIXED PATH: A1B

The A1B scenarios ('balanced technology') assume that all technologies will develop uniformly. Lower path dependency is therefore assumed than in the other A1 scenarios: A coordinated global strategy of research, development and application of technologies leads to regionally differentiated specialization in different technologies. To secure carbon dioxide stabilization, A1B scenarios focus on carbon storage and on increased development of non-fossil energy carriers and conversion technologies, particularly new nuclear reactors, hydrogen fuel cells for the transport sector and additional hydroelectric plants. Hydrogen is mainly produced from renewable energy sources.

STRONG DEVELOPMENT OF NON-FOSSIL ENERGY CARRIERS: A1T

The A1T scenarios are characterized by a rapid development of solar and nuclear technologies and by the large-scale application of hydrogen technology. Prerequisites are very large and dedicated investment in research, development and application of these technologies, for example in new, 'inherently safe' nuclear energy technologies (e.g. high temperature reactor) and renewables. Higher investments in energy efficiency are also assumed, so that the demand for final energy is lower than in the other A1 scenarios (with identical demand for energy services). On average, energy productivity increases by 1.4 per cent per year. In 2100, renewables and nuclear energy together have a primary energy carrier share of 86 per cent.

Due to the already low CO_2 emissions, the A1T scenarios require only few reduction measures to achieve the 450ppm stabilization target. Carbon storage is therefore only used moderately. Technological progress is similar to the A1B scenarios, although the departure from the fossil path is even clearer. The A1T-450 scenario thus illustrates development towards a hydrogen economy. Hydrogen is produced in nuclear reactors and with renewable energy sources (e.g. solar thermal technology). Coal utilization ceases at the end of this century. The utilization of nuclear energy (high temperature reactors) is increased slightly, and so is the utilization of hydrogen fuel cells for transport.

4.2.5.2
The role of carbon storage

Since CO_2 storage is cost-intensive (Section 3.6.1), it is not used in the reference scenarios without climate policy measures, except for re-injection into oil and gas fields. It is, however, used in all A1 post-SRES stabilization scenarios, albeit to a very different extent (Table 4.2-1). The figures are comparable with the potential estimates for carbon storage in oil and gas fields (200–500GtC) and in deep aquifers (100 to more than 1,000GtC) (Section 3.6), although these figures are subject to a high degree of uncertainty. The fossil-intensive paths A1C-450 and A1G-450, but also the 'medium' path A1B-450, far exceed the guard rail of 300GtC set by the WBGU as the upper limit for carbon storage during the 21st century (Section 4.3). In scenarios A1C-450 and A1G-450, the required CO_2 sequestration is greater than the potential estimated to be available for geological storage, so that storage in deep ocean water would be required. The WBGU regards anthropogenic carbon storage in the ocean as non-sustainable (Section 3.6.3).

4.2.5.3
Cost comparison

Figure 4.2-1 shows the dependency of energy system costs on the technological development in the reference scenarios and on the stabilization target (Roehrl and Riahi, 2000). The energy system costs are defined as the sum of investment, operation and maintenance costs, including the costs for distribution and for environmental technology. Whilst the MESSAGE model discounts energy system costs at 5

Table 4.2-1
Total stored CO_2 quantity for the period 1990–2100 in selected A1 scenarios (reference and 450ppm CO_2 stabilization scenarios). *EOR* Enhanced Oil Recovery, *EGR* Enhanced Gas Recovery (Section 3.6.1). *A1C* coal-intensive path, *A1G* oil- and gas-intensive path, *A1B* mixed path, *A1T* strong development of non-fossil technologies, *450* stabilization at 450ppm CO_2. Source: Roehrl and Riahi, 2000

Scenario	A1B	A1B-450	A1G	A1G-450	A1C	A1C-450	A1T	A1T-450
				[GT C]				
EOR + EGR	28	98	171	366	0	63	29	69
Other storage	0	762	0	1,148	0	1,492	0	148
Total	*28*	*860*	*171*	*1,514*	*0*	*1,555*	*29*	*217*

per cent (assuming minimization), they are shown as non-discounted in Figure 4.2-1.

It is evident that the cost differences between the reference scenarios and the associated stabilization scenarios are usually lower than the cost differences between the individual reference scenarios. For example, the difference between the costs for the fossil-intensive paths (A1G, A1C) and the A1T path, which strongly depends on non-fossil technologies, is far greater than the costs for stabilization at 450ppm, for example for the balanced A1B path or the A1T path. The fossil path is therefore an inherently expensive path. The main reason for the high costs is the application of partly outdated energy structures (path dependency), in which comparatively small learning effects can be expected. An additional factor is the increasing cost of resource extraction for non-conventional resources (Section 3.2.1). In the coal-intensive path, a large proportion of the future demand for liquid fuels has to be supplied through coal liquefaction or methanol production, both of which are very expensive.

For the coal-intensive path, whose reference scenario already shows the highest costs, the additional costs for a stabilization of the carbon dioxide concentration to 450ppm are particularly high. Fixation on a coal-intensive path therefore not only leads to an expensive energy system in the long term, but also causes very high CO_2 reduction costs with a stabilization target of 450ppm.

However, research and development costs and additional macro-economic adaptation costs or expenditure for the development and procurement of end-use devices (e.g. vehicles, production plants, domestic appliances) are not included in the energy system costs shown here. Expenditure for research and development is higher in the A1T scenario than in the other scenarios. The research intensity of the energy sector (i.e. the proportion of expenditure for research and development relative to turnover) in the A1T scenario between 1990 and 2050 reaches globally averaged values between 4 and 13 per cent, depending on how strongly research and development expenditure influence the technology costs (Riahi, personal communication). For comparison,

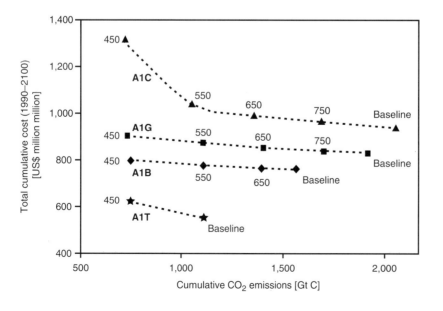

Figure 4.2-1
Total (non-discounted) energy system costs (1990–2100) and cumulative CO_2 emissions for the reference and stabilization scenarios (stabilization levels 750, 650, 550 and 450ppm of CO_2). Each point represents a scenario. A1C: Coal-intensive path. A1G: Oil- and gas-intensive path. A1B: Mixed path. A1T: Strong development of non-fossil technologies. Source: Roehrl and Riahi, 2000

the extremely research-intensive pharmaceutical industry currently has a research intensity of approximately 10 per cent (Section 2.3.1). By contrast, in the coal and nuclear-intensive scenario A1C, the globally averaged research intensity only reaches a value of approximately 0.3 per cent during the same interval. This corresponds to the current research intensity of the energy sector in the OECD countries. The massive increase in research and development expenditure (against the current trend) in the energy sector to US$$_{1990}$8–25 million million for the period 1990–2050 is a prerequisite for the cost reductions achieved in the A1T scenarios for technologies to exploit renewable energy carriers. The additional expenditure for research and development compared with the A1C scenarios, where only approximately US$$_{1990}$1 million million are spent, is at least compensated by the cost reductions and the much lower investment costs. During the same period, the investment costs for the A1T scenarios are approximately US$51 million million, for the A1C scenarios approximately US$73 million million. Nevertheless, even in the A1C scenarios the investment costs amount to a maximum of 1.7 per cent of GDP. The difference in the total energy system costs is even greater: For the A1C scenarios, total energy system costs amount to approximately US$230 million million during the same period, in the A1T scenarios they are approximately US$190 million million. In addition to strong market growth, higher expenditure for research and development over the next decades is a prerequisite for the realization of the comparatively very high assumed learning rates (26 per cent for solar photovoltaic, 11 per cent for wind energy, 10 per cent for electricity from biomass, 8 per cent for nuclear energy, 10 per cent for natural gas fuel cells; Riahi, 2002).

For the transformation of energy systems not only the cumulative costs are significant, but also their development over time (Fig. 4.2-2). For example, the cost advantages of the non-fossil A1T path vis-à-vis other paths only become apparent after more than 20 years when comparing the non-discounted energy-specific energy system costs. The cost advantage of the non-fossil climate change mitigation path compared with the coal-intensive reference path only becomes apparent from about 2040. However, the coal-intensive climate protection path A1C-450 is more expensive than the other paths from the outset, since expensive measures are required, e.g. for sequestration, due to the high emissions assumed in the reference path (Roehrl and Riahi, 2000).

This highlights the path-dependency risk, with the fossil path, and in particular the coal-intensive path only being able to be maintained at very high costs. The high costs of the coal-intensive climate protection path A1C-450 that are incurred from the outset make the political enforceability of ambitious climate policy measures in a world taking such a path appear doubtful.

4.2.5.4
Environmental effects

Figure 4.2-3 shows the annual emissions of carbon dioxide, methane and sulphur dioxide, as well as temperature change for the coal-intensive and non-fossil paths, both for the respective reference scenarios and for the 450ppm stabilization scenarios.

It can be seen that the non-fossil climate protection path causes far lower sulphur dioxide emissions than the carbon-intensive path. In the carbon-intensive path, they increase despite the significant invest-

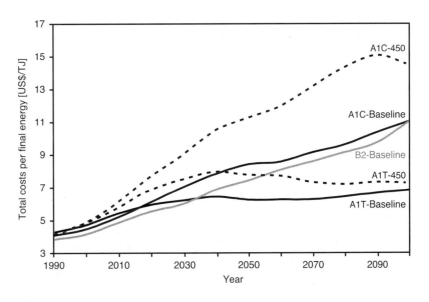

Figure 4.2-2
Specific (non-discounted) system costs (based on final energy) for the A1C and A1T reference scenarios and for the A1C-450 and A1T-450 stabilization scenarios. The cost path for the B2 reference scenario is shown for comparison. For the investment decisions in MESSAGE, future costs are discounted with 5 per cent. No discounting was carried out for this graph, in order to avoid a distortion of cost development over time. Sources: Roehrl and Riahi, 2000; Riahi, 2002

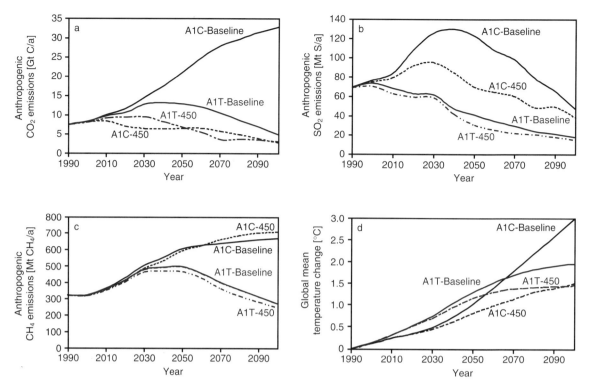

Figure 4.2-3
Environmental effects for a path with strong expansion of non-fossil technologies (A1T, red) and a coal-intensive path (A1C, black) with identical assumptions for population and economic development. In each case, the effects for the reference path (continuous line) and the stabilization scenario (stabilization of the carbon dioxide concentration at 450ppm, dashed line) are shown. Stabilization in the coal-intensive path A1C is only possible with extensive sequestration.
a) anthropogenic carbon dioxide emissions,
b) anthropogenic sulphur dioxide emissions,
c) anthropogenic methane emissions,
d) mean global warming (from 1990 baseline) with an assumed climate sensitivity of 2.5°C.
Source: Riahi, 2002

ments in low-polluting coal technologies over the next decades to nearly twice the current value before a reduction can be observed (Fig. 4.2-3b). Due to the cooling effect of sulphate aerosols, initially this leads to lower warming in the fossil path with the same CO_2 stabilization target, despite the higher methane emissions.

4.2.6
Selection of a scenario for developing an exemplary path

The WBGU considers it difficult to imagine that the development described by the A2 scenarios can be steered into the sustainable realm. The combination of a lack of global convergence, the associated slow technological development and low efficiency improvement and decarbonization, together with the absence of a general environmental orientation, make the attainment of climate protection targets

within the WBGU climate window extremely difficult and expensive, if not impossible.

A B2 scenario could be considered for the development of an exemplary path towards sustainability, even if it reaches the sustainable realm later in terms of the socio-economic guard rails (Section 4.3) than the A1 and B1 scenarios, which are characterized by strong convergence. However, since no B2 scenario is available with an energy system model containing the required technological detail, with stabilization at a CO_2 concentration of 450ppm (Morita et al., 2000), for pragmatic reasons the WBGU has not selected the B2 scenarios for further examination.

Because a world of global convergence (A1 and B1 scenarios) leads more quickly into the sustainable realm defined by the WBGU's socio-economic guard rails (Section 4.3), the B1-450 stabilization scenario would lend itself as a basis for a path modified according to WBGU criteria, in view of its orientation to both social and environmental compatibility. However, the WBGU regards it to be more advisable to demonstrate the option of an energy system transformation towards sustainability on the basis of a scenario that assumes strong growth in primary energy demand. The options for structural change towards less energy-intensive products and services, or changes in preferences, consumer habits and lifestyles are thus estimated more cautiously.

The WBGU regards the fossil and nuclear energy-intensive paths of the A1 scenarios (global convergence, high growth, rapid technological development) as non-sustainable, even if the minimum socio-economic requirements (Section 4.3.2) and a stabilization of the carbon dioxide concentration at 450ppm could be achieved. This stabilization is only possible with extensive sequestration, so that at the end of the 21st century the limits to storage in geological formations would be reached. Storage in deep ocean water may even be required, which the WBGU regards as non-sustainable. These scenarios repeatedly overstep the guard rail for the maximum tolerable utilization of carbon storage (Section 4.3.1.2). Furthermore, these paths hinder the transformation of the energy system for future generations, since they are unlikely to be able to be continued during the 22nd century. The WBGU also regards the high proportion of nuclear energy as non-sustainable. Finally, the energy system costs for these fossil-nuclear scenarios are very high. Stabilization of the CO_2 concentration at 450ppm only appears possible at very high cost. Moreover – with the same stabilization level of atmospheric carbon dioxide concentration – due to emissions of other pollutants (e.g. SO_2), the environmental effects of the fossil paths are significantly greater than those of the non-fossil path, despite the fact that significant investments in the development of power plants producing lower emissions are assumed.

Since even the A1B path with a balanced technology mix relies for stabilization at 450ppm on geological storage to an extent that is not acceptable to the WBGU, the WBGU uses post-SRES scenario A1T with a 450ppm stabilization target for the development of an exemplary path, because with certain modifications it can comply with all guard rails. It avoids a deepening of dependency on fossil technologies, has low emission values and assumes sustained economic growth and economic convergence among countries, as well as strong technological development.

This choice does not imply a statement about which of the SRES worlds the WBGU regards as the most probable one. Quite the opposite: Since the A1 world is based on optimistic assumptions, particularly in terms of the rapid reduction in divergence between rich and poor countries as well as low growth in population, the realization of a (modified)

A1T-450 path requires not only energy policy measures, but also economic and development policy measures. To reach robust conclusions, it would therefore be sensible to also examine an appropriate path from the B2 world. However, this would require the development of a corresponding climate change mitigation scenario in the B2 world with a detailed consideration of technologies.

4.3
Guard rails for energy system transformation

Guard rails are quantifiable limits to damage whose transgression would entail intolerable consequences today or in the future. Even major utility gains would not offset this damage (WBGU, 2001a). In the following, the Council defines guard rails for the protection of natural life-support systems, and for the operationalization of socio-ethical goals guided by the vision of sustainable development (ecological guard rails, Section 4.3.1; socio-economic guard rails, Section 4.3.2). These guard rails establish concrete limits in relation to energy use that must be observed in order to live sustainably (Box 4.3-1), and can be applied to test the sustainability of future energy scenarios. Guard rails should by no means be understood as goals. They do not establish desirable values or states, but rather absolute minimum requirements which have to be complied with if the principle of sustainability is to be observed. Nonetheless, the guard rail approach can indeed be used to deduce concrete goals (Chapter 5).

Even if development trajectories remain within the guard rails, this need not mean that all socio-economic grievances or ecological damage can be averted. Nor do global guard rails reflect the significant regional and sectoral disparities that may arise in relation to the impacts of global change. Finally, the guard rails stated by the WBGU can be no more than proposals, for the determination of non-tolerable impacts cannot be left up to the academic community but needs to take place – supported by scientific expertise – within a worldwide, democratic decision-making process (WBGU, 1997b; Box 4.3-2).

4.3.1
Ecological guard rails

4.3.1.1
Protection of the biosphere

In earlier reports, the WBGU has developed five 'biological imperatives' as general principles

Box 4.3-1

Guard rails for sustainable energy policy

Ecological guard rails

CLIMATE PROTECTION
A rate of temperature change exceeding 0.2°C per decade and a mean global temperature rise of more than 2°C compared to pre-industrial levels are intolerable parameters of global climate change.

SUSTAINABLE LAND USE
10–20 per cent of the global land surface should be reserved for nature conservation. Not more than 3 per cent should be used for bioenergy crops or terrestrial CO_2 sequestration. As a fundamental matter of principle, natural ecosystems should not be converted to bioenergy cultivation. Where conflicts arise between different types of land use, food security must have priority.

PROTECTION OF RIVERS AND THEIR CATCHMENT AREAS
In the same vein as terrestrial areas, about 10–20 per cent of riverine ecosystems, including their catchment areas, should be reserved for nature conservation. This is one reason why hydroelectricity – after necessary framework conditions have been met (investment in research, institutions, capacity building, etc.) – can only be expanded to a limited extent.

PROTECTION OF MARINE ECOSYSTEMS
It is the view of the Council that the use of the oceans to sequester carbon is not tolerable, because the ecological damage can be major and knowledge about biological consequences is too fragmentary.

PREVENTION OF ATMOSPHERIC AIR POLLUTION
Critical levels of air pollution are not tolerable. As a preliminary quantitative guard rail, it could be determined that pollution levels should nowhere be higher than they are today in the European Union, even though the situation there is not yet satisfactory for all types of pollutants. A final guard rail would need to be defined and implemented by national environmental standards and multilateral environmental agreements.

Socio-economic guard rails

ACCESS TO ADVANCED ENERGY FOR ALL
It is essential to ensure that everyone has access to advanced energy. This involves ensuring access to electricity, and substituting health-endangering biomass use by advanced fuels.

MEETING THE INDIVIDUAL MINIMUM REQUIREMENT FOR ADVANCED ENERGY
The Council considers the following final energy quantities to be the minimum requirement for elementary individual needs: By the year 2020 at the latest, everyone should have at least 500 kWh final energy per person and year and by 2050 at least 700 kWh. By 2100 the level should reach 1,000 kWh.

LIMITING THE PROPORTION OF INCOME EXPENDED FOR ENERGY
Poor households should not need to spend more than one tenth of their income to meet elementary individual energy requirements.

MINIMUM MACROECONOMIC DEVELOPMENT
To meet the macroeconomic minimum per-capita energy requirement (for energy services utilized indirectly) all countries should be able to deploy a per-capita gross domestic product of at least about US$$_{1999}$3,000.

KEEPING RISKS WITHIN A NORMAL RANGE
A sustainable energy system needs to build upon technologies whose operation remains within the 'normal range' of environmental risk. Nuclear energy fails to meet this requirement, particularly because of its intolerable accident risks and unresolved waste management, but also because of the risks of proliferation and terrorism.

PREVENTING DISEASE CAUSED BY ENERGY USE
Indoor air pollution resulting from the burning of biomass and air pollution in towns and cities resulting from the use of fossil energy sources causes severe health damage worldwide. The overall health impact caused by this should, in all WHO regions, not exceed 0.5 per cent of the total health impact in each region (measured in DALYs, disability adjusted life years).

(WBGU, 2001a). These form the basis for a sustainable use of the biosphere, and thus also of energy. They include:
– sustaining the integrity of bioregions,
– securing biological resources,
– conserving biopotential for the future,
– protecting the global natural heritage, and
– preserving the system control functions of the biosphere.
The definition of ecological guard rails presented in the following proceeds from this normative basis.

4.3.1.2
Tolerable climate window

Present energy systems are a prime driver of global climate change. This climate change will have significant impacts upon ecosystems as well as human civilization. The following discussion therefore develops a climate protection guard rail designed to exclude intolerable ecological and socio-economic climate change effects.

DEFINING THE GUARD RAIL
The WBGU already developed the climate window in earlier reports (WBGU, 1995, 1997b). The determination of what is deemed to be intolerable warming is oriented to the stress limits of human society

Box 4.3-2

Defining guard rails in terms of international law?

A binding definition of sustainable energy policy guard rails at international level can be established primarily through international law, starting with the binding principles and objectives enshrined in international treaties and conventions. However, it may be possible to derive guard rails from customary international law as well. Political declarations, resolutions and other 'soft law' agreements, although non-binding, can contribute towards the definition of international legal principles.

Until now, the key instrument in the development of sustainable energy policy has been the United Nations Framework Convention on Climate Change (UNFCCC), together with the Kyoto Protocol to the Convention. The Kyoto Protocol commits the industrialized and transition countries to reducing greenhouse gas emissions by 5 per cent on average by the 2008–2012 period against the baseline year of 1990. However, this must be regarded merely as a first step towards defining binding long-term climate guard rails.

For this reason too, there is increasing debate about whether states' obligations can already be derived from customary international law, particularly the principle which prohibits transboundary environmental damage. Tuvalu, the Maledives and Kiribati, whose survival is threatened by climate-related sea-level rise, are planning legal action against major industrialized countries before the International Court of Justice. For this reason, there is now more intensive debate on the issue of the industrialized nations' liability under international law for damage which can be expected to occur in developing countries in future, especially in the small island states which are most at risk. Until now, states have tended to be reluctant to enforce claims against other countries. The general preference has been to seek diplomatic solutions or develop international civil liability instruments. However, given the extent of climate damage and the completely inadequate resources available to the Adaptation Fund established under the Kyoto Protocol, the worst-affected states may well attempt to pursue claims based on the principle of state responsibility more vigorously in future. An increase in tort claims against major emitters can also be expected at the same time.

As well as many practical problems, enforcing claims in accordance with the principle of state responsibility creates a number of legal problems to which there are no clear solutions. They include, for example, the scope and application of the principle prohibiting transboundary environmental damage, the question whether the UNFCCC and the Kyoto Protocol may potentially exclude any possibility of inter-state claims, and which actions could establish any state responsibility for climate damage in the first place. Here, as with tort liability, far more research is required.

Although the Kyoto Protocol has given rise to a definition of the 'climate protection' guard rail for the first commitment period from 2008 to 2012, socio-economic guard rails (Section 4.3.2) have rarely been integrated into binding legal instruments until now.

The failure to develop, harmonize or define sustainable global energy policy guard rails or integrate energy-relevant development goals fully into climate protection policy is a distinct shortcoming in international climate policy at present. The WBGU recommends that a World Energy Charter be developed as a first step towards implementing a sustainable energy policy. In order to overcome resistance to this – initially non-binding – agreement, the key task will be to convince the international community that this type of global energy strategy offers added value (Section 5.3.2.2).

Sources: WBGU, 2001c; Tol and Verheyen, 2001

and the observed range of fluctuation in the recent Quaternary period, during which humankind and its environment have co-evolved. The aim is to preserve the natural environment as the life-support system for humankind and all other organisms. Defining this tolerable window thus establishes an important basis for implementing the UN Framework Convention on Climate Change (UNFCCC), the ultimate objective of which is to achieve 'stabilisation of greenhouse gas concentrations in the atmosphere at a level that would prevent dangerous anthropogenic interference with the climate system. Such a level should be achieved within a time-frame sufficient to allow ecosystems to adapt naturally to climate change, to ensure that food production is not threatened and to enable economic development to proceed in a sustainable manner' (Art. 2, UNFCCC).

Recent IPCC assessments (2001a) confirm the WBGU's decision to adopt the climate window as a guard rail against intolerable climate change. The climate window is defined by two limits: Both a temperature change rate of more than 0.2°C per decade and a mean global temperature change of more than 2°C relative to pre-industrial levels are deemed unacceptable. The form of the WBGU climate window also needs to be noted (Fig. 4.3-1), for the maximum tolerable warming rate declines with increasing proximity to the maximum tolerable absolute warming level, i.e. the climate window is not rectangular. The arguments in support of this guard rail are substantiated in the following.

CLIMATE ZONE SHIFT
One aim of the UNFCCC is to allow ecosystems to adapt naturally to climate change (Art. 2 UNFCCC). The impacts of climate change upon the world's ecosystems are not yet fully understood. What is certain, however, is that the geographical extent of damage to ecosystems will grow in step with both the rate of climate change and its absolute level and will involve the loss of biological diversity (IPCC, 2001a). If climate zone shifts are too swift or too large, several of the biological imperatives will be infringed. Hazards to certain particularly sensitive ecosystems

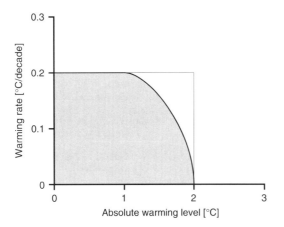

Figure 4.3-1
The WBGU climate protection window. The window shows
the tolerable rate of warming as a function of the absolute
warming level already reached. The acceptable rate of change
drops with increasing proximity to the maximum warming
level of 2°C relative to pre-industrial levels.
Source: WBGU, 1995

(e.g. coral reefs, tropical forests, wetlands) must be
expected from an absolute warming level of 1–2°C
and are in some instances already apparent today
(Box 4.3-3).

An absolute global warming level of more than
2°C may cause fundamental changes in many natural
systems. This would lead to major additional greenhouse gas emissions (IPCC, 2001a). The risk of
extreme climate events would then also rise sharply.
From a warming level of 3.5–4°C onwards, adverse
effects must also be expected for anthropogenic
ecosystems in most regions of the world (IPCC,
2001a).

Climate zone shifts trigger the migration of animals and plants. Transportation routes and human
land uses (agriculture, settlement, etc.) stand in the
way of this form of adaptation (WBGU, 2001a), so
that the loss of entire ecosystem types and species
must be feared.

THERMOHALINE CIRCULATION
The thermohaline circulation of the world ocean has
great importance for the global water and heat balance. It is driven by the downward flow of cold and
salty water, above all in the Labrador Sea and Greenland Sea, but also in the Weddell Sea. The cold water
flows as a deep current southwards through the
Atlantic, while in exchange warm upper water is con-
veyed from tropical regions to the north by, among
other things, the Gulf Stream. If certain climate
warming thresholds are overstepped, both a shut-
down of regional components (e.g. in the Labrador
Sea or in the Greenland Sea) and a complete collapse
of the circulation are conceivable. The changes thus
triggered could set in suddenly and would be irreversible over a period of centuries (IPCC, 2001a).
Within a few decades, the climate of western and
northern Europe could cool by about 4°C – with
immense impacts upon Europe's economy and ecology. As long as the climate guard rail defined above
is not overstepped, a collapse of the thermohaline
circulation is highly improbable (WBGU, 2000).

SEA-LEVEL RISE
The rising mean sea level is a result of warming driven by greenhouse gas emissions. Over the 20th century, the absolute mean sea-level rise already
amounted to 10–20cm (IPCC, 2001a). This is about

Box 4.3-3

Corals under threat through climate change

Even today, coral reefs are under threat due to increased
temperatures in the upper layer of the tropical seas. At temperatures above 30°C corals increasingly lose their sym-
biontic algae (zooxanthellae). This impairs the corals'
skeleton growth, reproductive capacity and stress resistance. Reefs are hotspots of biological diversity. Although
they cover less than 1 per cent of the seafloor, they host a
third of all known marine species. Moreover, they have
important resource functions, e.g. for coastal protection,
fisheries and tourism. As many reef-building corals live
close to their upper temperature limit, even a slight rise in
water temperature leads to increased bleaching. This phe-
nomenon will intensify with climate change and is already
observable today during an El Niño. If intervals between El
Niño events are short, it is scarcely possible for corals to
recover. The outcome is a loss of irreplaceable natural her-

itage together with its biopotential. Model computations
indicate that the temperature tolerance of reef-building
corals will be exceeded within the next decades.

Coral reefs are also threatened by sea-level rise.
Healthy reefs can achieve vertical growth of up to 100 mm
per decade; this is at the upper limit of the sea-level rise of
20–90 mm per decade estimated by the IPCC. However, in
view of the many additional anthropogenic stress factors it
is dubious whether present coral growth will suffice to keep
up with the anticipated sea-level rise. One reason is that the
increased CO_2 concentration in the atmosphere and in the
upper layer of the ocean is reducing the calcium carbonate
content of seawater. This is reducing the growth rates of
coral reefs and is thus hampering their adaptation to rising
sea levels.

A guard rail of at most 2°C mean global warming is
probably already too high to safeguard the survival of many
coral reefs.

Sources: Gattuso et al., 1999; Hoegh-Guldberg, 1999; IPCC,
2001a; Coles, 2001

ten times the mean rise over the previous 3,000 years. It is important in this context to take into consideration the great inertia of the climate system: The anthropogenic emissions already caused today and the resultant thermal expansion of seawater will only let sea levels stabilize at a new value in centuries. The time constants are even substantially longer if we consider the melting of inland glaciers such as in Greenland.

In the WBGU's opinion it is essential to prevent the melting of the Greenland ice, as such a process would cause the mean sea level to rise by several metres over many millennia (IPCC, 2001a). Model computations indicate that for this to happen the critical warming over Greenland is around 3°C. Local warming over Greenland is higher than global warming by a factor of 1.3–3.1 (IPCC, 2001a). If we assume an amplification factor of 2, then a global warming by about 1.5°C could already lead to an irreversible melting of the Greenland ice in its entirety.

The mean global warming level of 2°C relative to pre-industrial levels deemed unacceptable by the WBGU translates into an absolute sea-level rise of 25–100cm (depending upon model) over the next 100 years (IPCC, 2001a). It is thus probable that the upper limit of 15–25cm mean sea-level rise discussed by the WBGU in an earlier special report will already be exceeded significantly in this century (WBGU, 1997b). In scenario A1T-450 (Section 4.2.5), with a climate sensitivity (being the magnitude of global warming that follows from a doubling of the CO_2 concentration in the atmosphere relative to pre-industrial levels) of 2.5°C, sea-level rise will be around 50cm by 2100. Even such a rise will involve severe social impacts, major costs for the adaptation of coastal infrastructure and serious losses of valuable coastal ecosystems (IPCC, 2001a).

These more recent findings affirm the WBGU climate window. The WBGU wishes to stress that this guard rail cannot entirely exclude the risk of the Greenland ice sheet melting. Moreover, the existence of a number of small island states would be jeopardized, and the number of people threatened by storm surges would rise substantially.

Besides the absolute rise, the rate of sea-level rise is a key parameter, as the adaptive capacity of societies and ecosystems declines if the rise happens faster. Major socio-economic and ecological damage is anticipated in coastal regions in the event of swift sea-level rise. The curve of sea-level rise in the 21st century is very similar with different CO_2 stabilization concentrations and across different models; the rate of sea-level rise can thus be taken to be a function of the absolute rise (IPCC, 2001a). As the ratio of the maximum to average gradient of sea-level rise appears to be constant, a guard rail for the speed of

sea-level rise can be translated into a guard rail for the absolute warming level. If we were to postulate a guard rail of 5cm per decade, then this would only just be transgressed in scenario A1T-450. A slightly lower stabilization concentration would presumably ensure compliance with the guard rail. Thus, as the WBGU temperature window already limits the rate of sea-level rise, a separate guard rail does not appear necessary.

ECONOMIC ADAPTABILITY
A first-order analysis (WBGU, 1995, 1997b) has shown that from a temperature change of 0.2°C per decade onwards such high climate change costs would result that the adaptive capacity of national economies would be exceeded, resulting in intolerable economic and social disruption. It is thus also essential for economic reasons to remain within the WBGU climate window.

FOOD SECURITY
Climate change impacts greatly upon agricultural ecosystems and thus upon food security (IPCC, 2001b). In a global perspective, climate change will not impact negatively upon food production (if the increase of global mean temperature does not exceed 2°C), as both the increased temperature and the increasing precipitation in some 'winner regions' may be able to compensate for losses in 'loser regions'. Absolute warming of up to 2°C is even expected to lead to global growth in cereal production by 3–6 per cent (Fischer et al., 2001). The winners are predominantly industrialized or transition countries which, due to their location in colder regions, would profit from the increased temperatures (such as Canada and Russia). The losers would be developing countries, above all in sub-Saharan Africa and in Latin America (Fischer et al., 2002). However, this assessment does not yet take into consideration effects such as increasing extreme weather events or mounting soil degradation, so that the 'winners' may merely become the 'less affected'. Even if global climate change causes increased temperature extremes of only brief duration, this could lead to broad-scale crop failures (IPCC, 2001d). Climate change at an absolute level of 2°C would not trigger an acute global food crisis, but must be expected to heighten to an unacceptable degree the already prevailing global imbalance of food supply.

HUMAN HEALTH
It is to be expected that the negative effects of climate change upon human health will outweigh the positive ones (IPCC, 2001b). Increased extreme events such as heat waves and extreme storms or floods would amplify the risk of infectious diseases,

particularly in developing countries. For instance, rising temperatures could cause infectious diseases to spread into regions where peoples' immune systems are not adapted to these diseases (IPCC, 2001b). However, the presently available data do not permit realistic modelling of climate change impacts upon health. It is therefore not possible at present to state a tolerable limit of climate change in relation to human health.

TESTING THE GUARD RAIL

The A1T-450 scenario selected by the WBGU as the reference for its transformation pathway (Section 4.2.6) runs slightly outside the climate window if a climate sensitivity of 2.5°C is assumed (Fig. 4.3-2). Stabilization at 450ppm will thus not suffice to remain within the guard rail under all possible climate sensitivity values.

Further uncertainties attach to this finding:

- *Window shape:* The rounded right upper corner of the climate window has not yet been defined quantitatively in scientific terms; it merely reflects general systems theory considerations (WBGU, 1995).
- *Absolute versus relative:* Assuming a climate sensitivity of 2.5°C, the maximum absolute warming level of 2°C is only just complied with in this century. During the period between 2010 and 2030, the warming rate exceeds the permissible maximum of 0.2°C per decade.
- *Climate sensitivity:* Climate sensitivity is a key variable for the further debate (Fig. 4.3-2; Section

4.5.2.1) but is very hard to estimate. Building on climate model computations using seven different coupled atmosphere-ocean-land models, the IPCC (2001a) reports possible climate sensitivity values ranging from 1.7 to 4.2°C. The IPCC refrains from determining a most probable value and stresses that climate sensitivity can also be outside of the stated range (IPCC, 2001a).

- *Climate modelling:* There are a number of further factors that need to be considered: The response of carbon reservoirs to climate change, the climate effects of aerosols (direct and indirect effects; Section 4.5) and tropospheric ozone, condensation trails and other aviation-related ice clouds, as well as the effects of the Kyoto Protocol. The first factors relate to the natural sciences and amplify or diminish (in the case of aerosols) mean global warming. The Kyoto Protocol is contained indirectly in scenario A1T-450 and causes a reduction of the greenhouse gas emissions of industrialized countries. However, given the withdrawal of the USA and the accounting of carbon sinks under the Protocol, it must be assumed at present that industrialized country emissions will merely stabilize. Thus this uncertainty, too, leads rather to an underestimation of warming.

Therefore, with the assumed climate sensitivity of 2.5°C, scenario A1T-450 lies at the margin of the climate protection window and for a certain period of time is partly outside of it. The scenario is consequently by no means secure within the meaning of UNFCCC Article 2 at a stabilization concentration of 450ppm. The uncertainty attaching to climate sensitivity has a much greater effect than a few gigatonnes of additionally emitted CO_2. It follows that, to adhere to the precautionary principle, scenario A1T-450 must be complemented by substantially more vigorous climate protection policies that embrace activities beyond the energy sector. This will be essential to develop a sustainable scenario.

GEOLOGICAL CARBON STORAGE

Geological CO_2 sequestration in depleted or still worked oil and gas fields has a sustainable potential. This potential cannot be quantified precisely at present, but in total it can be assumed to be in the order of several hundred GT C. To test the present scenarios, a guard rail of 300GtC is assumed as a first approximation (Section 3.6). At 226GtC, total geological storage of carbon between 1990 and 2100 is within this guard rail in scenario A1T-450. The potential of geological storage in deep saline aquifers is substantially larger, but so too are the uncertainties relating to feasibility and environmental consequences. In view of the inadequate state of knowledge and potentially substantial ecosystem impacts,

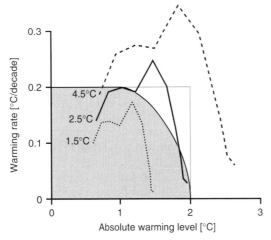

Figure 4.3-2
Scenario A1T-450 in the climate window for a broad range of climate system sensitivities (1.5°C, 2.5°C and 4.5°C climate sensitivity). Climate sensitivity expresses the warming level that follows from a doubling of the CO_2 concentration in the atmosphere relative to pre-industrial levels.
Source: WBGU, using IIASA data

the WBGU deems an application of this technology inappropriate before an in-depth technology assessment has taken place (Section 3.6).

4.3.1.3
Sustainable land use

In an earlier report, the WBGU provided a first-order calculation of how the sustainable use of the biosphere can be ensured for present and future generations alike (WBGU, 2001a). This indicated that 10–20 per cent of the worldwide terrestrial biosphere should be safeguarded by a global network of protected areas. As this network needs to be established in a manner differentiated according to biomes, countries, regions, etc., there can also be regions for which a far larger priority conservation area percentage is appropriate. For other regions, 2–5 per cent may suffice.

Population growth additionally exacerbates the land-use conflict between agriculture or forestry and nature conservation. In the densely populated regions of South Asia, 85 per cent of the potential arable area is already used for food production, but the agricultural area per capita is still smaller than food security would require (WBGU, 2001a). The IPCC assumes a worldwide growth of utilizable agricultural area by some 30 per cent in the 21st century. The range is from 7.5 per cent in developed countries to 96 per cent in Africa (IPCC, 2001b).

DEFINING THE GUARD RAIL

To prevent land-use conflicts, it is essential to define limits for the cultivation of bioenergy crops and for terrestrial CO_2 storage. The two following principles need to be observed in this context:

- Bioenergy carrier production and terrestrial CO_2 storage must not jeopardize implementation of the WBGU conservation area target of 10–20 per cent. As the present worldwide total of conserva-

tion areas only figures 8.8 per cent (category I-VI areas; Green and Paine, 1997), the conversion of natural ecosystems into land cultivated for bioenergy crops is rejected as a matter of principle.
- The production of food must have priority over the production of replenishable energy carriers and over storage.

These principles provide a basis for estimating the maximum area devoted to cultivating bioenergy crops worldwide and in certain regions. As a global guard rail, the WBGU recommends allocating at most 3 per cent of the terrestrial area to such energy purposes. Cultivation of bioenergy crops on this area could yield some 45EJ primary energy. However, in view of disparate local conditions, it is essential to carry out a detailed examination of the individual continents in order to avoid land-use conflicts with food and timber production, as well as with the conservation of natural ecosystems. Table 4.3-1 lists the WBGU's proposals for regional guard rails. A precondition to their implementation is that the required worldwide protected area network has been realized beforehand (WBGU, 2001a).

QUANTIFYING THE GUARD RAIL

For the European Union, Kaltschmitt et al. (2002) state 10 per cent of the arable area (7.4 million ha) as the potential area for cultivating energy crops; the figure is mainly a result of taking agricultural areas out of production (Section 3.2.4.2). If we assume for the whole of Europe that in future 10 per cent of arable land and 10 per cent of pasture land will be available for energy crops, then we receive an area of approx. 22 million ha or 4.5 per cent of the terrestrial area as a guard rail (Table 4.3-1).

In Asia, it has been shown that biomass resources are already over-exploited in some areas (Kaltschmitt et al., 1999). Consequently, the available area is small. No appraisals are available for Australia, but in view of the large desert and semi-desert

Table 4.3-1
Potential area for energy crops and its regional distribution, and energy quantity that can be produced annually from these areas. The energy quantity is calculated from a mean yield of 6.5 t/ha/y and a calorific value biomass of 17.6MJ/kg. The percentages relate to the total areas of the various continents.
Source: WBGU compilation

Region	Potential area		Source	WBGU guard rails		
	[million ha]	[%]		[million ha]	[%]	[EJ/a]
Europe	22	4.5	Kaltschmitt et al., 2002	22	4.5	2.5
Asia and Australia	37	0.7	IPCC, 2001c	29	0.5	3.3
Africa	111	3.8	Marrison and Larson, 1996	111	3.8	12.7
Latin America	323	16	Schneider et al., 2001; only Brazil	165	8	18.8
North America	101	5.9	Cook et al., 1991; only USA	67	3.6	7.7
World	*595*	*4.6*		*394*	*3.0*	*45.0*

areas the opportunities to cultivate energy crops are very limited.

An appraisal conducted by Marrison and Larson (1996) for Africa assumes that arable land will grow 2.4 fold and that the areas of forests and natural landscapes will remain unaltered. Schneider et al. (2001) estimate for South America that 16 per cent of the terrestrial area would come into question for cultivating bioenergy crops (extensive grassland, degraded soils) without needing to convert natural ecosystems. However, their study area in north-east Brazil, in the state of Maranhao, had only a small proportion of tropical rainforest. As that proportion is higher in the rest of Latin America, the guard rail for the whole subcontinent is set at 8 per cent of the area in order to safeguard the conservation of tropical primary forest remnants.

For the USA, 40 per cent of the bioenergy cultivation area reported by Cook et al. (1991) is made up of agricultural areas that are no longer utilized, while the other 60 per cent are made up in roughly equal measure of meadows, pastures and forests. If we exclude the conversion of forests, and reduce because of prevailing demand uncertainties the area of meadows and pastures coming into question by half, then we arrive at a potential cultivation area of 61 million ha. By adding 10 per cent to this value to represent Canada, we arrive at a maximum utilizable area for North America of 67 million ha or 3.6 per cent of the terrestrial area.

PRINCIPLES PRESENTING LIMITS TO BIOMASS USE
Energy crops: On all areas which, after balancing different uses, are deemed suitable for cultivating biomass as an energy carrier or for storing carbon, such cultivation must be sustainable and ecologically sound. Fertilizer and pesticide inputs must therefore be minimized and tillage must be low-impact in order to keep erosion to a minimum. These requirements are simpler to implement when cultivating perennial grasses and rapidly growing trees than when engaging in the intensive cultivation of annual energy crops (Graham et al., 1996; Paine et al., 1996; Zan et al., 2001). Moreover, when cultivating energy crops in plantations, a minimum of species, genetic and structural diversity must be maintained within the areas. Furthermore, cultivation should be integrated into the surrounding landscape.

Use of residues: In agriculture, care needs to be taken that the use of straw and other residues does not jeopardize over the long term the conservation of soil structure, the recycling of nutrients and the proportion of organic matter in the soil. The quantities that can be extracted vary depending upon nutrient supply and soil structure. They are close to zero on poor tropical soils (e.g. oxisols, ultisols), but on richer

soils they can quite well be around 1–2t per hectare and year. Assuming a global mean of 0.7t per hectare and year and a calorific value of 17.6MJ per kilogram, a utilizable agricultural area of 1,500 million hectares worldwide yields a potential of 18EJ per year.

The use of forestry residues, too, must not impair nutrient recycling. Care further needs to be taken that a sufficient quantity of dead wood remains in the forests. Under European conditions, it can be assumed that approx. 1.5t of forest timber can be used sustainably for energy purposes per hectare and year. In view of the inaccessibility and conservation value of large parts of the boreal and tropical forests, it appears purposeful to proceed at a global level from about one-third of the European value, i.e. to extract no more than 0.5t per hectare and year. Given a global forest area of 4,170 million ha and a calorific value for wood of 18.6MJ per kilogram, this results in a sustainable potential of 39EJ per year.

Thus, in total, the cultivation of energy crops (approx. 45EJ per year), the use of agricultural residues (18EJ per year) and of forestry residues (39EJ per year) yields a sustainably utilizable potential of modern biomass amounting to approx. 100EJ per year. This is joined by a further 5–7EJ per year from the traditional use of cattle dung to produce energy (Section 3.2.4.2).

CO2 STORAGE IN TERRESTRIAL ECOSYSTEMS
For the storage of carbon in biological sinks, too, it needs to be ensured that the appropriation of areas is sustainable. Food production and nature conservation goals must not be jeopardized. The intensified use of biomass as a renewable energy source competes with carbon storage on the same area. Because raising the share of renewable energy sources is a key goal of the WBGU, using the areas coming into question to produce energy from biomass has preference over pure carbon storage through afforestation of 'Kyoto forests'. In existing forests, however, the conservation of stocks has priority, for the rise in atmospheric CO_2 concentration that would result from the destruction of their carbon stocks is greater than the CO_2 savings that can be achieved through measures accountable in the first commitment period under the Kyoto Protocol. Thought therefore needs to be given to an appropriate commitment to also conserve natural stocks in future commitment periods.

TESTING THE GUARD RAIL
In scenario A1T-450 modern bioenergy provides about 205EJ primary energy in the year 2050. To produce the required biomass, the scenario assumes large areas for the cultivation of bioenergy carriers. These areas exceed both globally and on the individ-

Table 4.3-2
Scenario A1T-450:
Estimated proportions of
total terrestrial area of areas
cultivated for bioenergy
crops in 2050, compared to
WBGU guard rails.
Source: WBGU

Scenario region	Area and proportion of total area				WBGU guard rail
	High estimate		Low estimate		
	[million ha]	[%]	[million ha]	[%]	[%]
Europe	144	29.4	129	26.2	4.5
Asia	589	12.1	494	10.2	0.5
Australia	186	21.9	147	17.4	0.5
Africa	288	9.7	241	8.1	3.8
Latin America	266	13.2	225	11.2	8
North America	353	18.9	294	15.7	3.6
World	*1,826*	*14.0*	*1,529*	11.7	*3*

ual continents the WBGU guard rail for sustainable biomass use (Table 4.3-2). The scenario assumes an energy yield of 112GJ per hectare and year (corresponding, at a calorific value of 17.6MJ per kilogram, to about 6.6t per hectare and year), which certainly seems realistic.

The origin of the areas on which to cultivate energy crops remains unclear in this scenario. Over the period considered, the areas for forests, grassland and arable farming change very slightly, while the areas aggregated under the 'other' category decline by the year 2100 from 3,805 to 3,253 million hectares. If we assume that, besides deserts, mountains and built-up areas, the areas in question are mainly arid and semi-arid, then the question arises whether the yields which the scenario assumes for bioenergy production are possible on these areas. A sustainable scenario will need to manage with much smaller amounts of energy from biomass.

4.3.1.4
Biosphere conservation in rivers and their catchment areas

As for terrestrial land uses, so also are there limits to sustainable use for freshwater ecosystems (lakes, rivers) and their catchment areas. In relation to the energy sector, hydropower use is the aspect of particular relevance, for dams have numerous ecological impacts that also extend beyond the immediate location of the hydropower plant (Section 3.2.3.3). Of the 106 large catchment areas of the world, 46 per cent have been modified by at least one dam (Revenga et al., 1998). Even today, the cumulative negative impact on aquatic ecosystems is considerable. In many cases, people who had adapted their lifestyles and traditions to their river are also adversely affected (for instance by forced resettlement). For these reasons, the principle of sustainable development requires limits to the expansion of hydropower.

SUSTAINABILITY OF HYDROPOWER
Because many factors interact, it is not possible to state a simple absolute and global guard rail for the sustainability of hydropower projects (Section 3.2.3.3). Nonetheless, compliance with certain framework conditions is essential when hydropower plants are built (Section 3.2.3.4):

- *Nature conservation:* As for terrestrial areas, a certain proportion (about 10–20 per cent) of riverine ecosystems including their catchment areas should be reserved for nature conservation. Especially in the catchments of potential future hydropower projects, areas of particular conservation value need to be protected swiftly in a precautionary manner (Section 3.2.3.4).

- *Principles for major hydraulic engineering projects:* The existing international guidelines for sustainability (World Bank, OECD) need to be applied to all hydraulic engineering projects. An important basis has also been elaborated by the World Commission on Dams (WCD, 2000) in a worldwide discussion process. Implementation of the guidelines at national level presupposes technical and institutional capacity-building as well as long-term responsibilities. To elaborate sustainability analyses, the scientific basis first needs to be created. Research needs to be conducted for the specific catchment areas by independent regional centres, in a manner independent of concrete projects (Section 6.3.1). Such centres can also establish the basis for comparing regional site alternatives, and can keep indirect and cumulative effects (e.g. of a series of projects on one river) in focus (Section 3.2.3.3).

DEFINING THE GUARD RAIL
If the necessary framework conditions (investment in research, institutions, capacity building, etc., Section 3.2.3) are created over the next 10–20 years, then, with a sufficiently circumspect approach, an additional third of the presently utilized potential can be made accessible in a step by step process until 2030 (power production then totalling approx. 12EJ per year). Only if the above preconditions are met

could the value be raised by 2100 to approx. 15EJ per year.

Scenario A1T-450 envisages hydropower being expanded from its present level of approx. 9.5EJ to 35EJ in 2100, i.e. more than three-fold growth. This value oversteps by far the guard rail set by the WBGU.

4.3.1.5
Marine ecosystem protection

The marine biosphere is already impaired by conventional energy systems, for instance through oil pollution, the warming of estuaries and coastal waters, and the dumping of nuclear waste. New energy technologies are being debated in connection with the restructuring of global energy systems that may have further major impacts upon the marine environment. Decisions thus need to be taken on which of these technologies are non-sustainable. This appraisal is difficult because marine ecosystems are comparatively poorly researched and the consequences of interventions are thus hard to evaluate. Consequently there is a particular need here to observe the precautionary principle. As it is not possible to define a universal guard rail for marine conservation – this would have to remain too general – we discuss here the tolerable limits for specific technologies.

DEFINING THE GUARD RAIL
Two technology options are currently under debate for carbon storage in the oceans: Dissolution in seawater and storage in marine ecosystems (Section 3.6). Deep-sea injection of carbon dioxide elevates the partial pressure of the CO_2 and at the same time lowers the pH value of the seawater. There is inadequate scientific understanding of the biological consequences. Similarly, iron fertilization, for instance in the South Seas, is feared to entail severe impacts upon marine ecosystems. For both options, major uncertainties attach to the permanence of storage. With due regard to the precautionary principle, the WBGU therefore recommends using neither of the options within a sustainable energy system.

USING OFFSHORE WIND POWER
In principle, wind power is a form of energy production that is both renewable and has low environmental impact. The development of offshore technology has given wind power a major new potential that is expected to further accelerate the advance of this form of energy generation. However, the establish-

ment of large-scale offshore wind farms may have adverse effects upon the marine biosphere (e.g. in terms of bird conservation); research is currently under way to resolve these issues (Section 3.2.5). A need remains to develop guidelines for the handling of this technology on a firm scientific basis, in order to minimize environmental effects. Areas that already have a designated nature conservation status must be excluded from areas designated for offshore wind farms, as must areas that may fall within the scope of the European Union Habitats Directive, as well as important bird breeding or migration areas. Offshore wind power use competes with other uses of the areas in question: The demands of shipping, oil industry, fishery, nature conservation and other interests need to be reconciled when planning the uses of areas. The available data does not suffice to define a universally valid guard rail.

4.3.1.6
Protection against atmospheric air pollution

The ecological effects of atmospheric pollution are diverse. The nitrogen oxide (NO_X) and sulphur oxide (SO_X) emissions formed when fossil fuels and biomass are burnt play a key role in the human modification of biogeochemical cycles. They impact upon soils, terrestrial ecosystems and water bodies and are a cause of forest dieback. Ground-level ozone formed from NO_X and hydrocarbon emissions in a smog reaction under sunlight elevates plant respiration while at the same time reducing biomass formation and increasing plants' susceptibility to pests and disease (Percy et al., 2002). Reduced biomass formation lessens the sink effect of the biosphere and amplifies human-induced global warming. Emissions of volatile organic compounds, soot and other suspended particulates as well as of heavy metals and persistent organic compounds from combustion processes produce direct toxic, but also ecotoxic effects when these substances enter ecosystems and accumulate in organisms.

TENTATIVE GUARD RAIL DEFINITION
'Critical loads and levels' are science-based maximum limits for specific pollutants and receptors of varying sensitivity (ecosystems, sub-ecosystems, organisms through to materials). Critical loads and levels need to be formulated by preference at the receptor and in an impact-related manner (UBA, 1996; SRU, 1994). They are determined numerically as the rate of deposition which, if not exceeded, is not expected to damage receptors according to the current state of knowledge. Defining and reviewing these ceilings is resource-intensive and complex

because doing so requires high-resolution spatial mapping of the different receptors (e.g. ecosystem or soil types) and of pollutant loads for each individual pollutant. This concept is implemented in the shape of the 1979 Geneva Convention on Long-Range Transboundary Air Pollution. However, until now the Convention has remained limited to Europe and North America. It thus does not provide a basis for deriving a global guard rail.

As a proxy guard rail for worldwide emissions, we could take the criterion that pollutant loads elsewhere must not be higher than those in Europe today. Given that the situation in Europe today is not satisfactory yet for all pollutants, this can only be an absolute minimum requirement. It also entails a series of problematic assumptions, e.g. that the regional distribution of pollutants is similar, that pollutant imports and exports are negligible, and that ecosystem and soil types are all similarly sensitive. Regional guard rails could be set and implemented on the basis of the critical loads concept by adopting national environmental standards or multilateral environmental agreements. If we assume an equal quantity of utilized energy services, we can expect that vigorous application of the state of the art in power plants, households and transportation would allow compliance with the guard rail.

TESTING THE GUARD RAIL

To produce a preliminary assessment, the WBGU has calculated the SO_x emissions per unit area in the different regions. This provides a very rough measure of the environmental impacts of emissions. Considerable sources of error are tolerated in this calculation; for instance, it does not consider the transboundary or seaborne transport of pollutants. The test shows that, in scenario A1T-450, mainly East Asia (China, Korea and neighbouring countries) as well as eastern Europe suffer high loads. In this scenario the guard rail is complied with everywhere in the second half of the 21st century, because the technology by which to prevent these emissions is already available and is increasingly also deployed in the 'critical' regions where strong growth in energy demand is to be expected. A more rapid transformation towards renewable energy sources would hasten compliance with the guard rail.

4.3.2
Socio-economic guard rails

4.3.2.1
Human rights protection

When formulating strategies for the transformation of energy policy, the WBGU is also guided by human rights imperatives, i.e. by universal principles which apply to all social systems. In order to define more precisely and implement the socio-ethical objectives of the sustainable development model, reference can be made to norms codified in international law, such as the human rights and labour conventions, and principles of universal justice, such as the equal distribution of global environmental space.

MEETING THE INDIVIDUAL MINIMUM REQUIREMENT FOR ENERGY

The WBGU considers that one of the objectives of a sustainable transformation of energy systems must be to enable all households worldwide to gain access to modern energy. This objective is underpinned by the International Covenant on Economic, Social, and Cultural Rights. In Article 11, the States Parties to the Covenant recognize the right of everyone to an adequate standard of living for himself and his family and to the continuous improvement of living conditions.

This right includes adequate housing, which encompasses, among other things, access to energy for cooking, heating and lighting (CESCR, 1991). At the same time, it follows from this article, in conjunction with Article 12 which recognizes the right to health, that adequate housing must protect its residents from health risks. This means that the fuel use of traditional biomass is incompatible with the Covenant whenever it produces indoor air pollution to an extent that endangers health (Sections 3.2.3, 4.3.2.7). The energy supply must be developed in a way which ensures that burdens are shared equally and reasonably between women and men, and that the rights of children to special protection (Convention on the Rights of the Child) are guaranteed. Specific references to an obligatory timeframe within which a basic energy supply must be guaranteed cannot be derived from the Covenant.

SAFEGUARDING THE ENERGY-RELATED BASES OF THE RIGHT TO DEVELOPMENT

As an adequate energy supply is a key prerequisite for economic and social development (Section 2.2), a collective entitlement to the amount of energy required to facilitate and promote development could be derived from the 'right to development'. In 1986, the UN General Assembly adopted its Declaration on the Right to Development, which established the right to development 'as a universal and inalienable right and an integral part of fundamental human rights'. At the World Conference on Human Rights in Vienna in 1993, the Western states finally dropped their reservations about this Declaration. Article 4 (1) of the 1986 Declaration, which has no binding force under international law, imposes the following obligation on States: "... to take steps, individually and collectively, to formulate international development policies with a view to facilitating the full realization of the right to development". It remains unclear, however, precisely what is meant by development on an individual basis, whether any specific provision of services can be derived from this 'right to development', and if so, which services they are. Only Article 8 sets out a catalogue of objectives for 'equality of opportunity for all in their access to basic resources, education, health services, food, housing, employment and the fair distribution of income'. A legal entitlement to an adequate energy supply could be derived, at most, from the demand for 'equality of opportunity for all in their access to basic resources'.

A 'right to development' which is unenforceable in law is of little value. Nonetheless, it could – in conjunction with Article 11 of the International Covenant on Economic, Social, and Cultural Rights – underpin a claim to an adequate energy supply on human rights grounds, since this is necessary not only for the development of agriculture and industry but also for the 'continuous improvement of living conditions'. And this right is undisputed in international law as well.

4.3.2.2
Access to modern energy for everyone

Basic energy services include lighting, cooked food, comfortable interior temperatures, refrigeration and transportation (UNDP et al., 2000), but also access to information and communications and to motive power for simple industrial and agricultural processes. The WBGU considers that access to modern forms of energy is essential to safeguard these energy services, because traditional biomass use, especially for cooking and heating, is both an impediment to development and health-endangering and must therefore be substituted (Sections 3.2.4.2, 4.3.2.7). Electricity supply is also of great importance. Electricity not only provides lighting and refrigeration and facilitates domestic and industrial processes; it also offers access to communications, opening up

educational opportunities and expanding the scope for participation.

Section 2.2 makes it clear that the current situation is still far removed from this objective. If present trends continue, it will take more than 40 years for access to electricity to be available to all households in South Asia, for example, and around twice as long for sub-Saharan Africa (IEA, 2002c). If anticipated demographic development is also taken into account, the timescale is likely to be even longer. The WBGU considers this prospect to be intolerable. Additional efforts must be made to ensure access to modern energy.

DEFINING THE GUARD RAIL
As a minimum requirement for a sustainable transformation of energy systems, the WBGU recommends that access to modern energy should be progressively secured for the entire global population. This applies especially to the switch from health-endangering biomass use for cooking and heating to modern energy carriers (Section 3.2.4.2) and to electricity-dependent energy services. Based on this guard rail, the WBGU derives various objectives which are set out in more detail in Section 7.3.1.

TESTING THE GUARD RAIL
The A1T-450 test scenario does not supply the necessary data to allow this guard rail to be tested in quantitative terms. However, since this is a scenario with high economic growth, achieving the guard rail within a matter of decades seems entirely feasible. Indeed, the World Energy Council defines the provision of electricity to all households which currently have no electricity supply as operationally feasible by 2020 (WEC, 2000). Given China's recent success in connecting an average of 6 million people in remoter rural areas to the electricity grid each year (Chen et al., 2002), it should be possible to replicate this achievement ten-fold across the world, especially if stand-alone energy sources (e.g. village electricity supply systems) are established at first.

4.3.2.3
Individual minimum requirement for modern energy

DEFINING THE GUARD RAIL
In order to meet the world population's individual minimum requirement for modern energy, the WBGU recommends that:
• by the year 2020 at the latest, everyone should have at least 500kWh final energy per person and year;

Table 4.3-3
Minimum per capita final energy requirement. Failure to achieve this must be viewed as non-sustainable. The calculation is based on a 5-person household.
Sources: WBGU; G8 Renewable Energy Task Force, 2001

Energy service	Explanations	Final energy requirement [kWh per person and year]	
Potable water	Electric pump for 5l per person and day	2	
Lighting	5hrs a day with 20W per household	7	
Information, communications	Communication equipment (radio, TV, etc.) 5hrs at 50W per household	18	
Refrigeration	0.4kWh per day per household, primarily for food	29	
Total Interim		56	*(electricity)*
Cooking	1.5 cooked meals per day	400	(fuel)
Total		456	

- by 2050, everyone should have at least 700kWh final energy per person and year;
- by 2100, the level should reach 1,000kWh.

The WBGU considers the absolute minimum requirement for individual needs to be approximately 450kWh per person and year (in a 5-person household; Table 4.3-3) or 500kWh per person and year (in a 2-person household). These figures are within the range of 300 to 700kWh per person and year generally postulated in the literature. Indeed, some WEC authors consider 1,000kWh per person and year to be the adequate minimum requirement (WEC, 2000). A level of 450 to 500kWh per person and year must be the absolute minimum, since this takes no account of heating, transportation and supporting domestic and subsistence-economy activities. Furthermore, while the efficiency level assumed for the purposes of the calculations is desirable and feasible in principle, downward adjustments may be required. On the other hand, technological advances may mean that the basic requirement, defined above, can in future be met with less primary energy. The following discussion provides a more detailed quantification of the guard rail.

QUANTIFYING THE GUARD RAIL
Defining a minimum per capita energy requirement poses significant normative, methodological and technical problems. Among other things, climatic and geographical aspects must be taken into account, along with cultural, demographic and socio-economic factors. Furthermore, when converting energy services into the energy amounts required, assumptions must be made about the technologies used. For this reason, the literature contains very little detailed data about such a minimum requirement.

Despite these problems, the calculation seems reasonable (Table 4.3-3), as this minimum requirement is not defined as an objective but as an absolute minimum, and failing to achieve it must be regarded as incompatible with sustainability. Table 4.3-3 assumes

that efficient technologies are used in line with the current state of technology.

REQUIREMENT FOR ELECTRICITY
The individual minimum requirement for electricity to safeguard a basic supply for lighting, refrigeration and communications is around 60kWh per person and year, assuming a 5-person household. For 2-person households, whose numbers are increasing in developing countries as well, it must be assumed that the household requirement is somewhat lower – a smaller refrigerator, etc. Their per capita requirement is thus estimated at around 100kWh per person and year.

The report by the G8 Renewable Energy Task Force (2001) is one of the few analyses to break down the basic electricity requirement on a quantitative basis. It arrives at similar conclusions. A Chinese study distinguishes between low, medium and high basic requirements for a 4-person household and defines the requirement for China as the equivalent of 37, 94 and 668kWh per person and year in each category (Chen et al., 2002). Unlike Table 4.3-3, however, refrigeration is not calculated into the low and medium basic requirement, while a washing machine and freezer are factored into the high basic requirement.

The World Energy Council (WEC) estimates current electricity consumption by the people in developing countries who have access to electricity to be, on average, 1,300kWh per person and year. In the lowest income quintile with access to electricity, the average is 340kWh per person and year (WEC, 2000). If the technologies used were replaced by more efficient ones, this electricity consumption could more than cover the minimum requirement.

ENERGY REQUIREMENT FOR COOKING AND HEATING
The energy requirement for cooking and heating must be viewed separately from the other energy services as no electricity has to be used here. In the

developing countries, around 0.15EJ from biomass is used for cooking (IEA, 2001b; G8 Renewable Energy Task Force, 2001). In Table 4.3-3, it is assumed that the per capita basic requirement is for 1.5 cooked meals per day on average (Grupp et al., 2002). If highly efficient gas cookers are used, this yields an energy requirement of 700–750Wh per cooked meal and thus a requirement of around 400kWh per person and year.

The WBGU has not included space heating in its calculation of the individual basic requirement in Table 4.3-3. There are two main reasons for this: Firstly, the need for heating and the amount of energy used for this purpose depend on local climatic and building conditions which vary so much, both internationally and regionally, that statements about average heating requirements are meaningless. Secondly, in the majority of countries with a low electrification rate and a high proportion of traditional biomass use, indoor heating is rarely required. In individual cases, however, the energy requirement for heating may be very high. For example, to heat a room of 20m² requires around 800kWh per year even in a low-energy house in Germany.

ENERGY REQUIREMENT FOR TRANSPORT
There is a minimum requirement for mobility because schools, medical facilities and markets must be accessible for everyone under acceptable conditions. This minimum requirement probably varies even more substantially than the domains examined above, because infrastructure and distances, for example, also vary widely. It is almost impossible to convert the basic requirement for transport services into units of energy, since no general assumptions can be made about the mode of transport used (truck, ferry, bicycle, animal, etc.). For this reason, energy for mobility requirements is not included in Table 4.3-3. Nonetheless, a 'soft', i.e. purely qualitative mobility guard rail should be considered.

REDUCING DISPARITIES IN ENERGY SUPPLY
The disparities in meeting the minimum per capita energy requirement can be very substantial, both between countries and between population groups within a single country. The disparities between the best- and worst-supplied groups are often so great that the latter is unable to participate satisfactorily in economic, political or cultural life. For this reason, the WBGU recommends that the absolute 'Individual minimum requirement for modern energy' guard rail be supplemented by relative sub-limits. This applies, not least, because – as in the context of poverty – it is not only the actual, but also the perceived availability of energy services which is significant. It is impossible, however, to derive an absolute

quantitative guard rail for the disparities which continue to be acceptable at national and international level. Aside from normative/ethical and technical/methodological difficulties, climatic and socio-cultural differences between regions and countries stand in the way of such a calculation. However, analyses show that the current disparities between countries or regions are far from being what the WBGU classifies as sustainable. This applies even if extreme values are not included in the calculation. The WBGU therefore introduces a tangible reduction of current disparities as a realistic minimum condition upon the scenario.

'DYNAMIZING' THE GUARD RAIL
The minimum requirement for energy, like the sociocultural minimum subsistence figure (minimum household income), is not a value which is independent of the system state. For this reason, it would seem appropriate to 'dynamize' the 'Individual minimum requirement for modern energy' guard rail. This also implicitly takes account of the disparity aspect of the guard rail. If income and energy use in the developing countries increase ('Economic minimum energy requirement per person' guard rail; Section 4.3.2.5), the underlying distribution norm requires a relative lower limit so that the amount of energy available to the energy-poorest households increases.

TESTING THE GUARD RAIL
Taking the amount of energy available per capita to the bottom 10 per cent of the population, it is possible, in scenario A1T-450, to obey the guard rails almost across the board. Exceptions are, in 2020, the Middle East/North Africa and South Asia, where the values are missed by a narrow margin.

4.3.2.4
Limiting the proportion of income expended for energy

The WBGU considers that poor households should not need to spend more than one-tenth of their income to meet their elementary energy requirements (500kWh per person and year). This would be a significant improvement on the current situation in poor developing countries. Nonetheless, the proportion of income expended for energy would still be six times higher than in industrialized countries.

DEFINING THE GUARD RAIL
As a guard rail, the WBGU proposes that expenditure to cover elementary individual energy needs

should be a maximum of 10 per cent of household income.

QUANTIFYING THE GUARD RAIL
Defining a quantitative guard rail which describes the maximum proportion of household income which should be used to cover the minimum requirement for energy services also poses major normative and methodological problems. For example, by defining a percentage, direct assumptions are made about the share of income which is appropriate to cover other needs. The variations in household incomes must also be taken into account.

Assessments of the current proportion of expenditure bring us closer to a quantification of the guard rail. Various studies agree that in the OECD, i.e. the industrialized countries, this proportion is around 2 per cent (G8 Renewable Energy Task Force, 2001; World Bank, 2002a). Few case studies are available for the developing countries (LSMS, 2002; World Bank, 2002a). However, the general impression is that people in countries without access to 'modern' energy spend a far higher share of their income on energy than people in countries or regions with secure access. The estimates on the share of disposable income spent on energy by poor population groups in developing countries vary between 10 and 33 per cent, depending on the energy services and the type and size of the group observed (ESMAP, 1998, 1999; World Bank, 2002a). Expenditure on energy for cooking is often not included in these estimates. If fuel is gathered at no cost, this expenditure may be very low in many rural areas, but is nonetheless associated with sometimes high 'costs', such as the time spent and the damage sustained to health (Sections 2.6, 3.2.4).

With an estimated 1,200 million people currently living below the poverty line of US$1 per day, the 10 per cent guard rail means that these people must be able to meet their elementary energy requirement at a cost of at most US$37 per year. For a further 1,600 million people who live on between US$1 and US$2 a day, the tolerable amount ranges between US$ 37 and US$73 per year. On the very simplistic assumption that income and income distribution remain constant and all of the 2,800 million poorest people have a disposable annual income of precisely US$365, the first 500kWh per year should not cost them more than 7.3 US cents per kilowatt hour on average (electricity or fuel). The necessary cross-subsidizing or social transfers ('heating and electricity benefit') decrease as the income of the poorest groups increases. Here, the income-generating effects of access to modern and affordable energy should act as an accelerator.

TESTING THE GUARD RAIL
Whereas scenario A1T-450 does not supply the data necessary to test this guard rail (e.g. electricity prices, income/consumption), scenario B1-450 can be used in this instance. It is assumed that average private consumption among the poor population groups is equivalent to private income. Then, via the distribution of income in the poorest developing countries, an estimated value for the income of the poorest 10 per cent of the population is calculated. Since a maximum of 10 per cent of this value is available for 500kWh per person and year, a still tolerable electricity price can be calculated, which is then compared with the price in the scenario. This calculation poses many uncertainties. For example, electricity prices within a country may vary widely: In rural regions, where electricity is produced with diesel generators, the price is far higher than in the cities. Different subsidy practices further distort the values. The result of the calculation for scenario B1-450 shows that compliance with the guard rail can be guaranteed from the middle of this century. However, as scenario A1T-450 has higher economic growth rates and therefore income increases, the situation is likely to be better here. For these reasons, the WBGU considers that compliance with the guard rail from 2050 at the latest is a realistic prospect.

4.3.2.5
Minimum per capita level of economic development

Total per capita energy requirement must also include the energy services used indirectly in the manufacture and distribution of all public and private goods which the person consumes. This includes, for example, transportation services which, for methodological reasons, are not taken into account when calculating individual energy requirement ('Individual minimum requirement for modern energy' guard rail, Section 4.3.2.3). A useful indicator of the sum of the goods and services produced is gross domestic product (GDP), although it has various shortcomings and does not fully reflect the informal sector or family/voluntary work, for example. The WBGU is aware of the normative problems associated with a minimum value for the GDP indicator per person and year. However, as this threshold is defined not as a goal but a guard rail, transgression of which is viewed as unsustainable in social and economic terms, the WBGU nonetheless proposes the following definition.

DEFINING THE GUARD RAIL

All countries should be able to deploy a per capita gross domestic product of at least US$2,900 in 1999 values.

QUANTIFYING THE GUARD RAIL

The guard rail was determined as follows: Out of the 70 poorest countries, the ten were identified which combine a relatively high value on the Human Development Index (HDI) and income-adjusted HDI with a low value on the Human Poverty Index (Table 4.3-4).

The ten countries selected have an adjusted HDI of 0.7–0.8 and an HPI of 11–29. Despite their relatively low GDP, they are countries which UNDP has classified as falling within the medium range of human development, and they are also included in the 50 per cent of developing countries with an HPI lower than 30 (UNDP, 2002a). They include Latin American and Asian but not African countries. The arithmetic mean of the ten countries' annual per capita GDP is US$2,900 per person and year, which the WBGU considers to be the lower limit for a life in human dignity. Sixty countries with a total popula-

tion of 2,200 million did not achieve this threshold in 1999. In 21 countries with a total population of 375 million, the indicator was actually lower than US$1,000.

In principle, a macroeconomic minimum energy requirement per person and year could be derived from the primary energy consumption of the ten countries selected. If Jamaica is excluded due to its extremely high consumption, per capita consumption of commercial energy in the remaining nine countries stands at between 4,500 and 10,500kWh per person and year, with a mean of 7,500kWh per person and year (Table 4.3-4).

Alternatively, the lower limit of US$2,900 per person and year could also be used directly to derive the economic minimum energy requirement. Taking mean primary energy required by all countries with a GDP of between US$2,600 and US$3,200 per person and year to manufacture a product with a value of US$1 in 1998, this results in a economic minimum primary energy requirement of 7250kWh per person and year. Finally, reference could also be made to the average energy intensity of the OECD countries

Table 4.3-4
Indicators of selected low-income countries with acceptable successes both in the area of development and in poverty avoidance. *HDI* Human Development Index, *HPI* Human Poverty Index, *GDP* Gross Domestic Product. The average values were calculated as an unweighted arithmetic mean.
Sources: UNDP, 2002a; World Bank, 2002a

Country	Per capita GDP (1999) [PPP US$/ per capita/a]	HDI (1999)	Income-adjusted HDI	HPI (1999)	Population (1999) [Million]	Commercial energy use (1997) [kWh/ per capita]	Commercial energy use/GDP (1997) [kWh/ PPP US$]	Traditional energy (1997) [% of total use]
Vietnam	1,860	0.68	0.78	29.1	77.1	6,044	2.9	37.8
Nicaragua	2,279	0.64	0.69	23.3	4.9	6,394	2.9	42.2
Honduras	2,340	0.63	0.69	20.8	6.3	6,171	2.6	54.8
Bolivia	2,355	0.65	0.71	16.4	8.1	6,354	3.0	14
Indonesia	2,857	0.68	0.74	21.3	209.3	8,039	2.5	29.3
Ecuador	2,994	0.73	0.81	16.8	12.4	8,271	2.7	17.5
Sri Lanka	3,279	0.74	0.81	18.0	18.7	4,478	1.5	46.5
Jamaica	3,561	0.74	0.81	13.6	2.6	18,003	5.3	6
China	3,617	0.72	0.78	15.1	1,264.8	10,521	3.0	5.7
Guayana	3,640	0.70	0.76	11.4	0.8	–	–	–
Average	*2,878*	*0.69*	*0.76*	*18.6*		*8,253*	*2.9*	*28.2*
FOR COMPARISON:								
Developing countries	3,530	0.65	0.68		4,609.8		2.7	16.7
Poland	8,450	0.83	0.87		38.6	31,564	3.6	0.8
Portugal	16,064	0.87	0.89		10	23,792	1.7	0.9
Germany	23,742	0.92	0.93		82	49,080	2.1	1.3
USA	31,872	0.93	0.92		280.4	93,682	3.0	3.8
Eastern Europe and CIS	6,290	0.78	0.82		398.3		5.6	1.2
OECD	22,020	0.90	0.90		1,122		2.5	3.3

(Table 4.3-4). As this also stands at 2.5kWh per US dollar, the identical result is achieved.

However, due to a lack of available data, this value does not fully reflect energy consumption from traditional biomass use, which plays a significant role in almost all developing countries (Table 4.3-4). Taking this fully into account is likely to increase the average value by at least 1,000kWh per person and year.

In view of the wide range of energy intensity, the difficulties associated with sectoral and geographical comparison between economies, and the variations in the amount of traditional energy used, the WBGU has decided not to set a 'hard' or quantitative guard rail for minimum energy consumption.

The WBGU has simply tested whether, and from which point in time, the various scenarios enable an energy requirement of 7,250kWh per person and year to be achieved. Assuming an annual efficiency increase of 1.4 per cent (to 2040) and 1.6 per cent (from 2040), the threshold in 2020 would be around 5,400kWh per person and year, and in 2050 only around 3,500kWh per person and year.

TESTING THE GUARD RAIL
In order to obey this guard rail, all countries must exceed a per capita income of US$$_{1999}$2900. This is already the case in scenario A1T-450 from 2020, but these data are only available for four regions of the world. In a country review, supplying the amount of energy stated above by 2050 could pose a problem. However, due to the lack of available data in this scenario, this could not be tested.

Based on this guard rail – and depending on increases in energy efficiency – with around 7,600 million people in 2020, for example, a global primary energy consumption of 104–137EJ is derived. As 400EJ are already used worldwide today and primary energy consumption will increase to 650EJ by 2020 in scenario A1T-450, the guard rail is unlikely to pose any fundamental problems with the quantity of energy, but at most with its distribution.

4.3.2.6
Technology risks

DEFINING THE GUARD RAIL
A sustainable energy system needs to build upon technologies whose operation remains within the 'normal range' of environmental risk across the entire supply chain, from the various primary energy carriers to the end consumer and possible waste. Here, the 'normal range' – as opposed to the borderline and prohibited ranges – is defined in line with the WBGU's Report, 'Strategies for Managing Global Environmental Risks' (WBGU, 2000).

In the production and transportation of fossil fuels and the operation of fossil-fuelled power stations, accidents may occur due either to error or sabotage. Such accidents have a limited impact in spatial and temporal terms, and can therefore be included in the normal range of environmental risk (WBGU, 2000). The risk of emissions is restricted by other guard rails (CO_2: 'Climate protection' guard rail; other emissions: 'Prevention of atmospheric air pollution' guard rail). Other renewable energy carriers (small-scale hydropower, wind, various forms of solar energy, biomass, geothermal energy, etc.) are non-hazardous and thus fall within the normal range of environmental risk, well within the bounds of the guard rail. Even when operating normally, hydropower plants with large reservoirs fall within the borderline range of environmental risk and may therefore collide with the guard rail (WBGU, 2000; Section 3.2.3.3). This applies especially to the threat posed by terrorism.

NUCLEAR POWER
The present use of nuclear energy (from uranium extraction to reprocessing) involves the release of radiation and thus poses an environmental risk. In the context of nuclear power, there are two main domains which collide with the risk guard rail: Firstly, the risks associated with normal operation and waste management, and secondly, those associated with proliferation and terrorism (Section 3.2.2).

Normal operation and waste management: The risks associated with the normal operation of nuclear power plants fall within the borderline range of environmental risk (WBGU, 2000). The OSPAR limit values are an example of internationally defined thresholds for the discharge of radioactive substances into the sea. The objective of OSPAR is to achieve concentrations in the environment close to zero for synthetic substances. The discharge of radioactive liquid waste from the nuclear reprocessing plants at Sellafield and Cap de La Hague has, in both cases, exceeded the limit values regionally (EU Parliament, 2001). Then there is the standard set by the International Commission on Radiological Protection, which defines acceptable doses of radiation exposure per person and year (ICRP, 1991). This limit value has also been exceeded by many times in the area around the nuclear reprocessing plants (EU Parliament, 2001). The nuclear reprocessing currently taking place in Europe exceeds the limit values agreed at international level. In light of the situation concerning the storage of nuclear waste, discussed in Section 3.2.2, the WBGU considers that the management of waste from the nuclear industry must also be classified in the borderline range.

Proliferation and terrorism: Due to the unresolved problems (Section 3.2.2), the WBGU classifies both proliferation and nuclear terrorism in the low to medium probability range with the extent of damage being substantial. This falls between the borderline and prohibited ranges and thus conflicts with the risk guard rail.

As there is currently no immediate prospect of being able to guarantee worldwide the safe operation of nuclear power plants, the non-hazardous long-term storage of nuclear waste, and the non-proliferation and avoidance of any illicit use of radioactive materials for terrorist purposes, the WBGU recommends abstaining from the use of nuclear power in the long term.

TESTING THE GUARD RAIL
As scenario A1T-450 contains a substantial proportion of nuclear energy, it conflicts with this guard rail.

4.3.2.7
Health impacts of energy use

The International Covenant on Economic, Social, and Cultural Rights defines health as a fundamental human right (Article 12). It also recognizes the right to an adequate standard of living (Article 11), which includes access to energy, e.g. for cooking and heating ('Access to modern energy' guard rail). In many countries and regions of the world, these rights remain unfulfilled because no 'clean' or fit-for-purpose energy is available. The energy carriers used in these areas can significantly impact on human health. In all, around 25–35 per cent of health impairment can be attributed to environmental risk factors (Smith et al., 1999), but it is difficult to establish clear causal chains between energy production and use, on the one hand, and health damage, on the other. Alongside the fundamental risks which cannot be avoided when dealing with energy, there are two domains in particular which have a proven impact on health and which appear, in the WBGU's view, to be globally relevant for the definition of guard rails:

Local urban and indoor air pollution is identified worldwide as one of the major risk factors for health damage and mortality (especially in relation to acute respiratory illness; Michaud et al., 2001; WHO, 2002b). This is caused by fumes from the burning of fossil fuels or biomass (Sections 3.2.1, 3.2.4). The technology used for this purpose (which ranges from a 3-stone hearth to a modern low-emission power station) plays a key role in determining the extent of health impacts.

Radiation is health-damaging, so that the use of nuclear energy (from uranium extraction to reprocessing and storage) is invariably associated with health risks (Section 3.2.2).

DALYs (Disability Adjusted Life Years) can be used to formulate health guard rails with the aim of establishing tolerable limits for health impairment (morbidity) caused by energy production and use. DALYs are a measure of burden of disease in populations by combining 'Years of Life Lost' (YLLs) and 'Years Lived with Disability' (YLDs) (Murray and López, 1996). However, this indicator has been criticized due to its weighting of age and specific diseases, which can under- or overestimate certain health impacts (e.g. UNDP, 2002b). Nonetheless, DALYs are currently the best available measure for standardized and comparative statements. In the World Health Report 2002, the WHO has already begun to attribute specific risk factors to specific health impacts and quantify the proportion of health damage caused in terms of DALYs, and this applies to urban and indoor air pollution as well (WHO, 2002b).

DEFINING THE GUARD RAIL
Burning fossil fuels and biomass produces air pollution in the form of gases and particulate matter which pose major health risks to the exposed population (Fig. 4.3-3).

Urban air pollution, especially in the rapidly growing megacities in developing and transition countries, causes major health impairment claiming around 0.8 million lives every year (Section 3.2.1.3). Almost half of the 7.9 million DALYs attributable to urban air pollution worldwide affect people in the western Pacific region and in South-East Asia (especially China).

Fumes in indoor areas resulting from the burning of solid fuels (especially biomass) in households poses an even greater hazard, claiming around 2 million lives every year, primarily in developing countries (Section 3.2.4.2; UNDP, 2002a). Africa and South-East Asia each account for one-third of DALYs caused by indoor air pollution. In India, the health impairment caused by indoor air pollution is even greater than that caused by smoking or malaria (Box 3.2-1).

Even today, values below 0.5 per cent as a share of regional DALYs are achieved for urban and indoor air pollution in much of the world (Fig. 4.3-3). The WBGU therefore proposes, as a guard rail, that the share of regional DALYs caused by these two risk factors should be reduced to below 0.5 per cent for all WHO regions and sub-regions.

To this end, the phasing out of health-endangering forms of traditional biomass use and the development and implementation of appropriate alternatives are essential. This poses a major challenge (availability of clean fuels, improved burning and ventilation technology; Box 2.4-1; Section 5.2.3.2). To

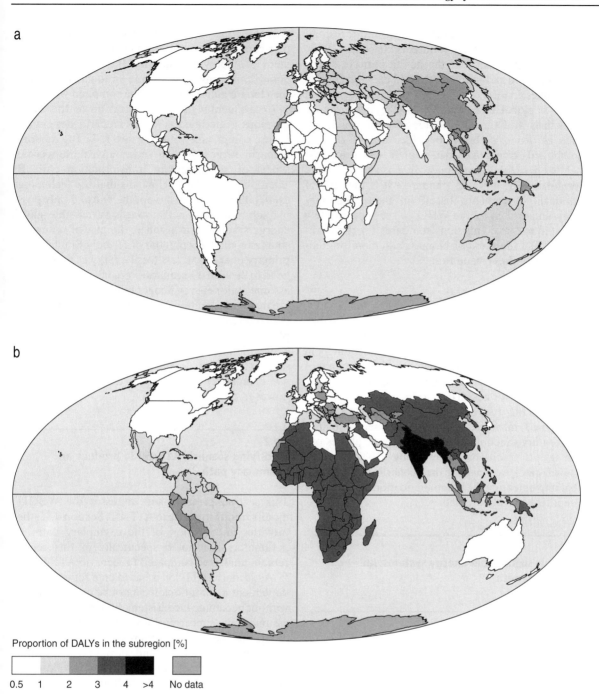

Proportion of DALYs in the subregion [%]

0.5   1   2   3   4   >4   No data

Figure 4.3-3
Health impairment attributed to local air pollution. The proportion of DALYs in the specific subregion is used as indicator.
Values above 0.5 per cent lie outside the guard rail proposed by the WBGU and are coloured red.
a) Health impairment caused by urban air pollution.
b) Health impairment caused by fumes in indoor areas.
Source: WHO, 2002b

promote compliance with the guard rail, threshold values can be set for air pollutants. Since the 1950s, the WHO has evaluated the health impacts of the emission of air pollutants. It drafted the 'Air Quality Guidelines for Europe' in 1987 and later extended them to global level (WHO, 1999). They propose guidelines and threshold values for air pollutants such as ozone, carbon monoxide, volatile organic compounds, nitrogen oxides, sulphur oxides and suspended particulates, and may serve as a basis for the formulation of national standards. It is the task of national governments, based on the preliminary work undertaken by the WHO, to set appropriately adjusted national emissions standards for the burning of fossil fuels and/or biomass and monitor compliance with these standards.

TESTING THE GUARD RAIL

The scenarios do not provide any values for DALYs, so that direct testing of the guard rail is impossible. Huynen and Martens (2002) observe, in an overview of 31 scenarios, that health is described adequately in only 14 scenarios, and just 4 scenarios take account of socio-cultural, economic and ecological factors as driving forces for health development. Scenario A1T-450 obeys the 'Prevention of atmospheric air pollution' guard rail across the board in the second half of the century (Section 4.3.1.6) and largely phases out the use of traditional biomass by 2100. The high growth rates assumed in the scenario thus suggest that implementing the recommendations set out here is entirely feasible by 2050.

4.4
Towards sustainable energy systems: An exemplary path

4.4.1
Approach and methodology

The previous section of this report tested scenario A1T-450 as to whether it is compatible with the guard rails set by the WBGU. It became apparent that this scenario oversteps various guard rails, such as the risk guard rail through the scenario's expansion of nuclear energy, or the ecological guard rails through its highly ambitious expansion of biomass use. The climate protection guard rail, too, is transgressed if a medium climate sensitivity is assumed. Nonetheless, scenario A1T-450 is valuable as a starting point for the WBGU's recommendations, as it combines the stabilization of atmospheric carbon dioxide concentrations with dynamic economic growth without requiring deep-seated changes in consumption pat-

terns (Section 4.2). The scenario has leeway for further climate protection through additional enhancement of energy productivity; this makes it appear possible to bring emissions down to values at which the climate guard rail would be complied with.

Consequently the present section of this report develops a modified A1T-450 scenario designed to ensure compliance with all guard rails. The modified scenario represents the technological, supply-side aspect of an exemplary transformation path. By 'exemplary', the WBGU means that the technological details of the path are not the one and only possible solution for a path towards sustainable global energy systems. For instance, the mix of renewable sources could be composed differently. Similarly, less primary energy and less fossil energy carriers would need to be used if a scenario were taken that does not assume major energy hunger but rather a path giving greater consideration to measures reducing energy demand (e.g. a B1 scenario, Section 4.2). The exemplary path thus provides 'evidence' that, even if energy demand continues to grow strongly, it is possible to transform the global energy system in a manner compatible with the guard rails. Other paths, however, could do this too.

4.4.2
Modifying scenario A1T-450 to produce an exemplary path

This section presents and discusses the WBGU's modifications to scenario A1T-450. Section 4.4.3 then provides an overview of the exemplary path. All parameters not noted specifically in this section remain unaltered compared to scenario A1T-450.

As scenario A1T-450 is based on a set of mutually dependent assumptions, it cannot be modified at will without becoming inconsistent. In particular, all basic assumptions concerning economic growth, investment, technological progress, the relationship between industrialized and developing countries, international cooperation, population development, etc. must be retained. Consequently, the WBGU has restricted its adjustments to a technological modification of the scenario. The energy sector in scenario A1T-450 receives by 2100 a strong hydrogen component supplying half of total global energy use. In order to ensure in this setting the compatibility of the exemplary path with scenario A1T-450, the electricity/heat/hydrogen and fossil/non-fossil energy supply ratios are kept identical as far as possible. As heat can be supplied efficiently at any time from hydrogen and electricity and conversion in the opposite direction entails high losses, it suffices to verify that, in the exemplary path, at all times at least as much electricity and hydrogen can be supplied as in scenario A1T-450. This verification is non-trivial and was provided through extensive calculations not reported in detail here.

The A1T scenarios start in 1997. Thus in scenario A1T-450 the values for 2000 are already projections. In contrast, the exemplary path uses observed values for 2000.

METHODOLOGY FOR QUANTIFYING PRIMARY ENERGY

The joint presentation of different forms and sources of energy within a global quantitative structure is a fundamental problem, because while some conversion paths supply high-grade final energy directly (e.g. power or hydrogen from solar energy), other conversion paths generate corresponding energy forms through an intermediate thermal stage (e.g. fossil power plants). We follow here the quantification approach taken by the world of the SRES scenarios by using the 'direct equivalent method': For nuclear energy and all renewables that deliver electricity or hydrogen directly as final energy (wind, hydro, photovoltaics, other renewables), the values stated correspond to high-grade final energy. For energy forms that can only be upgraded to electricity or hydrogen via the production of heat (e.g. fossil fuels, biomass, geothermal), energy figures correspond to thermal primary energy equivalent values. The scenario discussion summates the two energy quantifications without correction.

RENEWABLES AND NUCLEAR ENERGY

The WBGU made the following modifications to scenario A1T-450 regarding renewables and nuclear energy:

- *Nuclear energy:* In the exemplary path, the quantity of energy provided by nuclear sources for the year 2000 is based on real IEA figures. The A1T-450 value was adopted for the year 2010. However, in departure from A1T-450, nuclear energy use is phased out by 2050 in the exemplary path. Nuclear-based electricity or hydrogen is substituted by renewable sources and, for a limited period, by natural gas.
- *Hydropower:* The start value for 2000 was taken from IEA figures. Capacity is then expanded moderately, finally reaching 15EJ per year (compared to 35EJ per year in A1T-450).
- *Biomass:* The start value for 2000 is based on estimates (Kaltschmitt et al., 2002). Noteworthy uncertainties attach to this value, but it is close to the A1T-450 estimates. The breakdown in roughly equal shares of modern/traditional use is adopted following A1T-450. Traditional biomass use is reduced over the long term on a trajectory similar

to that in A1T-450; 5EJ per year remain permanently from 2050 onwards (compared to 0 in A1T-450). As it can be assumed that this energy quantity can then be used without causing indoor air pollution, this modification is compatible with the WBGU health guard rail. Levels of modern biomass use are raised in a manner similar to A1T-450, but expanded less strongly over the long term and finally limited to 100EJ per year (compared to 260EJ per year in A1T-450).

- *Wind energy:* The start value for wind energy was derived from real data on installed worldwide capacity (BTM Consult, 2001). An annual growth rate of 26 per cent is assumed until 2020 (tenfold growth per decade); this rate then slows. Over all, wind energy is expanded to a much greater degree than in A1T-450. Nonetheless, at 135EJ per year, the final level reached is still substantially below the technological wind energy potential identified in Chapter 3 of this report.
- *Solar electricity:* The start value for solar power generation in the year 2000 is based on measured data. In the following period, it is assumed that solar-based power generation (distributed photovoltaics, photovoltaic and solar thermal power generation) grows tenfold each decade until 2040. The aggregated growth curve of solar power generation meets the A1T-450 curve shortly before the year 2050; in the opinion of the WBGU, the A1T-450 curve is unrealistically high in the previous years. Towards 2100, the exemplary path slowly adopts a trajectory approaching that in A1T-450. The level reached in 2100 still falls far short of the maximum potential (Chapter 3).
- *Solar thermal:* In the exemplary path, the thermal use of solar energy ('solar thermal') is increased in a manner similar to that in A1T-450. The A1T-450 figure for the year 2000 is hard to substantiate, as global solar-based hot water production in 1998 was actually far lower. However, it is difficult to appraise precisely the contributions of active and particularly passive solar thermal use, so that, despite reservations, the A1T-450 value was adopted for the exemplary path.
- *Geothermal:* This form of energy is not quantified specifically in A1T-450. The WBGU considers the potential of geothermal energy to be so important – with regard to both thermal applications and electricity production – that it has set up a specific category. It was assumed regarding the power/heat ratio that half of the primary energy is used thermally (heating, cooling, process heat) and the other half is used to generate power. The respective efficiencies are lower than those of fossil power plants, because geothermal heat is usually available at a comparatively low temperature

level. The start value for 2000 was adopted from the World Energy Assessment (UNDP et al., 2000).

- *Other renewables:* Here primary energy contributions were assumed that are significantly more optimistic than those of A1T-450. The WBGU is of the opinion that the development of new technologies to harness renewable energy sources is far from concluded. Technologies already under debate today include solar chemical energy systems (producing storable energy carriers), tidal and wave energy, as well as energy conversion using artificial membrane systems employing processes similar to photosynthesis.

FOSSIL SOURCES
Fossil energy is key to CO_2 emissions and thus to the climate guard rail. Fossil energy use is modified only very slightly in the exemplary path, which applies real energy consumption levels for the year 2000, based on current US government statistics (US-DOE, 2002). From 2010 up to 2050, the exemplary path adopts the contributions of the individual fossil sources from A1T-450 almost unaltered. There is only one modification: The transient power bottleneck at the beginning of the century that results from the comparatively slighter use of biomass and hydro is compensated for by an additional, time-limited use of gas-fired power plants. This leads, over an interim period, to a slight increase in energy requirement compared to scenario A1T-450. The gas is used to supply sufficient electricity to compensate for the shortfalls arising among non-fossil energy sources compared to A1T-450. Assuming 50 per cent power plant efficiency, this results in some 17GtC additional CO_2 emissions over the period until 2050.

In contrast, the use of fossil sources is set somewhat lower than in A1T-450 throughout the second half of the century, as in that period intensified efficiency improvements lead to reduced energy demand. Initially demand is reduced equally in both the fossil and non-fossil sectors. As a result, energy-related CO_2 emissions drop by about 24GtC between 2050 and 2100. This more than compensates for the elevated level of gas consumption in the first half of the century.

ENERGY PRODUCTIVITY
It is assumed that energy productivity in the exemplary path outstrips energy productivity in scenario A1T-450 from 2040 onwards. While historically energy productivity has grown on average across the world by about 1 per cent annually since the onset of industrialization, the A1T scenarios assume approximately 1.3 per cent annually. Scenarios making more ambitious assumptions in this respect even take 2 per

cent per year (B1; Section 4.2). The exemplary path assumes an energy productivity growth of 1.6 per cent annually from 2040 onwards. This is still consistent with the assumptions of the A1 world, because scenario A1T-450 scarcely takes into consideration measures to reduce energy demand, for instance through price incentives. It thus leaves sufficient leeway to assume further improvements in energy productivity without – as in the B1 world – needing to presuppose a shift in values and structures towards less energy-intensive industrial products and services. As a result, energy usage is reduced by 22 per cent compared to scenario A1T-450 by the year 2100.

CARBON DIOXIDE STORAGE ('SEQUESTRATION')
A distinction can be made between carbon dioxide sequestration from fossil power plants and from biomass plants.

- *Carbon dioxide sequestration from fossil plants:* In scenario A1T-450, about 218GtC are sequestered in total by 2100; in that year, the rate of sequestration still figures 1.7GtC per year. The WBGU considers it important to limit the period during which carbon dioxide is sequestered because of the limited capacity of repositories. The exemplary path thus distributes storage across time in such a way that it is terminated by the end of the 21st century. However, as scenario A1T-450 already assumes that the greater part of carbon arising in power plants is sequestered, there is very little scope for such redistribution and the cumulative quantity of carbon stored had to be reduced slightly in the exemplary path compared to scenario A1T-450.
- *Carbon dioxide sequestration from biomass power plants:* In biomass-fired power plants, as well as in facilities for the production of synthesis gas (hydrogen) from biomass, the carbon contained in the biomass can be separated from the flue gas in the form of CO_2 and then consigned to storage. This leads to a net removal of carbon dioxide from the atmosphere. In the exemplary path, the corresponding technology is introduced and expanded from 2040, whereby energy production in biomass from facilities with storage amounts to 25EJ annually in the decades between 2060 and 2080. This technology can be phased out again towards the end of the century. With an assumed storage efficiency of 70 per cent, this development over time yields compared to A1T-450 additional CO_2 emissions savings amounting to some 21GtC. Figure 4.4-1 compares carbon sequestration in the exemplary path to that in A1T-450.

TOTAL CO_2 EMISSIONS
Figure 4.4-2 compares the energy-related CO_2 emissions of scenario A1T-450 with those of the exem-

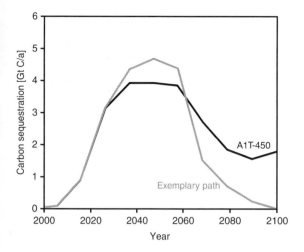

Figure 4.4-1
Carbon sequestration in scenario A1T-450 and in the
exemplary path.
Source: WBGU and Riahi, 2002

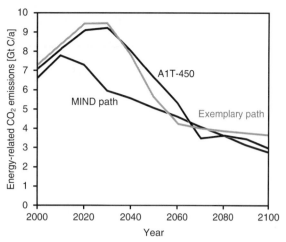

Figure 4.4-2
Energy-related CO_2 emissions in scenario A1T-450, in the
exemplary path and in the UmBAU path computed using the
MIND model.
Source: WBGU and Riahi, 2002; Edenhofer et al., 2002

plary path, as the outcome of all modifications discussed above. The differences are comparatively
small. Including non-energy-related CO_2 emissions,
which are the same in both scenarios, both result in
cumulative CO_2 emissions of just below 650GtC by
2100. The distribution of emissions over time is also
very similar in both scenarios. Even taking into consideration the uncertainties discussed in this section,
all statements regarding global warming that were
formulated when testing the climate guard rails for
A1T-450 thus also apply to the exemplary path. By
way of comparison, the figure also shows the CO_2
emissions of another path (MIND) derived from a
further model computation (Section 4.5).

4.4.3
The technology mix of the exemplary
transformation path: An overview

Table 4.4-1 shows the contributions of energy carriers to meeting energy demand in the exemplary path,
assuming the modifications discussed above. Table
4.4-2 provides an overview of CO_2 emissions and carbon sequestration. Figure 4.4-3 shows the contributions of energy carriers in the exemplary path. Due to
the uncertainties attaching to long-term projections,
the figure only shows the last years of the 2050–2100
period. The projection for 2100 highlights the great
importance of solar energy in this scenario.

Not only the way energy demand is met by a certain energy carrier mix is crucial to the exemplary
path, but also the greater energy productivity
enhancement assumed compared to scenario
A1T-450. Such an increase can be achieved in many

ways, for instance through price-induced reduction of
energy demand, which leads to efficiency improvements in both energy conversion and final energy
use, as well as through sectoral structural change and
altered settlement and transportation structures, or
changed consumer behaviour. In sum, this can prevent so much energy use that energy productivity
enhancement becomes one of the main pillars of the
exemplary path (Fig. 4.4-4).

4.4.4
Conclusion: The sustainable transformation of
global energy systems can be done

In the previous sections, the WBGU performed a
consistent modification of reference scenario
A1T-450 in such a way that it now obeys all guard
rails (on the climate guard rail cf. Section 4.5.2). This
produces an exemplary transformation path demonstrating that, even in a world characterized by rapidly
growing energy consumption, it is possible to transform global energy systems such that they become
sustainable. A number of key properties of this path
warrant special mention.

The exemplary path is characterized by high
growth rates of both energy use (three-fold growth
by 2050) and economic development (six-fold
growth by 2050). It builds upon three pillars: Declining use of fossil sources, rising use of renewables and
growing energy productivity. From the middle of the
century onward, renewables account for the greater
proportion of energy provision. In the exemplary
path their share amounts to some 50 per cent in 2050
and almost 90 per cent in 2100. This goal will require

	2000	2010	2020	2030	2040	2050	2100
				[EJ]			
Oil	164	171	187	210	195	159	52
Coal	98	111	138	164	126	84	4
Gas	96	138	196	258	310	306	165
Nuclear	9	12	12	6	3	0	0
Hydropower	9	10	11	12	12	12	15
Biomass, traditional	20	17	12	8	7	5	5
Biomass, modern	20	48	75	87	100	100	100
Wind	0.13	1.3	13	70	135	135	135
Solar electricity	0.01	0.06	0.6	6	63	288	1,040
Solar thermal	3.8	9	17	25	42	43	45
Other renewables	0	0	2	4	10	15	30
Geothermal	0.3	1	3	10	20	22	30
Total	420	519	667	861	1,023	1,169	1,620

Table 4.4-1
Global energy demand in the exemplary path, broken down according to energy carriers. The figures were calculated using the direct equivalent method (Section 4.4.2).
Source: WBGU

massive growth rates of almost 30 per cent annually for several decades. This is feasible, as the growth rates of wind and solar energy achieved in a number of countries have recently demonstrated. The great growth of solar energy is a key property of the exemplary path. Figure 4.4-5 illustrates the surface area that would be needed to supply the solar power projected for Western Europe and North America in the exemplary path, if all solar power facilities were concentrated at one location. Such a concentration is by no means envisaged; in fact, facility use is highly distributed, even at middle latitudes and in industrialized countries. A global energy system building essentially upon solar power would not require unacceptably large areas compared to settlements and present infrastructural facilities, particularly considering that in arid regions the use of land for solar power production scarcely competes with other land-use forms, and dual uses are possible, for instance on roofs or transport infrastructure areas.

In the scenario underlying the exemplary path, the rapid technological development is attributable to swift economic growth worldwide and particularly in developing countries, providing sufficient financial resources for the transformation process. Less energy-hungry paths will presumably leave even more scope for transformation, if the economic and technology development setting changes accordingly. This confirms the key statement of the exemplary transformation path: The sustainable transformation of global energy systems can be done.

4.5
Discussion of the exemplary path

This section examines in more detail the exemplary path derived above. This includes in particular a discussion of the uncertainties and costs of this path. To provide a basis for this discussion, Section 4.5.1 uses an endogenous model to perform alternative model computations in order to explore the scope for action that is available when transforming energy systems while observing defined guard rails. Section 4.5.2 then discusses the exemplary path, particularly with

Table 4.4-2
CO_2 emissions and carbon sequestration in the exemplary path.
Source: WBGU

	2000	2010	2020	2030	2040	2050	2060	2070	2080	2090	2100
						[GT C]					
Annual energy-related CO_2 emissions	7.3	8.4	9.4	9.5	7.8	5.6	4.3	4.0	3.9	3.7	3.6
Annual energy-related carbon sequestration	0	0.1	0.9	3.0	4.1	4.5	4.2	1.5	0.7	0.3	0
Annual non-energy-related CO_2 emissions (e. g. logging)	1.1	1.1	0.3	0.2	0.2	0.1	0.1	0.1	0.1	0.1	0.1
Total	8.4	9.5	9.8	9.7	8.0	5.7	4.4	4.1	4.0	3.8	3.8

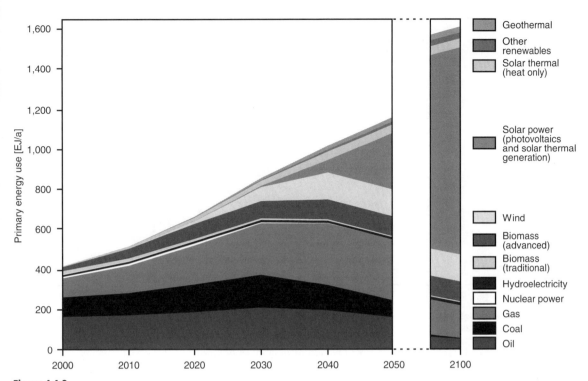

Figure 4.4-3
Contributions of energy carriers to energy demand for the exemplary transformation path. This path demonstrates that the sustainable transformation of global energy systems is technologically viable. A different renewable technology mix could also produce the same outcome.
Source: WBGU

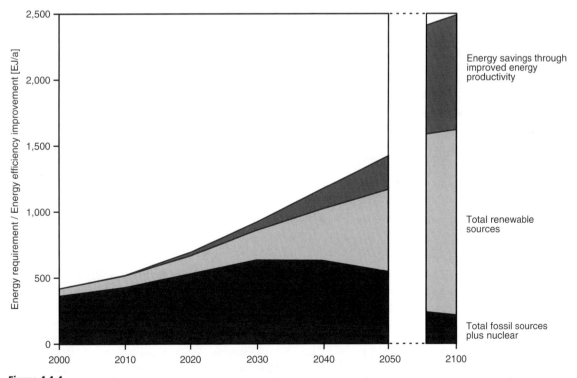

Figure 4.4-4
Energy efficiency enhancement in the exemplary path. This path assumes from 2040 onwards a 1.6 per cent annual increase in energy productivity, compared to the historical figure of 1 per cent annually.
Source: WBGU

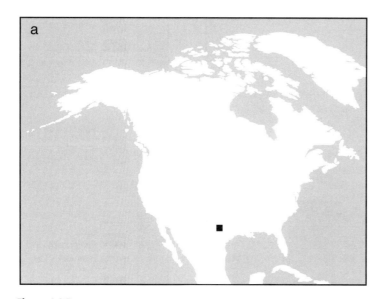
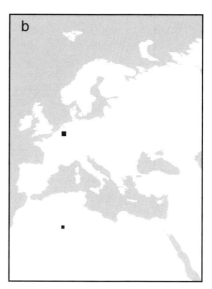

Figure 4.4-5
Visualization of the surface area required for solar electricity. The squares represent the surface areas that would be needed to produce the solar power assumed in the exemplary path for the year 2050. (a) Areas required for North America, assuming generation in Texas (100 per cent in solar power plants). (b) Areas required for Western Europe, whereby two-thirds of the solar power are generated in central Europe (upper square; insolation values of Belgium, 25 per cent generation in solar power plants, 75 per cent distributed) and one-third in the Sahara (lower square; insolation values of Algeria, 100 per cent generation in solar power plants). The regions of Western Europe (WEU) and North America (NAM) are defined as in Nakicenovic et al., 1998. The calculation is based on the technological potential. Transmission losses are assumed to be 10 per cent throughout.
Source: WBGU

regard to the aspect of compliance with the climate guard rail.

4.5.1
The MIND model

To underpin economically the findings of Section 4.4 above, the WBGU deploys an innovative modelling approach that permits a consistency check of the exemplary path. This is done by means of the MIND model (Model of Investment and Technological Development; Edenhofer et al., 2002). MIND is an endogenous energy system model coupled to a climate model.

The exemplary path derived in Section 4.4 builds strongly upon the assumptions of scenario A1T-450 (Section 4.2). These concern among other things population development, economic growth, the development of energy requirements as well as technological progress, which are all predetermined exogenously.

An alternative approach is to determine key variables such as economic growth, energy requirements or efficiency and productivity improvements endogenously, i.e. within the model. In contrast to the MESSAGE model with which the computations for A1T scenarios were conducted, only few framework data and a set of plausible decision criteria for dynamic optimization are predetermined in MIND. The

model delivers neither a regional resolution nor a technology resolution. Nonetheless, this alternative approach permits further plausibility tests of the exemplary path. MIND is a global model allowing assessment of long-term options for climate change mitigation action with regard to the necessary investments and technological dynamics. It permits, in particular, a critical review of the hypothesis that the costs of transformation are far too high and are out of proportion to the benefits.

In the model, essentially only population development, technology learning curves and the availabilities of coal, oil and gas are predetermined. Future developments, such as demand for fossil and renewable energies and consumption levels, are computed endogenously by the model. As is common practice in the theory of economic growth, an investor is assumed who seeks to maximize per-capita consumption of products and services over time (Ramsey, 1928). Emission paths are determined on the basis of endogenized technological progress, technological development being a process co-determined decisively by economic activities. This applies particularly when scarcities, such as an oil crisis, generate innovations (Ruttan, 2000; Goulder and Mathai, 2000). MIND also takes into consideration effects induced by efficiency improvement measures and learning, e.g. through rising production volumes or through mounting resource scarcity. Furthermore, the

WBGU does not use in MIND any cost-benefit analyses such as are common in climate economics (e.g. Nordhaus and Boyer, 2000). This is because the WBGU considers it exceedingly difficult to monetarize climate damage appropriately.

MIND contains three different energy sectors, each of which is examined as an aggregated whole: renewable energies, energy production from fossil fuels, and the extraction of fossil fuels. The fossil energy mix is not modelled explicitly; rather, the development over time of the carbon content of the fossil energy mix is adopted from the exemplary path. Fossil primary energy carriers are converted to final energy in MIND in the fossil energy sector. Conversion efficiency can be modified endogenously, by substituting primary energy with capital. Energy derived from non-fossil sources is added to final energy from fossil sources. This approach corresponds to the direct equivalent method, which was also applied to generate the exemplary path. In MIND, renewable energy carriers are the 'new' renewables, i.e. those that have a potential for further technological development in the future (solar, wind, modern biomass, etc.). They, too, are modelled as an aggregated set. Conventional renewables such as traditional biomass and hydropower are not modelled explicitly in MIND. The quantities of energy generated from these sources and from nuclear power are adopted from the exemplary path (Section 4.4). The same applies to emissions of other greenhouse gases (methane, nitrous oxide, CO_2 from land-use change and fluorinated gases), and to carbon storage in geological formations. This ensures that all factors not modelled by MIND agree with those of the exemplary path. MIND computes different scenarios for the BAU case (business as usual) and the UmBAU case ('UmBAU' meaning 'transformation' in German). In the BAU case, unconstrained cost-effective paths are computed, while in the UmBAU case the climate guard rail defined by the WBGU is introduced in addition. Finally, corridors of future emission paths are computed – these corridors are compatible with the climate guard rail and are economically acceptable in international development terms.

MIND FINDINGS
Figure 4.5-1 shows the findings of the MIND simulations for the development of primary energy use, and its breakdown between renewable and fossil energies. In the BAU case, new renewables only become economically profitable at the beginning of the 22nd century, as it is then that the higher exploration and extraction costs of coal, oil and gas lead to massive investment in renewables (Fig. 4.5-1a). However, the model computations also show that under these conditions the global mean temperature must be expected to rise by more than 4°C (Fig. 4.5-2c). The climate problem is thus not solved solely by the mounting scarcity of fossil resources. Consequently, a second case is examined – UmBAU ('transformation') scenarios involving the introduction of the climate guard rail.

Despite the different approach as compared to scenario A1T-450, MIND arrives in the UmBAU scenario (Fig. 4.5-1b) at energy requirement trajectories for fossil and renewable energies that are similar to those of the exemplary path (Fig. 4.4-3). Furthermore, it is apparent that in such a scenario renew-

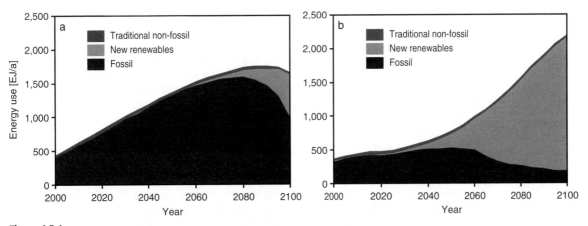

Figure 4.5-1
Energy use in the MIND model in the (a) BAU (business as usual) and (b) UmBAU ('transformation') cases, the latter involving compliance with the climate guard rail. The breakdown according to fossil, new renewable and traditional non-fossil energy sources is shown. In BAU, energy use dips slightly towards the end of the 21st century, because the increasing scarcity of fossil resources then comes into play. However, due to the massive introduction of renewables, energy use rises again after 2100 (not shown here). Because the MIND calculations start in 1995, the values of the UmBAU and BAU cases already diverge in the year 2000, as in the UmBAU case investors already take the climate guard rail into consideration in their decision-making. Source: Edenhofer et al., 2002

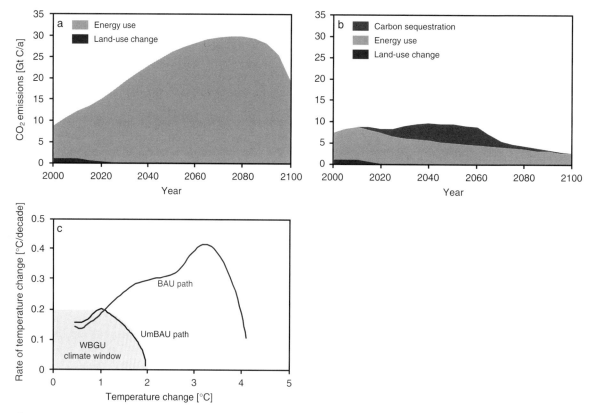

Figure 4.5-2
CO_2 emissions in the MIND model for the (a) BAU (business as usual) and (b) UmBAU ('transformation') cases. The UmBAU case shows in addition the CO_2 emissions prevented by sequestration. Figure (c) shows the WBGU climate window and the temperature development for the 2000–2100 period for the two scenarios (BAU red, UmBAU black). As MIND is an optimization model, in the UmBAU case part of the temperature development follows the edge of the climate window closely. A climate sensitivity (this is the warming that follows from a doubling of CO_2 concentrations compared to pre-industrial levels) of 2.5°C was assumed. The sum of red and light-red areas represents actual emissions. The sum of the black and light-red areas represents the resources extracted. Because the start year for MIND is 1995, the UmBAU and BAU values already diverge in 2000, as in the UmBAU scenario investors must already take the climate guard rail into consideration in their decisions. Source: Edenhofer et al., 2002

ables become macro-economically profitable far sooner. The model simulations thus verify the viability of the exemplary path.

The findings illustrate that in the UmBAU case CO_2 emissions can be cut massively compared to the BAU path (Fig. 4.5-2a,b). Nonetheless, the climate guard rail can only be complied with if over the next 100 years approx. 200GtC are sequestered in secure geological formations (Fig. 4.5-2b,c). Compared to the exemplary path, the UmBAU path entails about 100GtC less cumulative emissions over the 2000–2100 period. The atmospheric CO_2 concentration thus rises in the UmBAU scenario to no more than 410ppm (in the year 2100).

NECESSARY INVESTMENT
MIND confirms, as a coupled climate-energy system model, that emissions targets must be announced in a credible fashion if the expectations of investors are to

change in such a manner that they invest in transforming the energy system. If investors expect the policy arena to make entitlements for atmospheric emissions scarcer over the long term, e.g. through certificate trading or through environmental quality objectives, then this expectation already becomes a part of investment appraisals today.

For the UmBAU case, MIND computes cumulative investment into the global energy system from 2000 to 2100 amounting to US$330 million million, and for the BAU case US$300 million million. As major uncertainties attach to the appraisal of future global investments (UNDP et al., 2000), it is difficult to use empirical investment data of recent years to verify these figures.

The MIND simulations show in the BAU case for the year 2000 an investment of 3.9 per cent of global GDP (extraction, but without R&D) in the fossil energy sector. The World Energy Assessment Report

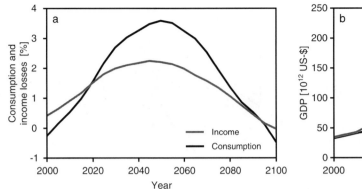
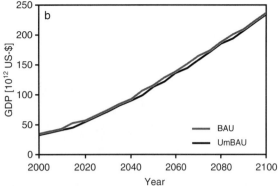

Figure 4.5-3
(a) Percentage losses of consumption and income for the UmBAU scenario compared to the BAU scenario: income (red), consumption (black). (b) Development of global GDP in the BAU (red) and UmBAU (black) cases.
Source: Edenhofer et al., 2002

produced by UNDP et al. (2000) states only 1–1.5 per cent. However, UNDP itself assumes this ratio to be too low (approx. 3 per cent is debated in the literature), because capital costs are significantly underestimated. MIND is initialized with a relatively high ratio of investment to GDP, because this is approximately consistent with the capital costs stated in that Report.

These uncertainties aside, the MIND computations show that, in the business-as-usual case, investment in renewables only starts when fossil resources dwindle. In the UmBAU case, there is an early change in investor behaviour; the climate guard rail can be complied with without massive macro-economic losses, and the transformation of global energy systems can be accelerated greatly. While slight consumption and income losses arise at the beginning of the transformation phase (amounting to less than 4 per cent change), after conclusion of the transformation phase (after 2100) welfare gains in fact ensue due to rising returns to scale (Fig. 4.5-3a). The underlying learning curves need not exceed historical rates. Over the relevant period (2000–2100) aggregated learning curves with a constant rate of learning are assumed for renewables. This involves the assumption that new technologies enter the market whose average costs can be lowered through learning by doing. Moreover, sensitivity analyses show that the findings would not change qualitatively even if learning curves have a declining rate of learning. Systems can be transformed sustainably without major macro-economic losses if the renewable energy sector succeeds in maintaining its historically achieved rates of learning (Fig. 4.5-3b).

PERMISSIBLE EMISSIONS CORRIDORS
Within the context of the scenario approach, as set out in Section 4.1, not only are cost-effective paths

examined with regard to compliance with a climate guard rail, but also an analysis is conducted of the scope that this constraint leaves to the economic system as defined by MIND to extract fossil resources, with the associated emissions. The Tolerable Windows Approach (WBGU, 1995; Toth et al., 1997; Petschel-Held et al., 1999; Bruckner et al., 1999) is a useful tool in this context. The Tolerable Windows Approach offers a procedure by which to identify emissions corridors (Leimbach and Bruckner, 2001). Emissions corridors represent the aggregate emissions values over time that an emissions path compatible with the guard rails can have. The Tolerable Windows Approach does not involve an optimization of the economic system, so that, in addition to the climate guard rail, socio-economic guard rails are introduced in order to demarcate the realm of permissible emissions futures (Section 4.3). To this end, a minimum growth of average per-capita consumption was stipulated in order to do justice to the development requirement of developing countries. As a further constraint, an excessive scarcity of energy as a factor of production was excluded.

Figure 4.5-4 shows corridors for emissions and resource extraction that are compatible with the guard rails. The width of the corridors results above all from the sequestration of a part of energy-related CO_2 emissions. Land-use-related emissions further widen the corridors. The bold lines within the corridors indicate the cost-effective UmBAU ('transformation') paths that obey the climate guard rail. Being within the corridors, these paths also comply with the other two guard rails. Moreover, the paths are located in the upper part of the corridors. This means that even more drastic emissions reductions are feasible within the space demarcated by the socio-economic guard rails. The economic system modelled by MIND is therefore flexible enough to transform

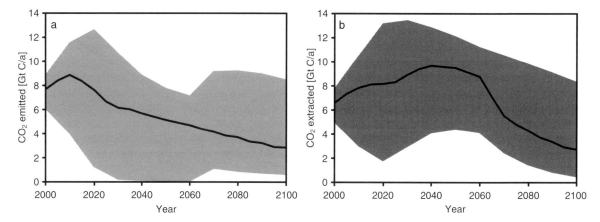

Figure 4.5-4
Corridors for (a) CO_2 emissions taking CO_2 sequestration into consideration, and (b) resource extraction. The corridors were computed under the requirement of complying with the WBGU climate window and the socio-economic guard rails. The non-sustainable realm is outside of the corridors. However, not all trajectories within the corridors are necessarily sustainable. For instance, a path remaining continuously at the upper limit of the corridor conflicts with the climate window. Compliance with the corridor is a necessary, but not sufficient condition for compliance with the guard rails.
Source: Edenhofer et al., 2002

energy systems early on and thus bring down the share of fossil energy carriers in energy production quickly enough to ensure climate change mitigation while at the same time doing justice to aspects of economic development.

MIND thus computes a climate change mitigation path which, for a medium climate sensitivity, remains within the WBGU climate window and still does not present politicians with unsolvable allocation problems. This reaffirms the exemplary path used by the WBGU. It does, however, require a massive increase in the level of investment in the renewable energy sector. The interplay of promotion measures – to accelerate learning – with long-term emissions limitation for climate change mitigation allows effective control of the transformation process towards a sustainable future.

4.5.2
The exemplary path: Relevance, uncertainties and costs

This section discusses the exemplary path developed in Section 4.4. The discussion concentrates on the uncertainties relating to permissible emissions, and on the question of the path's financeability.

4.5.2.1
Uncertainties relating to permissible emissions

The exemplary path entails about 100GtC more cumulative CO_2 emissions over the period from 2000 to 2100 than the path computed by the MIND model in the UmBAU scenario. However, the discussion in Section 4.4.2 shows that the exemplary path fully exploits the available scope to reduce CO_2 emissions. For instance, an assumed stronger increase of energy productivity would no longer be consistent with the underlying A1 world, more CO_2 sequestration would conflict with the requirement to phase out sequestration entirely by 2100, and a more rapid expansion of renewables is not possible. The following discussion illustrates how sensitive all scenarios are with respect to the uncertainties surrounding climate sensitivity values – an aspect already discussed in qualitative terms in Section 4.3.1.2.

CLIMATE SENSITIVITY OF THE EXEMPLARY PATH
Using an energy balance climate model, Kriegler and Bruckner (2003) have determined how much cumulative carbon emissions a pure CO_2 emissions scenario that neglects the net radiative forcing of aerosols and other greenhouse gases can contain at most over the period from 2000 to 2100 if it is not to generate global warming of more than 2°C relative to pre-industrial values. Their findings are shown in Table 4.5-1.

This shows clearly that permissible emissions can diverge by more than 1,500GtC between the two extreme values of climate sensitivity. This divergence is more than the total cumulative emissions of the

Table 4.5-1
Permissible cumulated CO_2 emissions over the 2000–2100 period for an absolute warming of at most 2°C relative to pre-industrial values, as a function of climate sensitivity. The IPCC states the possible range of climate sensitivity to be 1.5–4.5°C (IPCC, 2001a). By comparison, cumulative emissions within the exemplary path over the same period figure some 650GtC.
Source: after Kriegler and Bruckner, 2003

Assumed climate sensitivity [°C]	Permissible cumulative CO_2 emissions [GT C]
1.5	1,780–1,950
2.5	850–910
3.5	530–560
4.5	approx. 380

exemplary path (650GtC). To place these values approximately in relation to the exemplary path, which also gives consideration to other greenhouse gases and aerosols, the relative differences between the permissible emissions need to be transferred to the exemplary path's carbon dioxide emissions. An assessment has shown that the exemplary path would remain within the WBGU climate window at an assumed climate sensitivity of about 2.2°C.

The range of climate sensitivities assumed in Table 4.5-1 reflects the state of current knowledge (IPCC, 2001a). It must be noted in this context that, due to the major uncertainties, particularly with regard to the indirect and feedback effects in the climate system, the IPCC no longer states any value as being the most probable. Nonetheless, it can be said that a

value of 2.2°C is within the range stated by the IPCC and is thus a plausible assumption. Recent studies have attempted to reconstruct climate sensitivity from comparisons between model simulations and empirical data using probability density functions, and discuss climate sensitivities that significantly exceed 4.5°C (Andronova and Schlesinger, 2001; Forest et al., 2002; Knutti et al., 2002). There is an urgent need to engage in further research to provide a more accurate estimation of climate sensitivity (Section 6.1).

Table 4.5-1 yields a further important aspect: If the exemplary path would need to obey the absolute warming guard rail not at a climate sensitivity of 2.2°C, but about 3°C, then the volume of permissible cumulative emissions would shrink by about 200GtC. Viewed within the context of current debate, a climate sensitivity of 3°C is also a plausible assumption. Its consequences therefore need to be discussed. The question arises how the greater greenhouse gas emissions reduction that would then be required could be realized. Table 4.5-2 lists various options, showing how even more emissions could be prevented within the exemplary path.

The 200GtC could be saved through carbon sequestration, through emissions reductions in agriculture, and through a swifter increase of energy productivity. Savings will be difficult to tap in agriculture. Initial studies on methane emissions from rice fields and on the use of nitrogen fertilizers have shown that emissions reduction potentials either require interventions in cultivation techniques (Mitra et al., 1999; Bharati et al., 2001) or their implementation is limited or difficult and in some instances

Table 4.5-2
Climate sensitivity and potentials to reduce greenhouse gas emissions within the exemplary path.
Source: WBGU

Uncertainty factors and options within the exemplary path	Equivalent worldwide permissible greenhouse gas emissions, cumulative (2000–2100)	Notes
3°C climate sensitivity assumed instead of 2.2°C	Exemplary path would entail additional excessive emissions of about 200GtC$_{eq}$	The IPCC states the rate of climate sensitivity to be 1.5–4.5°C.
50% reduction of CH_4 and N_2O emissions in agriculture	Would yield about 200GtC$_{eq}$ savings	50% is a very high reduction rate. It can scarcely be assessed whether it will be possible to stabilize emissions in agriculture at all.
Storage up to the maximum amount permitted by the WBGU guard rail (300GtC instead of 200GtC)	Would yield about 100GtC$_{eq}$ savings	This would mean that CO_2 sequestration could not be brought down to zero by 2100 within the exemplary path.
Raise energy productivity growth rate from 1.3% per annum to 2% per annum	Would yield about 120GtC$_{eq}$ savings	Corresponds to a transition from the A1 world to the B1 world in the SRES scenarios.
Raise energy productivity growth rate from 1.3% per annum to 2.5% per annum	Would yield about 220GtC$_{eq}$ savings	2.5% per annum improvement may be the very maximum feasible.

also expensive (Scott et al., 2002). A major potential such as listed in Table 4.5-2 will therefore be very difficult to achieve.

If only little can be reduced in agriculture, then the missing 200GtC would need to be achieved mainly by raising energy productivity more swiftly, or by raising the volume of sequestration. About 100GtC could be sequestered additionally without overstepping the WBGU guard rail for secure storage in geological formations. This, however, would mean that in the exemplary path CO_2 sequestration could not be brought down to zero around 2100.

About 200GtC could be achieved by raising energy productivity more rapidly. The rate of 2.5 per cent enhancement per year stated in the table is viewed in the literature as the maximum rate feasible (Hoffert et al., 1998). It must be stressed that an energy productivity improvement rate of 2 per cent (B1 scenarios) or even 2.5 per cent can only be achieved if at the same time technological efficiency rises steeply and, as determined in scenario B1 (Section 4.2), structural change towards less energy-intensive products and services takes place, including changes in settlement and transportation structures as well as lifestyles. This additional productivity improvement of the energy sector requires strong incentives to reduce energy demand.

4.5.2.2
Costs of the exemplary transformation path, and financeability issues

Turning energy systems towards sustainability will require a transformation process extending over 100 years and more. To calculate the cumulative costs of this transformation, we would need to be able to predict not only the development of capital investment, research expenditure, operating input costs and system maintenance costs for the exemplary path, but also the costs of the BAU path and of other reference scenarios. It is obvious that price developments, e.g. for primary energy carriers, cannot be projected with sufficient reliability over a period of a century. The same applies to future fundamental innovations in energy conversion and use that are still unknown today, and to the costs or cost savings that may result from their implementation. These cannot be captured sufficiently by the commonly applied approaches of scale economies and learning curves.

This is further compounded by the major difficulties attaching to an identification and monetarization of the external costs of energy system transformation and alternative energy paths. Nonetheless, while no precise statements can be made about the level of external costs, comparisons are indeed possible. For instance, the exemplary path shows that a switch to renewable energy carriers reduces both local and global environmental damage. In contrast, the fossil-nuclear path leads to massive environmental impacts, particularly through climate change. Climate change, in turn, generates macro-economic damage and adaptation costs. Similarly, the external costs caused by health damage are also lower in the exemplary path. Furthermore, the exemplary path holds out, compared to the fossil path, the potential for a 'solar dividend'. This refers to savings in energy policy motivated defence expenditure, whose necessity declines through a switch to renewables, as import dependency upon fossil sources drops. It is estimated that the energy policy motivated defence expenditure of the USA alone amounts to approx. US$33,000 million annually (Hu, 1997). This would be joined by further avoided costs of combating nuclear terrorism.

In view of the uncertainties and imponderabilities attaching to cost assessments, the WBGU abstains from a quantitative estimate of total costs. However, as the question of the financeability of energy system transformation and thus the level of necessary investment is of immediate practical interest, the data of the model runs are taken to provide a rough picture of the investment that may be involved in implementing the exemplary path. The MIND model computes for the 2000–2100 period cumulative investments in the global energy system amounting to US$300 million million for the BAU (business as usual) case and US$330 million million for the UmBAU ('transformation') case. Scenario A1T-450, which is the reference scenario for the exemplary path, requires according to IIASA calculations a cumulative investment of about US$190 million million over the same period, while the coal-intensive and nuclear growth path of scenario A1C-450 would require some US$500 million million. Overall, the energy requirement of the exemplary transformation path is lower than that of scenario A1T-450. Fossil sources and carbon sequestration are almost identical in the two scenarios, but the exemplary path involves less nuclear, hydropower, solar thermal and biomass, and would therefore presumably require less cumulative investment for these energy sources. Cumulative investment only exceeds that of scenario A1T-450 in the case of wind power and solar electricity generated through photovoltaics. Taking all relevant costs into consideration, the Council assumes on the basis of its own estimates that the modifications to scenario A1T-450 entailed by the exemplary path may roughly offset each other in financial terms. These estimates relate to scenarios involving an immediate launch of the transformation process. If transformation is postponed by several decades, the investment costs and, all other conditions remaining equal, the transformation costs would rise steeply due to deepening path dependency. The WBGU is well aware that the exemplary path, involving strong initial support for new renewables, is more expensive over the short term than a path relying initially upon exploitation of cost-effective greenhouse gas reduction potentials. However, it can certainly be assumed that the exemplary path will be more cost-effective over the long term, as only this path makes available within a few decades the solar energy supply capacity needed to avert major global warming damage. It follows from all of the above that:

- Major uncertainties attach to statements on the costs of transforming the global energy system. Various models estimate cumulative investment costs for the period from 2000 to 2100 in the region of several hundred million million US dollars.
- Using modified model findings, a cumulative investment from 2000 to 2100 ranging from US$190 to 330 million million can be estimated for the exemplary path.
- Over the long term, the exemplary path entails up to 2100 much lower investment costs than a coal-intensive and nuclear path and, moreover, averts substantial macro-economic damage.

However, the financeability of transformation depends less upon the absolute level of cumulative investment over 100 years, than upon
1. the relative level of requisite energy investment (measured as e.g. the share of gross domestic product), and
2. the rate at which investment would have to grow over the short to medium term.

Neither the IIASA scenarios nor the MIND model require a strong rise in the ratio of energy investment to GDP over a long period. At no point does investment in the energy sector exceed twice its present share of GDP.

The second factor – the rate at which investment must grow over a given period – impacts more strongly upon the appraisal of financeability. Investment in the requisite transformation of energy systems must be carried out mainly by private-sector actors. Private-sector investment is based upon profitability considerations. This is why there can be no absolute guard rail for the level of investment in the energy system, e.g. in the form of a maximum share of GDP. If political decisions modify the investment setting, the profit expectations for investment in energy efficiency and renewables can improve greatly. On the other hand, changes in the setting redirect capital flows, which is problematic for those sectors of the economy from which capital is withdrawn. This can cause considerably adaptation difficulties in both

economic and social terms. The economic growth potential would be reduced, particularly if this process were to occur abruptly and within a short period. A doubling of investment in the energy sector within a few years could therefore exceed the adaptive capacity of national economies. If the process occurs over one or two decades, however, a doubling is possible without significant frictional losses. Historical experience shows that in other, similarly large sectors the investment ratios (ratio of sector investment to GDP) have even in some cases more than doubled within a decade without having caused significant macro-economic disruption.

In order to safeguard the financeability and thus economic feasibility of energy system transformation, it is essential to design a long-term transformation strategy (Chapter 5) that does not impair the adaptive potential of market-based mechanisms, but rather harnesses them for the transformation process. That economic actors have planning certainty is a precondition to this. It is essential that they can rely on certain energy policy framework conditions over a period of at least 10–20 years. If actors can orient their behaviour to interim targets and instruments set out in a transformation roadmap (Chapter 7), their investment behaviour will adjust accordingly. The WBGU thus considers it essential that the policy arena wastes no more time and gives clear signals pointing towards energy system transformation. This must take place at both the national and international levels. Under such conditions of a strategic transformation, the WBGU is convinced that the selected sustainable path is financeable and can be travelled without significant economic losses. Short-term adaptation costs will be unavoidable at certain points, but can be minimized by applying the right instrument mix (Chapter 5). Overall, the consistent, long-term implementation of energy system transformation will enhance societal welfare and will tap new welfare potential.

4.6
Conclusions

The analysis of scenarios for the long-term development of energy systems (Section 4.2) and the examination of an exemplary path (Sections 4.4 and 4.5) consistent with the WBGU guard rails (Section 4.3) produce a series of conclusions. These are also the basis for the recommendations for action developed in Chapter 5.

- Global cooperation and convergence – both economic and political – facilitate the rapid technology development and dissemination required for the transformation. High economic growth can then, in conjunction with a strong decrease in energy and carbon intensity, lead to sustainable energy supply. Present developing countries can profit from this: through catching up quickly with the development level of industrialized countries, through technology and capital transfer as well as through the opportunities resulting from the export of high-quality energy products. This can make it possible to attain, early on, the goal of providing all people with access to modern, clean forms of energy. This, however, will require not only energy policy measures, but also activities in the development and economic policy realms.

- Binding CO_2 reduction commitments and the associated price signals as well as other incentives will be essential in order to transform energy structures quickly enough for them to meet minimum climate protection requirements. Global CO_2 emissions will need to be reduced by at least 30 per cent by the year 2050 from a 1990 baseline, whereby industrialized countries need to reduce their emissions by about 80 per cent and developing countries need to limit their emissions growth to about 30 per cent.
- Energy policy activities need to be supported by further measures to reduce non-energy-related emissions (for instance from agriculture) and to preserve natural carbon stocks.
- While the exemplary path developed here by the WBGU is based upon a stabilization of atmospheric CO_2 concentrations at 450ppm, due to uncertainties attaching to driving forces and climate development this can not be taken as a safe stabilization concentration. With due regard to the precautionary principle, the WBGU therefore recommends retaining the option of aiming at lower CO_2 stabilization targets.
- Even if climate protection goals are met, a fossil-nuclear path entails substantially larger risks considered intolerable by the WBGU, as well as much higher environmental impacts. Moreover, it is significantly more expensive over the medium to long term than a path relying – as does the exemplary path developed by the WBGU – upon promoting renewables and improving energy efficiency. The exemplary path leads to the following recommendations: The share of renewables should be expanded by about 50 per cent worldwide by 2050, and should figure about 85 per cent by 2100. Energy productivity should rise over the long term by 1.6 per cent annually. The use of coal should be phased out by the end of the century, and the use of nuclear power already by 2050.
- Due to the long investment cycles, for instance of power plants or transport networks, the next 10–20 years are the decisive time window for

putting energy systems on a sustainable track. If this opportunity is used, it will be possible to transform systems in a way entailing only low income losses. This will prevent a deepening of path dependency upon present fossil-nuclear energy systems. Moreover, developing countries will be able to leapfrog non-sustainable technologies.

- The transformation will only succeed if the transfer of capital and technology from industrialized to developing countries is intensified. To this end, industrialized countries will need to strengthen technology development significantly in the fields of energy efficiency and renewable energy sources, for instance by raising and redirecting research and development expenditure, implementing market penetration strategies, providing price incentives and developing appropriate infrastructure. This can reduce the initially high costs of energy system transformation and can accelerate attainment of market maturity, thus in turn facilitating transfer to developing countries.
- Over the short and medium term, it is essential to swiftly tap those renewable energy sources which are already technologically manageable and relatively cost-effective today. These are in particular wind and biomass and also, within limits, hydropower. However, as their sustainable potential is limited (Chapter 3), their vigorous expansion will already meet its limits in the first half of the present century.
- Over the long term, the rising primary energy requirement can only be met through vigorous utilization of solar energy – this holds by far the largest long-term potential. To tap this potential, it is essential to ensure that installed capacity grows ten-fold every decade – now and over the long term.
- The utilization of fossil energy sources will continue to be necessary over the next decades. Wherever possible, this needs to be done in such a fashion that the efficiency potential is tapped and both the infrastructure and generating technology can be converted readily to renewable sources. In particular, the efficient use of gas, for instance in combined heat and power generation and in fuel cells, can perform an important bridging function on the path towards a hydrogen economy.
- To harness the worldwide potential of solar energy and balance regional fluctuations, it will be essential to establish global energy transmission networks over the long term ('global link').
- A certain volume of carbon sequestration in geological formations (oil and gas caverns), but not in the oceans, will be necessary as a transitional technology in this century, given sharply rising primary energy demand. The quantity of sequestration will

be moderate compared to fossil scenarios. The sustainable use of biomass through gasification, in conjunction with storage of the carbon dioxide arising in the process, even presents the opportunity to create a carbon sink.

- Besides modifying supply-side structures, it is important to engage in a strategy to enhance energy productivity far beyond historical trends – from a rate of improvement of about 1 per cent today, to at least 1.6 per cent annually as a long-term global mean value. This means that global energy productivity must improve at least four-fold within the next 60–70 years. Three-fold improvement should be aimed at by 2050. This will require measures (such as price incentives) to improve efficiency and reduce energy demand in both energy conversion and final energy use. It will also require further measures such as infrastructure policies aiming to modify transportation and settlement structures.

The WBGU transformation strategy: Paths towards globally sustainable energy systems

5.1
Key elements of a transformation strategy

The previous chapters of this report have set out the requirements that globally sustainable energy systems must meet. Present and future generations should command over the resources and goods necessary to meet their needs. This must be ensured in such a way that environmental changes do not jeopardize the natural life-support systems on which humankind depends, and in such a way that no unacceptable social developments occur. The non-sustainable realm is defined by the ecological and socio-economic guard rails set out in Section 4.3 of this report. If trajectories remain within the action space circumscribed by the various guard rails, there is a prospect that future generations have a similar scope for action as the present one. The WBGU transformation strategy towards globally sustainable energy systems thus builds upon two fundamental goals:

Goal 1: To protect natural life-support systems (compliance with ecological guard rails);

Goal 2: To secure access to modern energy forms worldwide for all (compliance with socio-economic guard rails).

The calculations presented in Chapter 4 show that not every development of energy systems is compatible with these requirements. The exemplary path set out in that chapter outlines a possible sustainable development trajectory and identifies the key elements of a global, sustainable energy strategy. The conclusions presented in Section 4.6 point to the following principal fields of action: engaging in climate policy; developing and applying new technologies; securing the engagement of developing countries; intensifying cooperation and convergence at the global level; and, finally, achieving greater policy integration. It must be kept in mind that the next 10 to 20 years represent the decisive window of opportunity for transforming energy systems – it is in this period that systems must be put on track towards transformation. The Council's recommendations thus concentrate upon this period.

The quantification of the exemplary transformation path underscores that by 2020 the share of renewables in the global energy mix should be raised from its current level of less than 13 per cent to at least 20 per cent, and should reach more than 50 per cent by 2050. A promising approach by which to achieve this goal is to set minimum quotas for renewable energies, which should be raised step by step (Box 5.2-1). In order that such a scheme covers the entire global energy mix, ideally all states should commit to binding quotas. Because the potential for sustainable expansion varies widely among the individual forms of renewable energy, it would be expedient to set sub-quotas differentiated according to energy carriers. In an economic perspective, it would be desirable to give the system greater flexibility over the long term by integrating agreed country quotas within a tradable quota system. Within such a system, a country would not need to fulfil its entire quota domestically, but would have the option of offsetting a part of its assigned quantity of 'green' energy against energy from countries which produce quantities of renewable energies exceeding their quota.

A further question is that of which concrete measures can be deployed to increase the share of renewables and transform present energy systems into globally sustainable systems. The following sections provide answers to this question. National-level measures (Section 5.2) must be supported and complemented by well-integrated policies and effective institutions at the international level (Section 5.3). The Council orients its selection of measures to certain guiding principles (Box 5.1-1).

5.2
Actions recommended at the national level

In the following recommendations, a distinction is made between industrialized, developing, newly industrializing, and transition countries in order to take account of the quite different settings in the various country groups.

Box 5.1-1

Guiding principles for the WBGU transformation strategy

- *Promote good governance*: For a global transformation of energy systems to get under way in Least Developed Countries, too, it is essential to strengthen their capabilities. Planning certainty, functioning state structures and functioning markets are basic preconditions for foreign direct investment and the sustainability and effectiveness of development policy measures.
- *Assume common but differentiated responsibility*: Climate change is essentially a consequence of present energy use in industrialized countries. It is the developing countries, however, which will suffer the greatest impacts. This places an obligation upon industrialized countries not only to initiate the transformation of energy systems themselves, but also to provide financial and technical support to developing countries in this process.
- *Obey the precautionary principle*: Global energy system transformation is a search process that needs to be re-adjusted continuously in line with the steadily growing body of knowledge and shifting framework conditions. To obey the precautionary principle in this process, it is essential that development trajectories do not cross the guard rails defining the non-sustainable realm.
- *Observe the subsidiarity principle*: The subsidiarity principle demands that competencies for tasks must in principle first be devolved to the lowest level. A shift to the next higher level is only legitimate if it can be proven that the higher level can implement and finance energy policy strategies more efficiently.
- *Pursue regional approaches*: Regional approaches (club solutions) can enhance the political enforceability of a transformation. Successful club solutions are an incen-

tive to other states or regional state groupings, and should therefore be promoted.
- *Create a level playing field for all energy carriers*: Fossil and nuclear energy generation still receive very large subsidies. Moreover, only a fraction of the external costs of fossil and nuclear energy systems have been internalized. The creation of a level playing field for all energy forms, particularly with regard to research and development, is therefore a basic precondition if market-based impulses for the transformation of energy systems are to unfold.
- *Shape liberalization sustainably*: The liberalization of energy markets creates, in many instances, the preconditions for harnessing economic potentials. However, the special conditions prevailing in the rural regions of Least Developed Countries also need to be kept in mind – here the initial concern is to secure supply. For new energy markets to develop according to sustainability criteria, liberalization needs to be combined with framework conditions set by the state.
- *Tap transformation potentials swiftly*: To enhance political enforceability, initially – and above all in developing, newly industrializing and transition countries – the most cost-effective transformation potentials should be tapped, such as efficiency improvements; concurrently those technologies that are not initially cost-effective should also be promoted. The financial resources thus saved can be used to provide targeted support to e.g. renewable energies.
- *Harness social and economic forces*: By integrating private-sector actors, catalysts can be gained for the transformation process. The energy industry commands over the necessary capital and to some degree over the requisite knowledge. The state must create an appropriate setting: at the international level by, e.g., opening markets and harmonizing international competition law, and at the national level by, e.g., preventing distortions of competition and removing market barriers.

5.2.1
Ecological financial reform

Ecological financial reform concerns the financial relationship between the state and its citizens, which should be restructured according to sustainability criteria. On the revenue side, the taxation of non-renewable energies has been the main focus of the debate until now. However, other environmental levies, tax credits and the general process of trawling the tax system to identify disincentives to sustainability are integral elements of an ecological reform of the revenue system. On the public expenditure side, it is the subsidies paid to industries and individual companies, as well as research subsidies but also transfers to private households, which must be reviewed and, if necessary, restructured according to environmental criteria. Not least, a general focus on the environment in all aspects of public spending (e.g. environmentally compatible procurement, environmental management of public institutions, etc.) is

an integral part of an ecological spending reform (Burger and Hanhoff, 2002). In the following sections, the WBGU focuses on taxation of non-renewable energy carriers and the systematic removal of subsidies which result in harmful effects on the environment as two key elements of an ecological financial reform.

5.2.1.1
Internalizing the external costs of fossil and nuclear energy

THE BASIC CONCEPT
The major obstacle to the establishment of globally sustainable energy systems is the inadequate internalization of the external effects of the fossil and nuclear energy chain from generation to use. As a result, fossil and nuclear energy is priced more attractively for the individual consumer than renewable energies, although their external effects are far lower.

This creates distortions of competition which put renewables at an disadvantage.

Full internalization of external costs worldwide would be the most significant contribution to establishing fair competition between the various types of energy. Fair competition is essential if renewable energy sources and efficiency increases are to become more profitable than current forms of energy. This would put in place the conditions for a swift energy reform towards sustainability.

Two primary effects justify an ecological tax reform:

1. *The environment-related incentive effect resulting from the taxation of non-renewable energy carriers:* A tax on fossil fuels increases the price of these fuels, leading to a fall in demand while market conditions stay the same, and resulting in their substitution by other energy carriers. There is also an incentive to increase energy efficiency and promote the technological development of renewable energies.

2. *The fiscal effect of revenue use:* Strictly speaking, revenue is merely a secondary effect which is largely irrelevant to the internalization approach. If the 'correct' rate of tax is selected, distortions of allocation are removed. However, the purpose of an ecological tax reform is not to introduce an individual pollution tax as an isolated measure, but to replace other revenues with a more efficient environmental levy.

This concept is based on the 'double dividend' hypothesis (Goulder, 1995): As well as producing behaviour-modifying effects in favour of environmental policy objectives (first dividend), the efficiency of national taxation systems can also be increased (second dividend). The second dividend is based on the assumption that through tax revenue, distorting – and therefore efficiency-reducing – levies, such as income tax or social contributions, can be reduced in a revenue-neutral way. If the distortions caused by the tax which is to be cut outweigh the distortions caused by an environmental levy (e.g. substitution effects in the area of intermediate inputs), a double dividend would be achieved. In practice, however, reducing the costs of the labour factor by cutting non-wage labour costs only produces a double dividend if specific assumptions are made about the labour and goods markets (SRW, 1998; Wissenschaftlicher Beirat beim Bundesministerium der Finanzen, 1997). Due to this uncertainty, the Council endorses the view, propounded by the German Council of Environmental Advisors (SRU), that taxing fossil energy carriers at national level is justified not in terms of the second dividend but solely in terms of its environment-related incentive effect (SRU, 2002).

As a general principle, tradable emissions rights (permits) – i.e. a quantitative solution – can achieve the same ecological objectives as taxes. But permits are not suitable for all pollutants or polluters. The national debate in particular has therefore generally been dominated by the issue of eco-taxes. However, in climate policy at global and European level, the focus is shifting towards the permit approach. So that the plethora of quantitative and tax-based solutions does not block the progress of environmental and energy policy at national or international level, international climate policy in particular must focus to a greater extent on ensuring compatibility between national and international instruments in future.

PRACTICAL STEPS

Against the background of the basic concept outlined above, the WBGU recommends the following measures to the German federal government:

- *Strengthening steering incentives by taxing non-renewable energy carriers:* In the interests of climate protection, lignite and hard coal should be subject to the highest taxes, followed by fuel oil, gasoline and natural gas. To produce dynamic innovation incentives, it is crucial always to raise the rates of tax in small steps so that actors can factor energy price rises into their long-term decision-making. Non-environmentally compatible exemptions for energy-intensive industries must be progressively dismantled. In highly integrated economic areas such as the EU, a joint approach is essential, and the long-term aim should be to adopt such an approach at global level as well.

- *Implementing an ecologically effective and economically efficient 'tool box':* The further development of ecological tax reforms should be guided by the CO_2 emissions trading established through the Kyoto process and planned on an EU-wide scale (Section 5.2.4.1). In this context, it is important to ensure that no double financial burden arises due to the co-existence of these two schemes. This means that stationary major polluters which join the emissions trading scheme should not be subject to eco-tax or other levies imposed in the interests of climate protection. However, the precondition for exemption from the eco-tax would be that the emissions reductions achieved through emissions trading should be at least equivalent to those achieved through the eco-tax. The WBGU views voluntary commitments as a supplementary climate protection tool which is compatible with, but not an alternative to, CO_2 taxes and permits.

- *Use of revenue:* The WBGU is in favour of allowing the revenue from national CO_2 taxes and other environmental levies to flow into the general budget, to be used – like other revenue – in line with the priorities set by Parliament. In the interests of a comprehensive ecological tax reform, a parallel reduction of purely fiscally motivated taxes – which produce high losses in terms of both allocation and growth – is recommended. However, temporarily ringfencing a proportion of the revenue accruing from the taxation of non-renewables to benefit research, development and commercialization of renewable energies is justified, in the Council's view, until external energy costs are fully covered by adequate tax or quantitative solutions and while subsidies on non-renewable energy carriers and forms of use remain in place.

The ecological financial reform is likely to be of primary importance for the industrialized countries in particular. Although internalizing the external costs of fossil and nuclear energy is important in principle for the developing, newly industrializing and transition countries as well, ecological financial reform is likely to fail in these countries in the short to medium term due to their often inadequate public fiscal systems. The WBGU therefore recommends working to ensure that in the long term, ecological financial reform can have a positive impact in these countries as well.

5.2.1.2
Removing subsidies on fossil and nuclear energy

THE BASIC CONCEPT
Unless energy subsidies are paid as compensation for an external benefit, they result in energy price distortions. The Council therefore regards the majority of overt and hidden energy subsidies as one of the most significant barriers to energy system transformation. Energy subsidies contribute substantially to the path dependence of the traditional energy system. With fossil energy, this applies especially to coal subsidies and price subsidies for oil and gas, which keep energy prices at an artificially low level. With nuclear energy, subsidies are generally indirect: For example, some states exempt their nuclear industry from 100 per cent liability and pay into a risk protection fund.

Removing these subsidies would have two positive effects: Firstly, there is an anticipated environment-related incentive effect. Price rises make fossil fuel use less attractive to energy producers and consumers. Secondly, a fiscal effect can be expected. Public funds are freed up and can be used for other – especially energy-related – purposes, e.g. promoting research into renewables and energy efficiency.

It is estimated that every year, around US$240 thousand million is paid in subsidies to the energy sector worldwide. This includes around US$80 thou-

sand million in the OECD countries (van Beers and de Moor, 2001). In the developing countries, state subsidies to the energy sector amount to US$50 thousand million – more than the total funds allocated to ODA (DFID, 2002). These figures still do not include non-internalized external effects (known as 'shadow subsidies') because they are extremely difficult to quantify precisely. Nonetheless, despite these quantification problems, the external costs should always be considered in qualitative analyses as they are highly relevant to energy policy decision-making.

Dismantling these subsidies offers great potential to make savings. In a study for a number of non-OECD countries, the IEA has calculated that by completely dismantling distorting energy subsidies, average efficiency gains of 0.7 per cent of GDP can be achieved. Furthermore, an anticipated reduction in energy use, amounting to 13 per cent on average, would cut CO_2 emissions by 16 per cent (IEA, 1999).

Removing subsidies and/or reforming subsidy policy also offers economic benefits. Nonetheless, opposition to these moves can be substantial because subsidies always have distributional effects. The economic sectors affected will therefore try to block any subsidy policy reforms which put them at a disadvantage. In order to reduce this opposition, subsidy policy reform should take place progressively in accordance with fixed timetables. This leaves adequate scope for adjustment. In the interests of political viability, some of the funds freed up could be used to cushion the social impacts of the resulting structural change.

Two developments at international level can benefit the reform of current subsidy practice in many developing and transition countries:
- *Accession to the WTO:* Accession to the World Trade Organization (WTO) is an important political objective for many of these countries, such as Russia. Domestic oil, gas and electricity prices, which are far lower than the world market price, are a major point of contention at the WTO accession negotiations. If the objective of WTO accession is pursued further, this is likely to drive forward subsidy policy reform as well.
- *International climate protection:* The CO_2 emissions reduction commitment may trigger the restructuring of the energy sector in line with market principles, which will also enhance energy productivity. The introduction of market prices for energy use would create incentives to improve productivity in many transition countries. In this respect, dismantling subsidies in line with climate protection criteria is doubly useful – benefiting public budgets and contributing to the fulfilment of reduction commitments.

PRACTICAL STEPS

The WBGU recommends the implementation of the following, mutually reinforcing measures to the German federal government:

- Drafting comprehensive documentation on environmentally harmful subsidies in general and the subsidies on fossil and nuclear energy in particular. This documentation could, for example, be prepared as part of the regular report on subsidies. It should list not only direct payments but also tax credits. Such a report on the ecological impacts of subsidies should also provide at least a qualitative impression of shadow subsidies (external effects).
- The German federal government should continue to press ahead with the removal of subsidies on fossil and nuclear energy at national level. It should also lobby, firstly at EU or OECD level and then at global level, for an internationally coordinated removal or reform of energy subsidies. In this context, the conclusion of a multilateral agreement on energy subsidies (MESA) should be given particular consideration (Section 5.3.5.1). In the interests of climate policy, it is especially important for the federal government to abandon its opposition to the phasing out of state subsidies for coal mining at EU level and support the target of removing coal subsidies by 2010. This would also be in line with the decision, adopted at COP7, which calls on countries named in Annex II of the UNFCCC (primarily the OECD countries) to phase out subsidies to sectors which emit greenhouse gases (UNFCCC, 2002).
- In the long term, all energy subsidies should be removed except for those which promote fundamental research into innovative energy technologies, renewables and rational energy use, because experience has shown that these services are not adequately delivered by the market. This should include research on the final storage of radioactive waste and, more generally, the closure of nuclear power plants.

It has already been pointed out that rigorously dismantling all state subsidies on fossil and nuclear energies is likely to make nuclear energy non-viable as an industry (Section 3.2.2). The removal of subsidies facilitates compliance with the guard rails outlined in Chapter 4. Germany has already begun to phase out nuclear energy. The Council recommends continuing along this path. Germany should also try to bring influence to bear on other industrialized countries, but also on developing, newly industrializing and transition countries, in order to achieve the progressive phasing out of nuclear energy worldwide.

For the latter groups of countries in particular, the success of such endeavours is likely to depend primarily on the availability of sustainable alternative supplies of energy. Measures to promote renewables, low-emission fossil energies, and especially greater efficiency in energy supply, distribution and use (Section 5.2.2), as well as to develop modern forms of energy and efficient energy use in the developing, newly industrializing and transition countries (Section 5.2.3), are therefore of key importance.

5.2.1.3
Conclusion

Ecological financial reform is a key component in the package of measures to transform energy systems towards global sustainability. In this context, the first priority must be to internalize the external costs of fossil and nuclear energies. This is essential to ensure that sustainable energies, which at present are generally characterized by far lower negative externalities but higher market prices, are able to achieve a breakthrough. Tools to compensate for these disadvantages include levies on fossil and nuclear energies due to their environment-related incentive effects, and if necessary, the temporary earmarked use of the accrued revenue. An ecological financial reform should be completed by 2020 in the OECD countries and be a long-term objective worldwide.

As well as internalizing external costs, it is essential to remove the existing subsidies on fossil and nuclear energy carriers. Fossil and nuclear energies should no longer be subsidized; indeed, they should be subject to fiscal charges equivalent to their negative externalities. Ecological financial reform will make fossil and nuclear energy carriers more expensive and thus reduce their share of the global energy mix. This will increase the share of renewables used. However, this will fall far short of the desired target of a 20 or 50 per cent increase. The WBGU therefore recommends the pro-active expansion of renewable energies. Ecological financial reform is likely to take place primarily in the industrialized countries. The measures outlined in the following section can also be implemented in developing, newly industrializing and transition countries.

5.2.2
Promotion

5.2.2.1
Promoting renewable energy

THE BASIC CONCEPT

Today, renewable energy only covers a small part of the worldwide primary energy requirement. In

OECD countries, its share is estimated at 4 per cent, most of which is generated from hydropower and wind power (Section 2.3). Direct support for renewable energy can come in the form of direct payments from the state to the relevant actors, tax exemptions, state financing of research and development, or measures promoting market penetration. In industrialized countries, many and varied supportive measures are in place (Table 5.2-1). The lowest level of support is currently found in the US, where expenditures for R&D projects to commercialize renewable energy are combined with tax credits for power plant operators who use renewable energy (IEA, 2002b). EU member states are intervening in their liberalized energy markets to a greater extent. In addition to promoting R&D, they mostly use price and quantity controls to this end.

Direct subsidies have primarily been used as price controls for renewable energy. In some EU countries, such as Germany, rates are also fixed for power fed into the grid. In these countries, the government first stipulates that grid operators have to purchase the electricity generated from renewable sources. Then, compensation for the power is agreed at a rate above the market price by law or in contracts freely negotiated between the power generators and the grid operators. Compensation rates vary according to the energy source and are primarily based on the cost drawbacks of the renewable energy sources compared to conventional energy. As the costs for the use of renewable energy will drop in the long term as technological progress is made and economies of scale realized, compensation generally tapers off over the years towards the market price.

The obligation to purchase and pay compensation for power fed to the grid basically constitute cross-subsidies within the energy sector. These obligations constitute additional expenses for grid operators and utility companies, which they then largely pass on to consumers. The result is higher consumer prices, which may lead to lower demand – and hence lessen the burden on resources. In Germany, Spain, Denmark, and France, the system of feed-in rates has led to steadily rising use of renewable energy. In the past few years, Belgium, Luxemburg, and Austria have also adopted the concept.

In contrast, the EU has long viewed price-based promotional measures for renewable energy sources critically (EU Commission, 1998). The European Commission argues that consumers in a completely liberalized European electricity market will choose the provider with the lowest prices, and that providers from countries without such costly obligations will be at an unfair advantage. In competition with foreign providers, utility companies that have to buy expensive renewable power would then no longer be able to pass on the higher costs to consumers. Hence, the European Commission advocates using quantity controls (EU Commission, 1999a) to promote renewable energy sources; at the same time, the EU is not vocal in its criticism of feed-in rates at present, and price controls remain in line with the EU's current subsidy regulations towards the harmonization of promotion mechanisms envisioned for 2005 (EU Commission, 2001d).

The simplest model of quantity-based promotion is the use of quotas for energy from renewable energy sources (Box 5.2-1). In this model, all utility companies and grid operators are obliged to cover a certain legally specified amount of the heat or power they sell with renewable energy that they either generate themselves or purchase. State tenders for cer-

Table 5.2-1
Overview of the political instruments available for environmental protection in specific industrialized countries. X in place, P planned.
Source: modified after Espey, 2001

	Price control			Quantity control				Financial incentives	
	Subsidies for renewables	Taxes on fossil sources	Fixed rate for power sold to the grid	Political targets for quantities	Emissions trading	Tenders	Quotas	State R&D	Market penetration
Germany	X	X	X	X				X	X
Denmark	X	X	X	X	X		P	X	
France	X	X	X	X		X		X	X
Great Britain	X	X		X	X	X	P	X	X
Netherlands	X	X	X	X			X	X	X
Sweden	X	X	X	X		P	P	X	X
Spain	X		X	X				X	
USA	X			X	X		P	X	
EU (framework-setting)	X			X	P			X	X

tain energy quotas for renewable energy that has to be sold to the grid are another possibility. In this procedure, the government sets quotas for the amount of energy from renewable sources that has to be sold to the grid, with the least expensive investor generally being awarded the contract. The tendering process ensures, through competition among power generators, that mainly those producers or generation technologies will advance which are able to overcome the cost barriers of market access quickest.

Recently, the EU has been discussing tradable quotas and 'Green Energy Certificates' for the use of renewable energy. In this model, which has been used in the Netherlands and more recently in Denmark, the government stipulates the minimum amount or share (quota) of electricity and heat from renewable energy sources, which can then be provided flexibly (Box 5.2-1). Tradeable quotas and Green Energy Certificates ensure that the state quotas can be fulfilled flexibly and at low cost. The power generation capacity is expanded where it is least expensive to do so. However, this model fails to promote specifically the very technologies that will be crucial in the future but are still expensive today. To do so, complementary state subsidies would be needed.

PRACTICAL STEPS

There is no ideal way to increase the share of renewable energy through direct support. There is a broad consensus that state expenditures for research and development have to be clearly increased, but this is the only uncontroversial approach. In light of the major role that renewable energy sources are to play in globally sustainable energy systems even in the next few decades and the very minor role they now play, the WBGU believes that support for public and private research and development in renewable

energy be increased quickly and comprehensively (Chapter 6).

The Council proposes that the German government should look into both quantity-based policies and price controls, especially a step-wise transition to quotas that takes account of each energy source and technology. Guaranteed rates for renewable energy sold to the grid is one of the options that makes a lot of sense as start-up financing, especially when it comes to promoting technologies that are far from competitive. In our experience, set rates for power fed to the grid and subsidies that provide different rates of compensation appear to be more effective than quotas in compensating for the higher costs of renewable energy sources. The example of wind power, which was uncompetitive at the beginning of the 1990s, underscores this fact (Table 5.2-2). While quotas could be set for each individual energy source (Deutsche Bank Research, 2001), the high costs of the technical implementation of emerging technologies make this difficult, and Green Energy Certificates are not easily tradable. Therefore, Germany should maintain its current practice of tapering feed-in rates for relatively uncompetitive energy sources and technologies.

However, the outcomes shown in Table 5.2-2 are less the result of a fundamental superiority of feed-in rates than of the specific design of the various instruments: quotas for renewable energy are often quite modest, while the rates for renewable energy sold to the grid are relatively high. Hence, it seems that incentives are easier to implement via price controls than quotas.

Once the market share of a technology for the use of renewable energy has grown significantly and the technology has become competitive, direct quotas should replace subsidies for renewable energy so that competitive processes can better promote the innov-

Table 5.2-2
Comparison of the expansion of wind energy capacity under various promotion models in 2000.
Source: modified after Gsänger, 2001

Incentive mechanisms	Country	Installed capacity end of 1999 [MW]	Installed capacity end of 2000 [MW]	Growth rate [%]	Rated power [W/per capita]	Rated power per surface [kW/km²]
Price control (fixed rate for power sold to the grid)	Germany	4,443	6,113	38	74.51	17.12
	Spain	1,542	2,535	64	64.39	5.02
	Denmark[a]	1,771	2,282	29	430.48	52.95
Quantity control (tenders, quotas)	Great Britain	344	406	18	6.88	1.67
	Ireland	73	93	27	25.10	1.32
	France[b]	22	60	173	1.02	0.11

[a] The great wind capacity in Denmark is mostly due to previous state minimum price controls.
[b] In 2000, France switched to a price-control model like that of the German *Einspeisevergütung* (fixed rate for power sold to the grid). The very high growth rates are the result of special state subsidies for wind energy (EOLE 2000 incentive programme).

Box 5.2-1

Quotas, tradable quotas, green energy certificates

QUANTITY TARGETS FOR THE USE OF RENEWABLE ENERGY SOURCES

The goal is a minimum share of the desired type of energy in overall energy consumption. The target can be an absolute value or a quota (such as 10 per cent of the electricity generated) to be fulfilled by a certain deadline.

NATIONAL AND INTERNATIONAL QUOTAS

The German government has set a goal of doubling the share of renewable energy in its electricity supply, which means the national target is 12.5 per cent by 2010. The guide value for the European Union is to raise the share of renewable energy in its energy supply from 6 per cent in 2001 to 12 per cent in 2010, and specifically for the supply of electricity to 22 per cent by 2010. The EU's overall quota is split up into sub-quotas for the member states. The various states can choose how they fulfil these quotas. Several states have chosen quantity targets (Table 5.2-1 for the European Union).

QUOTAS VERSUS TRADABLE QUOTAS

Within a country, power companies or energy providers are usually obliged to provide a quota of the power they sell from (certain) renewable forms of energy. Relatively widespread in industrialized countries, this concept is called a portfolio model. If the overall quantity target is broken down across the various parties obliged to meet the target, the target is quantified for each actor. In addition, it is generally not difficult to monitor the degree to which the target has been met. However, the economic efficiency of rigid quotas is limited.

As power companies and energy generators incur different costs for the fulfilment of the quota, the quotas should be made flexible by allowing trading to lower costs. If an electricity generator has to generate a certain amount of electricity from renewable energy within a certain period, it can chose to generate it itself or purchase 'green power' from another power generator. As a result, the generators of 'green power' who operate most inexpensively will produce power, while generators with higher production costs will purchase power from them. With tradable quotas, the overall quota would thus be reached at lower overall costs than in the portfolio model.

Up to now, tradable quota models have mostly been implemented at the national level. However, there are a few model projects to test international trading quotas for renewable energy.

GREEN ENERGY CERTIFICATES

Basic concept

The model of green energy certificates represents a further development of flexible, tradable quotas. Here, individual quotas or absolute amounts that individual parties have to fulfil are a prerequisite. Unlike with tradable quotas, the parties do not have to generate the energy themselves or buy green energy physically, however, but only prove they have fulfilled the quota by owning a relative amount of green energy certificates. These certificates are issued to generators of 'green power' by a state supervisory authority for a certain amount of energy (say, 1MWh). Not only power companies and producers, but also consumers can receive tradable green energy certificates. The free trading of certificates would create a new market – an 'environmental service market'.

Pricing and economic efficiency

Pure price competition will continue to determine the price for electricity on the traditional market. Generators of 'green electricity' can compensate for cost disadvantages with additional income from the functionally separate certificates market. Providers of certificates will demand at least the price based on the difference between the costs of generating 'green power' and the market price for electricity. The most efficient providers of 'green electricity' will be awarded contracts. No distorted costs due to collusion or abuse on the part of dominant power generators are expected because the purchasing power company can also produce 'green electricity' itself to procure these certificates.

The advantage over tradable quotas is mostly their greater transparency and lower contract costs. In addition, consumers such as voluntary participants (environmental protection organizations, environmentally concerned citizens) can be better integrated.

Strategies for not yet competitive new technologies

In practice, the advantage of economic efficiency can lead to a disadvantage: specializing in inexpensive generation and inexpensive forms of energy will only promote competitive energy sources, such as windpower and hydropower, in the current state of technology. Thus, not yet competitive, but promising energy sources would be left out of this system of incentives. As a result, quotas would have to be set for these technologies and certificates markets established – such as a certificates market for power sold to the grid from geothermal plants. This would not make much sense for the latest technologies as the market volume of certificates sold and bought would be too slight to ensure that the market would remain functional. Hence, start-up financing is indispensable for emerging, not yet competitive forms of energy (such as state subsidies, fixed rates for power fed to the grid, etc.). As soon as the technologies have matured enough, quotas can replace subsidies, and a green energy certificate market for that particular energy source can be envisioned.

ative and allocative functions of renewables. In the long term, a transition to a model of Green Energy Certificates for 'green' power generated with increasingly competitive renewable energy sources seems possible and appropriate, especially as greater efficiency will result from the international trading of such certificates. The time frame for the transition of price controls – such as compensation for power sold to the grid – towards tradable quotas will basically depend on the expected volume of the quota market. For instance, wind power and hydropower can be expected to cover a large part of electricity genera-

Box 5.2-2

Renewable Energy Certification System

Electricity providers (generators, traders, grid operators, etc.), state institutions, associations, service providers, and households can participate in a Renewable Energy Certification System (RECS). Up to now, some 170 organizations are members in the RECS. They include almost all EU countries, Norway, and Switzerland. There are also cooperative deals with Australia, Japan, New Zealand and the US. The participants agree to accept the 'basic commitments', which define trading institutions and the flow of trading. The national 'issuing bodies' play a crucial role; they issue certificates and monitor trading. Certificates are issued to accredited electricity generators if the power was not produced with fossil or nuclear energy sources. Details about generation are checked in the course of accreditation and kept in national databases. The certificates from the issuing body are based on units of 1MWh of electricity generated. To prevent electricity being certified twice, the certificates contain information about the location of the production

plant, the manner of generation, the date of issue for the certificate, and the identification of the certifier. The networking of the issuing bodies' databases provides the IT basis for the trading system. Each generator has an account in the databases, where it receives certificates for each MWh fed to the power grid. Generators can offer certificates on a market. The buyers in the market are generally providers of 'green' products, private households (in the Netherlands), or environmentally conscious power consumers who require proof of the origin or means of generation of their power. Finally, power companies that have to fulfil a certain quota of renewable energy (portfolio standards) could use certification to demonstrate that they have fulfilled their quotas. Transactions between market partners have to be reported to the issuing body, which checks and corroborates the information provided by the participants. A certificate is invalidated as soon as it has reached the customer of a 'green' product or – in the case of the Netherlands – has been turned in for tax credit. By mid-2002, certificates for more than one TWh of electricity had been traded within the RECS.

Sources: RECS, 2002; Groscurth et al., 2000

---

tion quickly, which is not the case for photovoltaics or geothermal power. At the same time, policy makers will not be able to do without subsidies completely in promoting innovative renewable energy technologies anytime soon. On the one hand, free-enterprise competition is not expected to produce sufficient research findings; on the other, promising new technologies will emerge in the period of transition – which will last at least 50 years – that will not be able to compete if only supported in quota systems. Hence, no single promotion mechanism will suffice for the successful market penetration of renewable energy; rather, a mix of various mechanisms will best take account of the different degrees of competitiveness of the respective types of renewable energy sources. Thus, the WBGU recommends:

- Continuing and expanding market penetration strategies (such as temporary subsidies, fixed rates for power sold to the grid, quotas). Until a considerable market volume has been attained (or a technology has proven to be non-viable), fixed, tapering rates for power fed to the grid is one of the best options.
- In the mid-term, quotas should be used more often, leading to a Europe-wide system of tradable quotas and Green Energy Certificates for renewable energy sources in the foreseeable future. Intensive research is necessary for the design and transferral of tradable quotas to the global level.

The incompatibility of the various national systems makes international trading of these certificates difficult. An initiative of European power companies created the certificate system RECS (Renewable Energy Certification System) to demonstrate that

Green Energy Certificates can be traded; this system was tested from 2001 to 2002 (Box 5.2-2). Participation in RECS is currently voluntary. The incentive for the companies is image benefits; for private households (such as in the Netherlands) tax credits for the certificates purchased. After the test phase, the RECS system is to be further institutionalized. Expanding the innovative initiative of certificates to promote international trading would be an appropriate first step towards a global solution. The WBGU recommends that more German institutions should join the RECS initiative to include their ideas and concerns in the trading system.

For specific countries, the time frames and designs of the transition from a price control system to a quantity control system for renewable energy have to be specified in accordance with their special settings. Social science research needs to be conducted to identify optimal strategies (Section 6.2).

To expand renewable energy beyond its current market volume in developing and newly industrializing countries, the proper technologies will have to be mass-marketed and promoted. The WBGU thus recommends that the German government design its development cooperation programmes both technically and financially to achieve this goal. In particular, training programmes and demonstration projects would be helpful in the field of energy to promote marketing strategies for renewable energy, as well as other measures designed to increase demand for renewable energy (Section 5.2.3).

The success of renewable energy in the transition countries will depend on whether market-based reform of the energy sector are pursued consistently.

Here, subsidies for fossil and nuclear power will have to be done away with and incorrect market signals corrected. A stable, reliable set of market rules and regulations for the energy sector and corresponding qualities of public administration play a key role in this context. The use of market-based instruments such as eco-taxes (green taxes) or certificate trading could provide incentives for greater use of renewable energy.

The eastern European EU accession states already meet important institutional prerequisites for the implementation of instruments to promote renewable energy. It is recommended to integrate them in the planned EU certificate trading as soon as possible to take optimal advantage of economic cost reduction potential. The situation regarding instruments in the CIS nations is quite different, however. In light of the continuing great need for reform and the insufficient capacity within companies and public administration, there is even a risk that implementing such instruments would have the opposite effect (Bell, 2002).

5.2.2.2
Promoting fossil energy with lower emissions

THE BASIC CONCEPT
In the short to mid-term, the world cannot do without fossil energy. Renewable energy will not be able to replace fossil fuels completely anytime soon. In order to ensure power supply even while reducing dependence on fossil fuels, two factors have to be kept in mind: on the one hand, new investments in fossil energy sources should be kept to a minimum; on the other, investments in the field of fossil fuels that seem to be indispensable for socio-economic reasons should be made in types of energy that produce lower emissions and can be implemented in a flexible infrastructure. A temporary expansion of fossil energy can thus be designed to allow for new facilities – such as power plants or grids – to be operated with renewable energy sources as well. Increasing the share of natural gas could provide the desired flexibility if, for instance, plants that initially run on natural gas can be later retrofitted to run on biogas or hydrogen.

PRACTICAL STEPS
The temporary expansion of fossil energy with lower emissions is especially important in developing, newly industrializing, and transition countries with no real alternatives for power supply in the short to mid-term, especially given their strong economic growth. For instance, Russia plans to cover the expected growing demand for energy by using its coal reserves to reduce its dependency on natural gas (IEA, 2002a). Such a development is less critical when the additional use of coal can be compensated for by technology that will reduce emissions. At least in the transition phase, modern technology for the use of fossil energy sources will have to be transferred to the transition and developing countries. Here, development cooperation will play an important role, either directly or indirectly (such as by means of export credit guarantees). The Kyoto mechanisms (joint implementation und clean development mechanism) could prove to be beneficial here: they provide industrialized countries with an incentive to transfer low-emission technology for fossil energy sources to developing and transition countries.

Problems can be expected mostly in developing countries if a country with great coal reserves and many coal-fired power plants wants to switch to natural gas, and hence import gas that it cannot pay for. Thus, financial and technical support should accompany the expansion of low-emission technologies for the use of fossil energy in these countries (Section 5.2.3).

Liquefied petroleum gas is most likely to replace traditional biomass for cooking in the mid-term. Even if all of the 2,400 million people currently living in energy poverty switch to liquefied petroleum gas, the emissions would only make up some 2 per cent of the world's emissions (Smith, 2002). Brazil has already replaced biomass with liquefied petroleum gas for cooking in 94 per cent of its households. While 40kg of liquefied petroleum gas are consumed there per capita annually, China and India only consume around 10kg, while sub-Saharan Africa uses less than 1kg per capita (Reddy, 2002). The Council finds liquefied petroleum gas to be an especially good substitute for the traditional use of biomass as the technology is readily available and can be implemented quickly; it also will allow for a later transition to renewable energy sources. In the long term, liquid gas should be made from biomass.

5.2.2.3
Promoting efficiency in the provision, distribution and use of energy

THE BASIC CONCEPT
There is great potential for enhanced efficiency in the provision, distribution and consumption of energy. The technologies required are already available for the efficient and inexpensive provision of energy services, the prevention of transport losses in grid-based energy transport across long distances, and the effi-

cient use of energy by consumers (Section 3.5). In industrialized countries, the untapped efficiency potential is around 60 per cent on the demand side (Enquete Commission, 2002). In the developing, newly industrializing, and transition countries, the potential efficiency gains are probably much greater.

EFFICIENT PROVISION OF FINAL ENERGY
The liberalization of energy markets will provide most of the incentives for more efficient energy provision. Here, the crucial elements are the abolishment of governmental supervision of investments, demarcation contracts and concession agreements, the separation of electricity generation and grid operation, and a limitation of the role of the government in setting rules. With grid-based power supply, customers can take advantage of the liberalization of electricity markets by choosing their power provider, thus influencing the structure and technology of generation. Liberalization can in principle be expected to lead to structural changes among the electricity providers. Only the providers who produce electricity with economic efficiency will be able to stay on the market in the mid to long term. Indeed, electricity prices had been falling in the EU until recently. However, critics of liberalization fear that a concentration of companies will eventually hamper competition on the electricity market (Kainer and Spielkamp, 1999). For instance, in Germany the four largest companies provide a majority of the power supply. But such a concentration is not a foregone conclusion. After all, liberalization means opening the market up to a potentially larger number of providers. For instance, more providers of 'green power' have entered the market since liberalization.

Liberalization can also be expected to produce long-term structural changes in the generation of electricity. For example, distributed power units may produce more electricity. For industrialized countries, this trend would not be a return to the spot solutions of the last century, but rather the inclusion of small, local power plants in the public grid. The development of information and communications technology makes it easier to bundle and coordinate a number of small power plants in one 'distributed power plant' (Section 3.4.3). As Germany is planning to revamp its network of power plants in the next few years from the ground up, this country offers an excellent opportunity to change the spatial supply structure by building new power plants. The Council feels that the German government should look into restructuring the supply structures and promoting pilot projects and market implementation programmes. Current campaigns within the EU and Germany should be expanded even further.

A more distributed power supply could increase the share of renewable energy and combined heat and power plants (CHP). Local distributors (municipal power companies) used to be the main source of investments in renewable energy and CHP on behalf of the communities. Now that complete liberalization has removed the special status of local distributors, however, a lot of CHP plants operated by municipal power authorities are being shut down in Germany (BMU, 2000). One of the reasons is the subsidies for fossil and nuclear energy and the insufficient internalization of external costs, which has distorted prices.

If the potential efficiency gains are to be realized on both the supply and demand sides, one will have to move beyond liberalization (IEA, 2000). The additional internalization of the external costs of fossil and nuclear power will ensure that the price for efficiently generated power is raised to the price level that corresponds to the true scarcity of the sources. Only then can the central criterion that consumers base their purchases of electricity on – the price – be used to create incentives for more efficient power demand. In addition, it would make sense to have mandatory labelling for electricity from renewable sources. The demand for electricity from regenerative energy sources would be easier to realize, and competition on liberalized markets would have a qualitative component in addition to prices. The labelling would also be an initial step towards a system of tradable green energy certificates (Batley et al., 2000; Section 5.2.2.1).

In addition, liberalization means opening national electricity markets to power imports, thus increasing competition on the national electricity market. However, such electricity imports will only increase the efficiency of supply and demand if the external costs of fossil and nuclear energy are internalized in foreign countries as well. Energy imports from countries where the external effects have not been internalized could lead to the suppression of domestically generated energy that is more expensive due to the greater degree of internalization (Section 5.3.5).

The WBGU holds that the state should intervene in the process of liberalization to help shape these events. Long-term, stable general frameworks should be created for the new markets, and the proper functioning of competition should be ensured. The increasing integration of the energy markets in Europe requires the creation of a supranational regulatory authority to protect competition, such as a European competition authority (Duijm, 1998). Nevertheless, the principle of subsidiarity stipulates that national and regional decision-makers should determine energy policy to the extent possible so that special national and regional needs and features can be taken into consideration.

The simultaneous supply of electricity and heating or cooling from CHP plants offers great potential efficiency gains (Section 3.3). But the price drops caused by liberalization, the ongoing subsidies for fossil and nuclear power, and the wholly insufficient internalization of external costs are weakening the current competitiveness of this comparatively environmentally friendly technology. Hence, the WBGU recommends that incentives be continued specifically for CHP plants and that quotas be implemented at the EU level. The WBGU also calls for a greater share of power from CHP than the level suggested in the EU's 6th Environment Action Programme: 20 per cent instead of 18 per cent by 2012. In addition to the proposal for a European CHP Directive, which calls for disclosure of CHP power by 2005 and requires that national subsidies should only be granted to CHP power when the heat is also used, the German government should set national, binding quotas as soon as possible. Tradeable quotas would be one way of providing power in the most economically efficient manner from CHP plants. These regulations should take the concept of distributed generation of electricity (Section 3.4.3) into account.

Efficiency increases in the provision of energy in developing, newly industrializing, and transition countries often require the prior transfer of better technologies to these countries. While some eastern European coal-fired plants only have 28 per cent efficiency and some in China only 20 per cent, modern gas turbines attain nearly 60 per cent efficiency. The WBGU thus recommends granting the transfer of energy technologies a higher status. On the one hand, more technology could then be transferred in the context of development aid; on the other, private technology transfers could receive more support. In addition to better credit conditions, tax exemptions and state risk guarantees would be possible. In addition, the German government should support the step-wise establishment of international standards for minimum efficiency levels for fossil power plants. Such standards should be based on the EU IPPC Directive and take effect no later than 2005.

EFFICIENT ENERGY TRANSPORT STRUCTURES
In the course of the liberalization of the energy markets, environmentally desirable increases in efficiency among operators of transit grids could stem from the separation of distributors and providers. As these companies concentrate on operating grids, better transmission technologies can be expected to provide efficiency gains, thus reducing transmission losses. However, grid operators that also own the grids only have an incentive for more efficient grid operation when the income from better transmission technologies is greater or the returns to scale for centralized power less than the costs for grid losses. This will probably not be the case very often. Long-term strategies have to be developed to make the grids ready to cover large areas of power generation.

In transition countries, considerable increases in energy efficiency have resulted from improvements in existing district heating systems. These systems have been suffering from high transit and distribution losses. Relatively simple measures – such as the introduction of consumption meters, variable-speed pump motors, and the revamping of the insulation used on piping – could provide major efficiency gains. Investments of this type usually pay for themselves within only two years (van Vurren and Bakkes, 1999). The WBGU recommends supporting transition countries in reducing their transit and distribution losses.

EFFICIENT ENERGY USE
The efficiency of energy consumption can also be increased for consumers who co-determine the energy demand in buildings or with machines, appliances, cars, transport services, etc. However, private households are rarely able to get information about the energy balance sheet for individual alternative products and houses or apartments and to assess it. Hence, labels and minimum efficiency standards have been established since the 1980s. However, these labels are often voluntary or limited to certain market segments. For instance, for a long time there were only efficiency classes and labels for 'white goods' (refrigerators, washing machines, etc.) in the consumer goods sector, while 'brown goods' (televisions, stereos, etc.) were rarely labelled.

The Council thus recommends expanding these labels to all consumer goods and making this development mandatory in the EU. In the long term, labels should be made mandatory for all energy-intensive consumer goods, buildings, industrial plants, and services to the extent possible. The EU's Energy Performance Directive (EU Commission, 2003) has already set the date for energy labelling for buildings at 2006.

At the same time, the requirements for labelling are to be adapted to the current state of technology in regular intervals. When revising the label features, more attention should be paid to consumption during stand-by for many consumer appliances. Stand-by consumption is estimated at around 10 per cent of the electricity demand in homes in the OECD countries, with potential savings averaging around 75 per cent (IEA, 2001c).

Box 5.2-3

EU-wide mandatory labelling of consumer goods

Since August 1999, there has been mandatory EU-wide labelling of 'white' appliances (cooling units, washing machines, etc.). The various EU states implement the mandatory labelling, monitor it, and ensure that the great variety of labels stays within certain bounds and consumers are informed about the labelling by means of campaigns.

The labelling is also beginning to apply to 'brown' goods (televisions, stereos, etc.). For instance, the EU adopted the US 'Energy Star' labelling for office and telecommunications equipment (PCs, screens, faxes, copiers, scanners, etc.) in 2001. The value of the transatlantic trading of such equipment – some US$40,000 million – is one main reason why the label was adopted at least up to 2005 instead of developing a European label. There are similar agreements between the US and other states such as Australia and New Zealand.

Sources: WTO, 2001; Energy Star Australia, 2002

The international harmonization of efficiency standards and labels would be helpful towards overcoming the confusion produced by a plethora of labels, especially for goods traded in large volumes internationally. Bilateral agreements within the EU could be a step in this direction (Box 5.2-3). International labels are usually only based on the lowest common denominator so that they should not replace, but rather be integrated into more ambitious labelling systems. If, for example, a 'single global energy star' were introduced for tradable consumer goods for the long term, regional and national labelling systems should continue to be implemented and further developed to provide information about far more ambitious efficiency standards, such as a European 'double energy star' or national 'triple star'.

Great efficiency gains could also be made for the use of heating and cooling energy if ordinances were passed to reduce heat loss in winter (insulation, heat recovery) and provide better protection from summer heat for buildings (Section 3.5.2). The construction sector could be made more environmentally friendly if the legal minimum standards contained ambitious targets and suitable incentives. For instance, the Energy Performance Directive stipulates that the energy consumption analysis of buildings is to be based on the consumption of primary energy (including energy for cooling and lighting) starting in 2006, which is already being worked on in standardization committees. In the process, essential elements of the German Building Energy Conservation Ordinance (Energieeinsparverordnung) of 2002 and some methods already established in Switzerland can be adopted. Furthermore, applied research is currently promoting demonstration projects that provide a comprehensive energy analysis of heating, ventilation, cooling, and lighting. There is also a great potential to lower CO_2 by supporting better supply technology, such as through low-interest loans (Enquete Commission, 2002).

Efficiency gains are also to be made on the demand side by means of 'demand control' in the strict sense of the term, i.e. demand side management. Load management would be one way of doing so; here, the maximum electricity consumption is shifted to times of lower consumption, thus helping to reduce the number and capacity of power plants needed. Variable rates and support for certain storage technology would provide incentives for this approach (Melchert, 1998). Though variable rate structures already exist today, consumers rarely take sufficient notice of them. Here, one can imagine having a display of the current rate in apartments on an electronic display or automatically controlling appliances based on the current rate.

In addition, 'contracting' is recommendable, especially for companies (Melchert, 1998). Here, a third party – the contractor – plans, conducts, and possibly finances energy projects. Whereas equipment contracting means that the construction and operation of a certain production plant is outsourced, the contractor in performance contracting looks for ways to save energy in the company and takes measures to save energy independently (Freund, 2002). Not only do the contractor and the customer profit from performance contracting, but also the environment. Hence, this model is a win-win-win instrument worth promoting.

For contracting to be established as a voluntary service on a liberalized market, the liberalization of the markets for grid-based energy supply is crucial and will have to be completed quickly. Only then will all customer groups be able to switch to energy providers with an especially favourable service offer. One can also imagine contracting offers that ensure that renewable energy or cogeneration (CHP) units can cover the energy services needed for a building or a plant. In the end, energy service providers can help make the current market for final energy (electricity, gas) into a market for services (lighting and warm/cool rooms, hot and cold meals, etc.).

To provide consumers with the relevant information, the WBGU recommends better consumer information about contracting and demand side management. Standardized sample contracts including environmental information could lower the information costs and help consumers, especially small ones, overcome their inhibitions.

INCENTIVE SYSTEMS FOR SPECIFIC TARGET GROUPS

Lifestyles influence the structures of energy consumption decisively, especially in industrialized countries (Section 2.2.3). Likewise, the possibilities of taking advantage of efficiency potential on the demand side or attaining a sustainable level of energy consumption differ according to lifestyles. Lifestyle research shows that environmentally alternative lifestyles like those of the 1970s and 1980s have not established themselves (Reusswig et al., 2002). In other words, for the foreseeable future strategies to save energy will probably only be successful if people feel that they will not have to lower their standard of living. The 2000 Watt Project in Switzerland can be considered a relatively successful attempt to propagate sustainable energy consumption, including efficient energy use, without leaving the impression of having to 'do without' (Spreng and Semadeni, 2001).

The incentive systems will have to be as varied as the target groups they address. The motivation to save energy can be the act of saving itself, an interest in modern technology/innovations, or an internalized responsibility for future generations. To reach all lifestyle groups, there will have to be a mixture of environmental policy instruments.

Communication concepts for the specific target groups appear suited to complementing and supporting political frameworks. For instance, the municipal utility of Kiel conducted a market study in order to allow it to respond to a highly differentiated buyer's market with marketing tailored to specific target groups. In the process, the utility found out that energy-saving behaviour not only differs from group to group, but also from sector to sector (heating, electricity, water). The various groups choose their general environmental behaviour and, in particular, their ways to save energy for all kinds of reasons that stem from their lifestyles (Reusswig, 1994).

The WBGU thus recommends integrating the discussion about sustainable lifestyles and environmental awareness in the ongoing negotiations on the implementation of Article 6 of the UNFCCC dealing with education, training, and public awareness about climate change. By 2005, a discourse about lifestyles should be part of school curricula in industrialized countries. Furthermore, campaigns addressing specific target groups should be carried out among the population. In the end, the extent to which consumers associate the transformation of energy systems towards sustainability with a vision that will personally benefit them will be decisive: a better standard of living, more freedom to choose, more jobs, and technological innovations.

5.2.2.4

Conclusion

Fossil and nuclear energy must be made more expensive than they currently are by means of fiscal measures so that they will become less attractive. On the other hand, the transformation of energy systems also requires special incentives to make both renewable energy and low-emission fossil energy more attractive in addition to directly increasing the efficiency of the production, distribution and consumption of energy.

In principle, price controls and quotas can both be used to promote renewable energy. Direct subsidies or set rates for power fed to the grid can be used very well as price controls. The various types of quotas for minimum shares of electricity or heat from renewable energy sources are essential for quantity control. These quotas can be designed so that certificates for the production of electricity from renewable energy sources can be purchased and sold in a market.

Quantity controls basically have the advantage of being more targeted. Tradeable quotas – with the addition of the system of green energy certificates – are also an attractive national and international instrument for the expansion of renewable energy in terms of efficiency. On the other hand, promising new technologies may not be competitive due to a lack of product maturity and small production volumes, problems that tradable quotas will not remedy. Here, the promotion through price instruments is more suitable. The selective transition from a price-based to a flexible, quantity-based system of incentives should be tailored to the special features of the technologies, markets, and countries concerned.

Efficiency gains can come from measures on both the supply and demand side. Here, state regulations play an important role, as do an increase in competition (due to liberalization) and better information for consumers from the propagation of mandatory labelling on all energy-intensive consumer goods, buildings, industrial plants, and services. The international harmonization of efficiency standards and labels is recommended for goods traded internationally in large volumes. In addition, the instruments of demand side management and contracting should be supported in liberalized energy markets and concepts for incentives and communications developed to address specific target groups.

Great potential efficiency gains in energy used for heating and cooling would result from ordinances regulating the thermal performance of buildings. Legal minimum standards combined with ambitious targets and appropriate incentive programmes are recommended.

The transfer of technology from industrialized to transition, newly industrializing, and developing countries is important to increase the efficiency of fossil power plants. In addition, the German government should support international standards for minimum efficiency levels for fossil power plants. The Council also recommends specifically promoting combined heat and power (CHP) plants. Here, the German government should work for the quick adoption of national target quotas in the ongoing negotiations on the EU's CHP Directive. By 2012, 20 per cent of the electricity in the EU should be generated by CHP. Tradeable quotas would be one way of providing power in the most economically efficient manner from CHP.

5.2.3
Modern forms of energy and more efficient energy use in developing, transition and newly industrializing countries

5.2.3.1
The basic concept

For developing and newly industrializing countries, population growth and the sometimes above-average growth of the gross domestic product are only some of the reasons why the demand for energy services can be expected to increase. The investment costs needed to meet this demand in the energy sector will amount to some US$180,000–215,000 million (in 1998 dollars) over the next twenty years, or 3–4 per cent of the annual GDP of these countries (UNDP et al., 2000; G8 Renewable Energy Task Force, 2001). The latest estimates of the IEA find that the transition countries will have similar demand. For instance, from 1999 to 2020 the investments needed for Russia's energy infrastructure (including investments in energy efficiency and promoting renewable energy) amount to US$550,000–700,000 million (IEA, 2002a). In light of the weak economies of these countries and the extent of the financing required, they will not be able to finance these investments on their own in the foreseeable future (Dunkerley, 1995). For these reasons, and because private investments can be expected to provide far more financing than funds from development aid programmes, the only way that the energy sector can be expanded as required is

if private investments are expanded, especially from foreign investors.

However, private investments will only be made when profits can be expected at least in the mid-term. In light of the current energy policies of developing countries, the incentives for private investments in the energy sector are very limited, both for the expansion of current electricity grids and improvements in access to modern energy in rural areas. Hence, in the mid-term expanding the energy supply to rural areas by means of distributed approaches and microgrids will primarily remain a task for the local governments and development aid agencies.

In addition to improvements in the supply of energy, better access to modern energy with low emissions also requires measures on the demand side. Expedient measures range from subsidies for efficient, low-emission energy consumption to the expansion of microfinancing for private households and aspects concerning the acceptance of different technologies and means of financing.

Various measures for supply and demand are now discussed in greater detail. It should be kept in mind here that the energy systems of developing, newly industrializing and transition countries all vary greatly, even within the respective groups. Hence, the recommendations for energy policies given in the following should not be considered a recipe for all cases. On the contrary, the great variety of demographic, geographic, cultural, social, economic and political distinctions in addition to the great variety of current energy systems make it clear that any proposals for a transformation of energy systems will have to be adapted to local conditions.

5.2.3.2
Practical steps on the supply side

CREATION OF ATTRACTIVE GENERAL FRAMEWORKS FOR PRIVATE INVESTORS
The initial experience of some developing countries with the partial privatization of the energy sector have shown that private investors increase the efficiency of the power supply considerably given proper regulations. At the same time, the countries benefit from access to international capital markets (Bond and Carter, 1995). This is not, however, true of the poorest developing countries, where not enough purchasing power can be created on the demand side to make great investments profitable. In many developing countries, especially the poorest, the grids of the public power companies are limited to large cities, in particular the inner cities.

Privatization creates incentives for expansion and efficiency increases in the provision of energy. The

WBGU thus recommends that one of the following privatization paths be taken when granting credit and projects to developing countries, depending on the setting in the specific country (Bond and Carter, 1995):

- Private energy companies are granted access to the existing grids of the state power company. After a transition period, the state power company (with or without previous restructuring) is incrementally or completely privatized.
- The state power company is spun off into smaller units, with the generation, transport and sale of power separated. Then, the individual units are privatized; at the same time, independent generators are granted access to the existing grids.
- The whole state energy monopoly is privatized. Only later is the market opened for other private providers.

In many developing countries, high import duties are charged for the import of industrial goods. This not only makes it more expensive to set up new plants, but also to purchase the necessary spare parts. There are special barriers for photovoltaics, for instance, which are seen as luxury items subject to great import duties. Locally manufactured energy units or spare parts, on the other hand, are often not reliable enough to ensure the smooth operation of technologically complex plants (UNEP-CCEE, 2002). The WBGU thus recommends that the German government work for an increase in the quality of locally produced units and components in its development policy. The more competitive locally produced components are, the less the domestic industry will have to be protected with import duties. Furthermore, industrialized countries could lower their trade tariffs and other trade barriers to give developing and newly industrializing countries an incentive to lower their own customs duties.

The prerequisites for private investments in the energy sectors of developing countries are often unfavourable due to the limited markets. Regional integration could enlarge the markets, allowing for a more efficient use of investments and even contributing to a greater diversification of commercial risks. Given proper competition, economies of scale would lead to lower prices for consumers. Therefore, technical standards should be unified and the planning of energy projects coordinated at the regional level to promote the integration of regional markets. In the process, the infrastructure needed for energy transport has to be treated as a primary goal of such investments. The expansion of trading organizations should be considerably stepped up, especially in Africa, where intraregional trade only makes up 6 per cent of the trade volume (Davidson and Sokona, 2001). The Council recommends that the German

government integrate these aspects in its planning of development policies.

The attractiveness of the developing, newly industrializing and transition countries for foreign investments in the energy sector can be increased with specific measures in the energy sector, but also by means of general economic and legal policies. Some examples are measures to increase legal security and lower political risks (Johnson et al., 1999). Such steps could set the foundation for the expansion of energy service companies in countries like Russia, where unclear legal relations and opaque permit procedures have kept foreign investors from committing to this potentially giant market (EBRD, 2001).

Foreign direct investments in the energy sectors of developing, newly industrializing and transition countries are especially attractive when the countries the investors come from provide favourable credit and export guarantees. To promote global, sustainable energy systems, it makes sense to grant such favourable terms to project categories that fulfil sustainability criteria. Hence, no investments or exports should be promoted for new plants to generate electricity from fossil fuels or nuclear power or to develop and market raw materials for fossil or atomic energy. Exceptions should be made if the following can be proven: the alternative with the lowest carbon emissions was selected; the project will fit the long-term, sustainable energy planning of the guest country; and renewable energy does not constitute a feasible or useful alternative. Promoting old plants for the generation of electricity from fossil fuels at least makes sense in the transitional period if the goal is only to modernize and take advantage of existing capacity so as to increase efficiency significantly.

The Council is aware that some governments and non-governmental organizations in developing countries, such as India's Centre for Science and Environment, reject such conditionality. The reason given is that the 'North' is setting the development path for the 'South' and wants to have the South shoulder the burden of the additional expenses for a sustainable energy path on its own (CSE, 2001). The WBGU nonetheless holds to its conviction that state investment and export incentives have to be subordinated to the overriding goals of sustainable energy policy. However, this also means that there must be a structural change in the energy policies of industrialized countries.

At the Economic Summit in Genoa in July 2001, a task force set up by the G8 made proposals of how export credit institutions could play a decisive role in the transformation of energy systems (G8 Renewable Energy Task Force, 2001). State export credit institutions should become active in expanding the OECD guidelines in this area. For renewable energy sources, the deadlines for repayment, interest rates, and the criteria for risk assessment have to be modified just as they have been in special sector agreements for nuclear energy, power plants, ships and planes.

The second proposal from the G8 task force concerned the OECD's environmental guidelines for export credit agencies. These guidelines are to specify universally applicable minimum standards for energy efficiency and carbon intensity as well as a unified reporting framework for the local and global environmental effects of a project. The Council supports these recommendations. To give the activities of export credit institutions the greatest possible leverage for a global transformation of energy systems, the WBGU supports further reform (Maurer and Bhandari, 2000):

- *Full-cost calculations for projects relevant to energy:* In order to prevent the use of fossil energy sources leading to social and environmental damage, export credit institutions have to insist on at least an approximate inclusion of external costs in the profitability calculations for specific projects.
- *Quotas for projects for the transformation of global energy systems:* The export credit institutions should specify quotas in their portfolios for renewable energy and for increases in energy efficiency in the context of international agreements. Starting in 2005, progressive minimal standards for the admissible level of carbon intensity should be specified for energy-generation projects.
- *Criteria for promotion:* In promoting large dams, the criteria formulated by the World Commission on Dams must be upheld. In the opinion of the Council, nuclear energy should not be promoted any longer.
- *Greater transparency:* The public should be given ample notice when a project is applied for, especially concerning its environmental and social effects, before a decision is reached to promote the project. The US Export-Import Bank is exemplary in this respect. It publishes environmental impact assessments for projects 30 days before a decision is made and lists all incentives and all of the companies involved, including the amounts, in its annual report.
- *Limits on state export incentives:* As general subsidies of export activities should be viewed critically in terms of market-based policies and such subsidies are the very purpose of export credit institutions, international negotiations to reduce the intervention of export credit insurance policies should be held with the aim of making sustainability an expressed prerequisite for the inclusion of exports in the support of export credit institutions.

REGULATION OF THE ENERGY SECTOR

In addition to the regulation of market frameworks in which, for instance, the conditions for access to power grids are specified, other regulations can make an essential contribution to increasing the environmentally friendliness and efficiency of the energy sector. To ensure this, rates and standards have to be specified. The WBGU recommends that the German government take this into account in its energy policy activities concerning developing countries as follows:

1. *Market supervision and competition for the market.* Competition in or for liberalized energy markets increases the efficiency of energy conversion and reduces the costs of conversion and power supply. When competition is fair, lower prices are passed on to customers, with the poorest thus gaining access to modern energy more easily. Often though, power companies keep the benefits of efficiency gains to themselves, especially for grid-based energy. In developing nations, which rarely have an effective policy to protect them from unfair competition, it thus makes sense to check power companies regularly to see if they have intentionally engaged in unfair competition and to ensure both that efficiency gains are passed on to consumers in an appropriate manner and that the promises made in tenders are kept. Another problem of liberalization is that customers in rural areas get short shrift. To accommodate for them, power companies should be obliged to provide power to all consumers within a certain region. To ensure the efficiency of the energy supply, the WBGU recommends that subsidies be granted in a transparent bidding process.

2. *Setting standards.* Energy systems in developing countries often suffer from low efficiency and great transport and distribution costs. Even if liberalization creates incentives to save and leads to the revamping of the power system to increase efficiency, quality standards should be set for the development of new energy projects. Such standards ensure the efficiency and proper functioning of energy systems, thus improving acceptance among the population that often suffers from the poor quality of the energy supply (UNEP-CCEE, 2002). In addition, technical standards in certain market segments can contribute to the opening of larger markets. This is the case for liquefied petroleum gas, which is only efficient if standardized containers are used in the whole country or at least large regions. To keep costs as low as possible, technical standards should not be based on the western model, but should instead take account of the often varying needs of consumers in developing countries. On the demand side, binding stan-

dards for consumer devices can contribute to efficiency increases in energy consumption, and hence to greater availability of energy (Davidson and Sokona, 2001).

5.2.3.3
Practical steps on the demand side

PROMOTING ACCESS TO MODERN FORMS OF ENERGY
The Council holds that access for all people to modern energy is the minimum requirement for the sustainability of the transformation of global energy systems (Section 4.3.2.2). This access is to be ensured for all people by 2020, with at least 500kWh of modern energy available per capita annually to everyone by 2020, 700kWh no later than 2050, and 1000kWh per capita by 2100 (Section 4.3.2.3). As a person's overall energy consumption consists both of individual energy needs and energy services used indirectly (manufacture and transport of goods), the overall per capita energy consumption will be even greater (Section 4.3.2.5).

Table 5.2-3 contains a selection of technological options for the development of sustainable energy systems in rural areas of developing countries. First, the use of traditional biomass is damaging to people's health and must be reduced (Sections 3.2.4; 4.3.2.7).

To this end, the WBGU recommends that at least 80 per cent of the world population should not have to use biomass in ways damaging to human health by 2020, with this figure increased to cover the world population no later than 2050. Second, liquefied petroleum gas could be used instead (Section 5.2.2.2). In addition, access to energy services based on electricity must be provided (lighting, cooling, support for household and commercial activities, and access to communication).

The WBGU recommends paying attention to disparities in all efforts to transform energy systems. For disparities within a country, disadvantaged groups have to be specifically promoted and attention has to be paid to cultural and gender differences. For disparities between countries, the most important challenge is that the per capita income in the poorest countries has to increase faster than elsewhere. In some cases, this may require cross-subsidies or social transfers (state support for electricity and heating).

Two important prerequisites must be met to improve access to modern energy services in developing and newly industrializing countries: on the one hand, the infrastructure for energy supply must be created or expanded; on the other, energy must be affordable for the whole population. The Council proposes that by no later than 2050 no household should be forced to spend more than 10 per cent of its

Table 5.2-3
Examples for selected technologies for the possible development of energy systems in rural areas of developing countries. Source: modified after Reddy, 2002

Activities	Currently	Short-term	Mid-term	Long-term
Cooking	Wood stoves	Liquefied petroleum gas Biogas		
			Biogenic liquefied petroleum gas	
Lighting	Candles, oil, kerosene, battery-powered light bulbs	Fluorescent lamps		
			LED lamps	
Shaftpower	Machines powered by humans and animals,	Electric motors		
			Highly efficient combustion engines, motors running on biofuels	
	combustion engines, micro-hydropower			
Water	Hand pumps, use of surface water and shallow wells	Electric pumps (powered by e.g. photovoltaics), purifcation technology for drinking water from conventional sources, activation of deep wells		
			Highly efficient irrigation technology, desalination plants running on renewable energy	
Telecommuni-cations	TV and radio			
		Mobile phones		
			Internet connections	Satellite-supported Internet connections
	Battery power	Photovoltaic power Wind power		
	Power from diesel generators		Power from advanced motors/generator systems	

income on its most elementary energy needs. In the long term, the figure should be much lower. Companies operating for profit will only make sure that access to modern forms of energy is expanded to areas with sufficient purchasing power so that the great investment costs both for the expansion of grid-based and distributed energy systems pay for themselves relatively quickly. Hence, access in poor, sparsely populated and remote regions – in the mountains or in shanty towns around cities – will depend on public funds, which will have to be paid for with development assistance. In addition, in such areas complete privatization of the energy supply and far-reaching liberalization of the markets would not appear to be suitable, at least in the transitional period. Slight margins and great investment risk would have to be made more attractive by means of temporary monopolies. Privatization and liberalization would be counterproductive here without a suitable regulatory framework. In projects promoted by development cooperation, public-private partnerships should be considered very important.

Distributed energy supply (such as with hybrid systems using diesel generators and photovoltaic systems) is often a better solution than grid-based power in sparsely populated areas (BMZ, 1999; Goldemberg, 2001). Expansions of grid-based energy systems mostly depend on the distance to the current grid, the number of households to be connected, and their demand (World Bank, 2000). The low demand due to the lack of purchasing power and low population density mean that the grid is only expanded when the distance to the current grid is below around 10km (ESMAP, 2001).

Poor sections of the population in developing and newly industrializing countries and consumers in transition countries have only been able to purchase energy services because state subsidies often lowered the prices far below the costs of generation. However, it is the better-off (urban) population that usually benefits from these subsidies because the rural population does not have access to the subsidized goods or the consumption of the poor remains low despite subsidies (UNDP et al., 2000). If poor sections of the population are to be explicitly granted access to modern energy services, subsidies for specific target groups make more sense in combination with rate structures determined by the market. The WBGU recommends that the German government work towards such structural changes in its development cooperation. To prevent the negative effects of such subsidies, four criteria should be met (UNEP and IEA, 2001):

1. The subsidies should be limited to a clearly defined target group to the extent possible. Beforehand, the economic, social and environ-

mental effects should be analysed to ensure that the positive goals desired can be attained.
2. The programme must make do with little administration.
3. The costs and operating principle of the programme have to be transparent. In particular, burdens on the public sector should be itemized in the state budget.
4. In designing the programmes, long-term incentives for the provision of energy services should be created (UNDP et al., 2000).

The Argentinean model is interesting in terms of the last point. In Argentina, there was a call for tenders for the expansion of the power supply to rural regions – at set rates, but without specifying the type of electricity supplied. The bidder who offered to expand the power supply with the least subsidies was awarded the contract (ESMAP, 2000). In South Africa, a similar model was used during the restructuring of the electricity market. There, concessions were auctioned for areas with a certain share of regions not connected to the grid and with the provision that power be provided either by expanding the grid or setting up microgrids. While the state covers the initial investment costs in this model, later there is a cross-subsidy between the various consumer groups (Clark, 2001). Although it is too early to assess the success of the two models, subsidies at the sales level seem especially suited to reaching large rural regions comparatively quickly. Given attractive market conditions, investors will be interested in such concessions so that the energy supply may increase in the mid to long term.

It is important that the subsidy mechanisms be designed for a transparent bidding process that not only power companies, but also consumers, village communities and project organizers have access to. In addition to explicitly addressing different target groups with the subsidies, the period for the subsidies is important towards improving access to energy. Here, a distinction can be made between short-term credit – especially microfinancing – and long-term financing.

In particular, long-term financing can be ensured by means of leasing contracts, consumer credit, or 'pay-for-service' contracts. In all of these cases, the seller of the energy ensures the required financing. Whereas the buyer generally retains ownership in leasing and consumer contracts once the financing phase has been completed, in 'pay-for-service' contracts the customer only pays for the energy service made available. The special advantage of this type of contract is the great incentive for sellers to ensure that the system is kept in working order and that users of the system are instructed about how to use it correctly. One problem in this type of financing is access to start-up capital for small and mid-size energy providers.

USE OF MICROFINANCE SYSTEMS

Microfinance systems were created in the 1980s as a counter-movement to the state development banks and subsidized credit programmes that were neither available to a lot of people nor economically viable. Instead of promoting individual state credit programmes, today the focus is on improving general institutional frameworks and setting up privately organized microfinance institutions (GTZ, 1998). In the meantime, they have proved useful in practice and are often conducted by NGOs. On the one hand, microfinance programmes are tailored to low-income sections of the population; on the other, local knowledge helps prevent wrong decisions being made and helps detect undesired developments. Microfinance is characterized by

- Low credit and savings volumes,
- Closeness to the customer, which simplifies and facilitates credit granting procedures,
- The relatively low importance of past economic data for credit approval in favour of prospects of future income, hence enabling the financing of innovative activities of small entrepreneurs,
- The acceptance of credit security uncommon for banks (jewellery, group liability, etc.).

The drop in public support requires a bundling of activities. Therefore, in the past few years a number of networks of microfinance institutions have sprung up (such as the Grameen Trust, or Banking with the Poor). At the same time, donor coordination committees were founded (such as Sustainable Banking with the Poor, or the Donor Working Group for Financial Sector Development), which are mainly controlled by the World Bank.

Microfinance projects can make an important contribution towards the financing of energy projects. The combination of micro-credit with investments in energy systems, especially photovoltaic units, does pose some difficulties, however. A photovoltaic system can cost around US$500, which is more than usually granted as microcredit. At the same time, most microfinance organizations have short terms of 6 months to no more than 2 years, while most potential purchasers would need terms of up to five years to pay off a loan for a photovoltaic system (Philips and Browne, no year indicated). The positive examples of the Grameen-Shakti photovoltaic programme in Bangladesh and Genesis in Guatemala prove that these difficulties can be overcome in some cases.

The WBGU is of the opinion that the importance of microfinance in improving access to modern energy services that are better for human health and the environment should not be underestimated, especially for private households, small businesses and micro-enterprises. The Council thus recommends using and expanding the microfinance systems to include energy projects as well. To this end, microfinance systems should continue to be supported from development cooperation funds and state subsidies taken into consideration for development measures.

CULTURE-SPECIFIC FRAMEWORKS

Even if a greater supply of modern energy can be provided in developing and newly industrializing countries and purchasing power can be created to take advantage of this availability, the efficient use of energy is still not ensured. Types of energy use handed down and accepted combined with a lack of knowledge about how to deal with new energy carriers or about their advantages can prove to be barriers towards the use of sustainable forms of energy in these countries. In developing countries, switching from a three-stone stove to a gas cooker can prove to be just as difficult as switching from a centrally controlled source of heat to a distributed supply with consumption-based billing in transition countries. In addition, the financing models for modern forms of energy, such as the purchase of Solar Home Systems or monthly invoicing for electricity, will probably also not be accepted without further ado.

To overcome such barriers, quantitative and qualitative improvements in training for energy systems are needed and knowledge about investments and saving disseminated in development cooperation. In addition, research about the acceptance of technical and financial systems has to be intensified with representatives of the countries, regions and communities and of ethnic and social groups (Section 6.2).

INCLUSION OF WOMEN

The use of modern energy also depends on how the technologies and their financing are tailored to potential users. One especially important group, particularly for the use of energy in private households in developing countries, is women. They are traditionally responsible for the procurement of energy sources for cooking, heating and drying (Section 2.4). More frequent use of modern forms of energy could lead to considerable improvements here. For this to happen, women will need to accept the modern forms of energy and their financing and have access to them. Here, too, the findings of acceptance research are just as indispensable as the general frameworks in creating incentives for women. These considerations should be integrated in development projects.

5.2.3.4
Conclusion

Measures on both the supply and demand side can lead to improvements in access to modern forms of energy with low emissions and increase the efficiency of energy consumption in developing, newly industrializing and transition countries.

On the supply side, privatization and liberalization need to be combined with state regulations. Depending on regional specifics, the mix of these three fields will vary. For liberalization and privatization, attractive general frameworks for private investors and the tapping of international sources of capital are important. Governments will have to become more heavily involved in the specification of standards and the expansion of public-private partnerships, supported by bilateral and multilateral development cooperation.

On the demand side, the goal must be to increase purchasing power for energy, especially among poor sections of the population. One way of doing this is to target subsidies to specific groups; another, to expand microfinance systems. Specific cultural and gender issues on the demand side have to be taken into account to ensure that not only purchasing power, but also the willingness to use energy in more sustainable ways will increase.

5.2.4
Related measures in other fields of policy

Energy policy measures have to be accompanied by related measures in other areas of policy. Key areas –

climate policy, transport policy and agricultural policy – are dealt with below.

5.2.4.1
Climate policy

National climate policies, especially those of the industrialized countries, have to support and step up international climate protection. Here, the Council would like to underscore the positive role the German government played both in the 5th national climate protection programme and in climate negotiations for the formulation of the Kyoto Protocol at the EU and international level. This pioneering role is an essential contribution to the further development of the Protocol. In two respects, the WBGU finds Germany's pioneering role to be especially interesting for the future: in international certificate trading models, and in the creation of a CDM standard. The Council refers the reader to the 2002 report of the German Council of Environmental Advisors (Sachverständigenrat für Umweltfragen, SRU) for recommendations concerning national-level measures to mitigate climate change (SRU, 2002).

PIONEERING AMBITIOUS INTERNATIONAL CERTIFICATE TRADING
Germany and the other EU countries could lead the way in designing the planned EU emissions trading directive (Box 5.2-4):
• The German government should develop criteria and strategies aimed at getting other member states to impose equally ambitious reduction targets on their industries as Germany has done. That

---

Box 5.2-4

Planned emissions trading in the EU

In December 2002, the Council of Environment Ministers agreed to introduce European trading of emissions rights starting in 2005. In the beginning, stationary generators of electricity and heat, the iron and steel industries, refineries, the paper and cellulose industry, and the mineral-processing industry (cement, glass, ceramics, etc.) will take part in the trading, with some states being able to opt certain industries out of the trading system up to 2007 or opt some in from 2008 onwards. The trading partners initially receive entitlements at no charge; starting in 2008, up to 10 per cent are to be auctioned according to the wishes of the European Parliament. After the first certificates are issued, trading with CO_2 emission entitlements can begin. In addition to the industrial companies involved, other actors such as non-governmental organizations can take part. To ensure that the transfer of entitlements among the emitters involved is also linked to equivalent adjustments in CO_2

emissions (reduction or increase only in the amount allowed), the proposed directive provides for comprehensive monitoring and reporting duties on the part of member states vis-à-vis the European Commission.

The further consultations in the European Parliament, which has the right of veto, should clarify the other details of the system. According to the Council's resolutions, sanctions in the amount of €40 per tonne of CO_2 (€100 starting in 2008) shall be imposed for any emissions not covered by certificates. The central and eastern European accession countries could be directly included in the trading system through their reduction commitments under the Kyoto Protocol. The Commission also aims to consider linking EU trading with other trading systems and is thinking about ways of expanding the EU system to the entire European Economic Area (EEA). If all EEA states, Canada and Japan can be included, more than three-quarters of the Annex B parties to the Kyoto Protocol would be involved in the trading system. The EU would then have a considerable influence on the ultimate design of international emissions trading in the first Kyoto Protocol commitment period.

should prevent industry playing governments off against each other to lower reduction targets or governments providing hidden subsidies by allocating too many emission rights.

- The absolute level of emissions reduction in the self-imposed declaration by German industry should serve as a starting point for the mandatory reduction goals for industry to be imposed by government. These goals should, however, be made stricter as industry will have a certain degree of flexibility in emissions trading. This is especially true when – as the European Commission is planning – parts of the reduction commitments are to be met via JI or CDM projects outside the EU.
- The Council thus recommends that the German government works towards ensuring that emitters meet at most half of their reduction commitments through international emissions trading with countries outside the EU. This is to ensure that emissions trading does not replace national measures.
- The system should then be expanded to cover the whole European Economic Area. If the EEA states can be included, three-quarters of the Annex B parties in the Kyoto Protocol would be involved in the trading system. The EU would then probably have a considerable influence on the further design of the international emissions trading system in the further Kyoto process.

CREATION OF A 'CDM STANDARD' FOR THE RECOGNITION OF EMISSION CREDITS

In the course of implementing the Kyoto Protocol at the national and EU level, the German government should work for the creation of a Germany-wide and – if possible – EU-wide standard for clean development mechanism projects. Generally, the standard should only permit CDM projects that promote renewable energy (except large hydro due to the unsolved sustainability problems; Section 3.2.3), increase the energy efficiency of existing plants (including those fired with fossil fuels), or concern demand-side management. Applicants for CDM projects promoting fossil energy should prove that the alternative producing the least amount of carbon was chosen, that the project will dovetail with the partner country's long-term sustainable energy planning, and that renewable energy does not represent a feasible alternative in the foreseeable future; financing cannot be the argument.

5.2.4.2
Transport and regional development

In addition to energy policy, national transport and regional development policy plays a crucial role in the transformation of global energy systems towards sustainability. As a detailed discussion of these policies would go beyond the scope of this report, a few approaches shall only be briefly described.

Motorized transport is continuing to grow in industrialized countries. In the past 30 years, the number of cars in the EU has tripled, and growth is expected to continue. This causes major consumption of fossil energy (Section 2.3.1). The EU's transport sector derives 98 per cent of its energy from mineral oil, some 70 per cent of which is imported (EU Commission, 2001c). Road transport causes a considerable amount of toxic emissions. Some 29 per cent of the total CO_2 emissions in the OECD are caused by the transport sector (IEA, 1997). Transport emissions have long been monitored by national and local environmental authorities and regulated by law. This has led to technological improvements (catalytic converters, optimization of engines, etc.), thus reducing emissions of some pollutants (such as lead, SO_2) dramatically (Section 3.7). Such national and local solutions should be stepped up even further and transferred to transition and developing countries to a greater extent.

To move the transport sector into a more sustainable direction, more renewable energy sources and forms of energy low in carbon and pollutants should be used as fuels to the extent possible, and as quickly as possible. To this end, taxation of fossil fuels (ECMT, 2001; Gröger, 2000) and (cross-)subsidies for alternative fuels in the form of price-controls and quotas are recommended. It is important that the market penetration of vehicle technologies – such as fuel cell cars – be promoted in research and development projects and that the necessary infrastructure be established.

Dynamic standards (such as the EU pollutant standards EURO 2, 3, 4) can step up the process of developing efficient technologies to facilitate market access for the new technologies (Johansson and Ahman, 2002). In addition to reducing emissions, incentives should basically be designed to: promote the abatement of pollutants that have local and regional impact, lower noise pollution, reduce the amount of land required for infrastructure, and lower the risk of accidents.

On the demand side, there are also various ways to increase the efficiency of and lower demand for transport services. To this end, programmes should be developed that improve capacity utilization in private car traffic (car sharing, etc.) and public passenger transport. Here, 'multimodality' should be the goal – a combination of the various means of transport: private cars, local public transport, trains and planes (WBCSD, 2001; Royal Commission on Environmental Pollution, 2000). Multimodal nodes – such as ports, railway stations and airports – should be equipped with telematic systems and their infrastructure modernized. In the long term, new regional planning concepts that bring residential areas and business areas closer should complement such measures. Research and social discourse on these concepts has to be further intensified (Section 6.3.3).

In the rail sector, the incremental liberalization in many industrialized countries has not increased the competitiveness of the sector compared to other transport modes to the extent desired. The Council is not of the opinion that the rail sector cannot benefit from liberalization; rather, the poor general setting for the sector was the cause of the failures. For instance, Deutsche Bahn is focusing on faster long-distance connections to the detriment of regional connections due to the one-sided profitability criteria of the privatization process. To remedy this, the Council recommends the following: restructuring the expansion of railway lines to improve coverage of rural areas; gradual reduction of mileage allowances; temporary subsidies for unprofitable lines; and greater competition. Here, a rail system that serves the aims of sustainability should not only compete with planes, but also cars, i.e. taking the train from and to mid-size and small towns should not entail changing trains often or require complicated planning.

In developing, newly industrializing and transition countries, a great increase in transport volume is expected in the next few decades (Sections 2.4 and 2.5). As standards of living rise, the number of cars per person and the general transport volume will grow. Hence, measures have to be taken in the short to mid-term to keep the growth of energy consumption and emissions in the transport sector below these other growth rates. Technological measures alone will not suffice to make transport sustainable; individual conduct will also have to change. Therefore, the transport sectors of developing, newly industrializing and transition countries require a comprehensive strategy. Some core elements could be the following (van Vurren and Bakkes, 1999):
- Maintaining and expanding public short-distance transport systems;
- Promoting rail freight transport;
- Promoting joint ventures and other kinds of collaboration between automobile manufacturers from western Europe and the transition countries to reduce fuel consumption;

- Promoting international research collaboration on energy efficiency and reducing emissions in urban passenger and freight transport (Section 6.3.3);
- Inclusion of external costs in prices for fuels.

The Council recommends that the German government support developing, newly industrializing and transition countries directly or indirectly (via conditionality) in developing and implementing comprehensive strategies aimed at promoting the sustainability of the transport sector.

5.2.4.3
Agriculture

The agriculture sector is responsible for a considerable share of greenhouse gas emissions: some 50 per cent of the methane, 70 per cent of N_2O, and 20 per cent of the CO_2 emissions (IPCC, 2001a). The main source of methane emissions is rice cultivation and the husbandry of ruminant livestock. Some 70 per cent of anthropogenic N_2O comes from fertilized fields (Beauchamp, 1997), even from fields taken out of food production.

Today, it would be relatively easy to reduce methane emissions from rice cultivation; for instance, irrigation strategies could be improved and better adapted rice varieties planted (Bharati et al., 2001). In contrast, reducing methane emissions from ruminating animals is more difficult as CH_4 is a product of the feed used. While these methane emissions could be greatly reduced if feed additives were used (methane oxidants, bacteria that hamper the creation of methane, or an increase in the content of starch/cellulose in the feed), this approach will probably be too expensive for developing countries.

The picture is much the same for reductions of N_2O emissions. In global terms, the use of fertilizers is increasing; only in industrialized nations is the level roughly constant (Scott et al., 2002). Highly technical approaches like 'precision farming' (the use of geographic information systems and global positioning systems to increase efficiency) are available but can only be implemented in industrial countries in the short term. For developing countries, they are too expensive, and there is a lack of local knowledge about such systems.

In conclusion, a sustainable reduction of emissions from the agricultural sector requires both specific infrastructure and specific knowledge, both of which are often lacking. Studies show that the agricultural sector will remain problematic in climate policy terms, even in industrialized countries (Kulshreshtha et al., 2000). In light of increasing food production, a reduction of emissions would require a vast effort.

5.2.4.4
Conclusion

A successful transformation of energy systems towards global sustainability is only realistic if other areas of policy support – or at least do not counteract – the measures taken in the field of energy. Above all, the goal must be to increase energy efficiency, use less fossil and nuclear energy and more renewable energy, and curb greenhouse gas emissions drastically.

Greater sustainability in the transport sector, less emissions from agriculture, and a determined commitment in climate policy are indispensable. These goals can be attained via a great variety of measures. The optimal combination of measures will vary between regions and target groups. Social science research in the areas mentioned can make an important contribution to providing consistency and coherence in the respective policy fields and in the field of energy.

5.3
Actions recommended at the global level

Implementing ecological financial reforms, regulating liberalized energy markets according to sustainability criteria, and promoting energy efficiency, renewables and access to modern forms of energy are all important elements of the WBGU's transformation strategy. It is beneficial, as a matter of principle, to have a great diversity of tools and ways in which they are shaped at country level. Different tools or tool mixes will be appropriate in the various countries depending upon their ecological and geographical, socio-economic, political and cultural settings. Efficiency considerations of location theory also speak in favour of open competition among different approaches.

Nonetheless, despite these arguments for a diversity of tools, there is also a need to act at global level, as already noted in connection with RECS or consumer labelling schemes (Sections 5.2.2.1 and 5.2.2.3). This has several reasons:

1. *Utilizing the benefits of 'rational' harmonization.* Transboundary and global environmental impacts triggered by state-level approaches to energy use, as well as the ever closer economic integration among states, jeopardize the ecological effectiveness and economic efficiency of national-level measures. For instance, levying a national-level CO_2 tax can lead to industries shifting their location, which reduces the ecological effectiveness of this measure (Copeland and Taylor, 2000). As concerns economic efficiency, it would be advantageous if the market-based instruments favoured over the long term by the WBGU – notably emission rights, tradable quotas and green energy certificates – were to be applied beyond national boundaries. This, however, requires compatibility among instruments; in individual cases, explicit harmonization may even be necessary. Moreover, 'rational harmonization' reduces the potential for conflict between the WTO and energy policy, as well as the resultant welfare losses. A further benefit is that of timely adaptation to international climate protection instruments. Within the climate regime, emissions trading presents the parties to the UNFCCC with a framework into which national-level instruments must be fitted. An internationally coordinated approach can do justice to future requirements ex ante and can thus reduce the cost of instrument adaptation that would otherwise be necessary.

2. *Inadequate financial resources at national level.* Many states, especially the poorer developing countries, do not command over sufficient financial resources to meet the initial extra costs of transforming their energy systems. Here the international community is called upon to support these countries financially and technologically – in accordance with the subsidiarity principle and the ability to pay principle – in reconfiguring and building their energy systems.

3. *Inadequate administrative capacities and capabilities.* Many developing countries, but also some newly industrializing and transition countries, lack sufficient state governance capabilities to shape the transformation process according to sustainability criteria. This concerns, for instance, the promotion of renewable energies and of efficiency technologies, the internalization of negative external effects, and the liberalization of energy markets. There is a great need for industrialized countries to provide supportive consultancy inputs in this respect.

4. *Overcoming barriers at national level.* Many domestic companies fear that they will suffer international competitive disadvantages as a result of national measures to curb emissions, internalize external costs and remove subsidies in the energy sector. An internationally coordinated approach reduces this threat and thus contributes to removing barriers. A further aspect is the incentive effect that the internationalization of an instrument can produce. For example, the resistance expressed among many developing countries against greenhouse gas emission caps will be reduced if they are integrated into a system of international emission rights trading and can then generate revenues. Conversely, for such a trading

system to function, the climate protection instruments deployed at national level need a certain degree of harmonization.

The following sections set out the measures required within a global energy policy context – but also climate, economic and development policy context – in order that a transformation of energy systems towards sustainability can succeed.

These recommendations reflect the Council's conviction that transboundary problems necessitate both international cooperation and a legal safeguarding and deepening of such cooperation through multilateral regimes. In the opinion of the WBGU, a multilateral approach at global level is the key to transforming energy systems.

5.3.1
Expansion of international structures for research and advice in the energy sphere

The worldwide transformation of energy systems towards sustainability requires a significant and swift intensification of research efforts (Section 5.2). In the view of the WBGU, the tasks of energy research institutions are above all:

1. *Assessment:* Analysing global energy trends and identifying options for action;
2. *Coordination:* Promoting the formation of networks and of complementary cooperative arrangements, and coordinating initiatives and organizations;
3. *Implementation:* Carrying out and financing research projects.

While Chapter 6 sets out the substantive focuses that future national and international research projects will need to pursue, the following section explores, proceeding from the above three functions, the institutional architecture of energy research at global level.

ASSESSMENT
In its previous reports, the Council has frequently noted the importance of independent scientific policy advice for global sustainability policy (WBGU, 1997a, 2000, 2001a). To be able to identify and resolve problems in situations frequently characterized by 'action under uncertainty', it is essential that scientific analyses are conducted at regular intervals, and that these are presented in a manner relevant to policy processes. Systematic dissemination of scientific findings and options for action creates a basis upon which policy governance bodies can adopt precautionary strategies and adapt existing strategies to new needs. The Intergovernmental Panel on Climate Change (IPCC), with its assessment reports on cli-

mate change, provides a model. The IPCC process shows how, through broad-based international participation of researchers, a widely recognized scientific basis for climate policy decisions can be built. In order to keep the influence of political interests upon the outcomes of scientific assessment processes as small as possible, it is essential to safeguard the independence of such advisory bodies.

As a classic cross-cutting theme, energy policy touches upon many policy areas, such as environmental, development, economic, trade and transport policy. The WBGU therefore wishes to stress that, when developing goals, strategies and instruments for promoting energy system transformation, it is essential to analyse systematically the interactions of energy policy with other policy arenas and to take these into consideration in recommendations for action. In particular, care needs to be taken to ensure the coherence of policy measures. A comprehensive stocktaking of the status and trends of global energy policy should proceed along the lines of the goals and content of a World Energy Charter that yet needs to be developed (Section 5.3.2.2). This is essential in order to be able to assess progress made in the process of transforming global energy systems. In this connection, the following issues need to be addressed in particular:

• Compilation and presentation of trends in primary energy use, together with statements on the range, energy productivity and shares of final consumption sectors in total primary energy use, presented wherever possible at a worldwide level and broken down according to regions;
• Development and continuous updating of energy scenarios;
• Detailed quantification of the sustainable potentials of renewable energy sources;
• Analysis of the environmental, social and health impacts of present energy systems at the local, regional and global levels;
• Presentation of developments relating to the basic supply of commercial energy services in developing and newly industrializing countries;
• Processing and presentation of policy-relevant data such as the levels of direct and indirect energy subsidies (worldwide and broken down according to regions or countries) and the levels of research and development expenditure (in the private and public sectors);
• Examination of the geopolitical aspects of globally sustainable energy systems;
• Presentation of successful best-practice national strategies to promote energy system transformation towards sustainability.

The World Energy Assessment published in 2000, involving the UNDP, UNDESA and the World

Energy Council (WEC), provides orientation for a way to institutionalize such a comprehensive stock-taking of global energy policy. The goal should be to report regularly – at least every 5 years – on the successes and deficits in implementing globally sustainable energy systems. The Council therefore recommends to the German federal government that it urges a continuation of the World Energy Assessment process, possibly in the form of 'assessment reports on global energy policy'. In view of the positive experience with the IPCC, the Council recommends establishing for this purpose an Intergovernmental Panel on Sustainable Energy (IPSE). This should involve maximum regional representation, whereby, in analogy to the IPCC, the participation of scientists from developing countries should be promoted by means of targeted support. Over the long term, a newly created International Sustainable Energy Agency (ISEA; Section 5.3.2.3) could assume responsibility for the World Energy Assessment process.

COORDINATION

To derive synergies from national energy research, it is recommendable to coordinate existing initiatives and programmes. The aim should not be to create a coordinating body seeking to establish international division of labour – for energy research profits from the competition between different research initiatives. However much cooperation and consultation there may be at global level, the subsidiarity principle should always be observed, and competition among national research bodies fostered. Thus, to coordinate energy research at global level, rather a worldwide networking of research centres should be promoted. Following the model of the World Climate Research Programme established in 1980, the Council recommends the establishment of a corresponding World Energy Research Coordination Programme (WERCP) under UN auspices. This programme could be managed scientifically by a joint scientific committee, representing relevant disciplines of the natural, engineering and social sciences.

IMPLEMENTATION

The Council argues in favour of pluralism and diversity in the international research landscape, on both the implementing and financing sides. Consequently the WERCP should not have a mandate to implement and finance international research projects. Its task should rather be to harness national-level grant funding for international research projects. This means the programme could serve as a clearinghouse for information on research funding worldwide. This would initiate competition among national research institutions for the award of grants. All research insti-

tutions coming potentially into question could thus be motivated to collaborate in solving priority research questions in the energy sphere. A prime goal in this context should be to stimulate transboundary cooperation.

For energy research and for the transfer of research findings into practice alike, a cooperation arrangement among national research promotion institutions modelled e.g. on the European Science Foundation (ESF) would be worth considering.

5.3.2
Institutionalizing global energy policy

At present, global energy policy institutions are highly fragmented (Section 2.7), characterised by duplicated work, overlaps and conflicting developments. The WBGU recommends that the German federal government work pro-actively for more energy policy coherence. To this end, the coordination of individual processes and actors is urgently required.

The link between energy policy and development/environmental issues has also been neglected at global level so far. There is an urgent need for more effective integration of sustainable energy policy into the international institutional architecture. The development aspect in particular means that the range of tasks must not be defined too narrowly here.

In a resolution as early as November 1990, the UN General Assembly noted with concern that the programme of action for the development and utilization of new and renewable energies was being implemented too slowly, thus failing to meet the pressing needs of the developing countries. It underlined the need for sustained commitment and action by the international community and called on Member States to consider further measures to promote new and renewable energies, including the establishment of an international institution. Yet despite repeated calls by the UN, the institutional weakness of global energy policy and the lack of coordination between the relevant actors were not addressed at the UN Summits in Rio de Janeiro in 1992 or Johannesburg in 2002. The German Advisory Council on Global Change (WBGU) has already criticized the poor institutionalization of global sustainable development issues and has proposed various options to remedy the situation (WBGU, 2001b).

5.3.2.1
The functions of international institutions

The first step in strengthening the institutional architecture is to identify the functions which are necessary for energy system transformation. They can be divided into groups, which are listed below in ascending order as the transfer of energy policy competences to global level increases:

1. *Advisory function: Providing continuous analysis and options for action by the scientific community*
   - Defining research gaps; initiating and coordinating research; establishing research networks;
   - Assesing the global situation: Status and trends in energy transformation (e.g. follow-up to the World Energy Assessment), organizing data collection and statistics;
   - Developing global scenarios for a sustainable energy future;
   - Devising options for action and strategies for energy system transformation.
2. *Clearing-house function: Organizing transfer of information and technology*
   - Establishing a clearing-house for sustainable energy systems; collating and circulating information;
   - Organizing non-commercial technology transfer; evaluating applied technologies and best practices.
3. *Coordination function: Coordinating activities between the international institutions and boosting cooperation; coordinating national transformation policies*
   - Ensuring a clear division of responsibilities between organizations and encouraging cooperation;
   - Through the approximation of national energy policy instruments, reducing the worldwide costs of transformation, dismantling economic distortions of competition, and anticipating future requirements (e.g. under the Kyoto Protocol).
4. *Implementation function: Implementing sustainable energy policy at national and regional level*
   - Providing strategic advice for governments: Devising and implementing (with support from partners) national programmes for energy system transformation and/or a sustainable energy infrastructure;
   - Promoting capacity-building in developing countries: Training and development (for civil servants, technicians, craftspersons, small and medium-sized enterprises, etc.), distribution of information, provision of advice and assistance with funding options;
   - Establishing regional research&development and centres of excellence.
5. *Management function: Establishing instrument platforms*
   - Providing management capacity, e.g. for the organization of a global trading scheme for Green Energy Certificates or similar global or regional instruments;
   - Involvement in defining environmental and social standards for the energy sector.
6. *Financing function*
   - See Section 5.3.3

These functions can either be integrated into existing institutions or carried out by newly established agencies. In principle, the Council is in favour of establishing a new global organization, as this would end the current fragmentation of energy policy activities. However, equipping a new institution with competencies and responsibilities for policy development would entail a loss of sovereignty for the nation-states, so that resistance to its establishment can be expected. The World Summit on Sustainable Development (WSSD) showed that many countries' willingness to commit to real targets and measures is very low. The WBGU therefore views a step-by-step approach as realistic: A new institution should not be demanded from the outset. Instead, existing and new initiatives should form a 'core' which should be integrated in the UN system and could be expanded further if necessary.

5.3.2.2
Developing a World Energy Charter

As a common basis for the work of the relevant institutions, the Council initially recommends the development of a global energy strategy which should be negotiated at international level without any binding legal force and which could take the form of a World Energy Charter, for example. It should set out the elements of global energy policy, including verifiable targets and time frames (Box 5.3-1). The Council has already proposed the launch of a World Energy Charter in its policy paper for the WSSD (WBGU, 2001c).

The option of its further development into a legally binding energy convention over the long term should not be ruled out, but should not be pursued as the primary objective. Implementing the World Energy Charter would remain the task of the individual states and international institutions operating in the relevant sectors.

Developing a World Energy Charter will only hold out any prospect of success if the current opposition to more intensive utilization of sustainable energies

Box 5.3-1

Elements of a World Energy Charter

1. *Protecting natural life-support systems*
   - Substantially reducing greenhouse gas emissions,
   - Removing subsidies on fossil energy carriers and nuclear power in the medium term,
   - Substantially increasing energy efficiency,
   - Significantly expanding renewables,
   - Phasing out nuclear power.

2. *Safeguarding access to modern forms of energy worldwide*
   - Attaining a minimum level of supply worldwide,

   - Targeting international cooperation towards sustainable development,
   - Mobilizing financial resources for global energy system transformation,
   - Enhancing the Least Developed Countries' capabilities,
   - Utilizing pilot projects as a strategic lever, and forging energy partnerships.

3. *Pressing ahead with targeted research and development*

4. *Pooling and strengthening global energy policy at institutional level*
   - Improving policy advice at international level,
   - Creating a coordination body.

can be dismantled successfully from the outset. Among the industrialized states, it is primarily the present governments of the USA and Australia who are opposed to global energy system transformation and wish to safeguard their energy supply by expanding fossil energy. Although the international community will probably have to develop a global energy strategy without any involvement by these countries initially, it is likely that the pace of technological development in the renewables sector and their long-term superiority over fossil development paths will ultimately lead to a shift in policy. In the long term, a sustainable energy policy will not be viable unless these two countries are involved.

The newly industrializing and developing countries are also sceptical about the efforts to promote renewable energy. They are concerned that this is an attempt to deny them opportunities for cost-effective development. To ensure that the developing countries are also involved in the development of the World Energy Charter, these concerns must be the starting point for the strategy. That means persuading the developing countries that a mix of sustainable energies and greater efficiency of fossil technologies is a future-proof development path, and that they can count on international support if they pursue this route.

The World Energy Charter could and should be an important outcome of the International Conference for Renewable Energies in Germany in 2004, which was announced by the German federal government at the WSSD.

5.3.2.3
Towards an International Sustainable Energy Agency

The existing institutional architecture should be concentrated and reinforced on the basis of the World Energy Charter (Fig. 5.3-1): Starting with the functions defined above, the first task is to focus and coordinate the work of existing organizations (Phase 1). Then, on this basis, the institutional foundations for a global energy policy should be enhanced through the pooling and strengthening of competencies (Phase 2). Finally, if necessary, the option of establishing a new overarching institution (International Sustainable Energy Agency – ISEA) should be explored (Phase 3).

PHASE 1: FOCUSSING AND COORDINATING THE FUNCTIONS OF EXISTING INSTITUTIONS

Boosting research and advice
Institutionalizing independent scientific advice on global energy policy and improving coordination between national research institutes and international initiatives were discussed in Section 5.3.1.

Organizing the transfer of information and technology
To support the dissemination and utilization of sustainable energy technologies at regional and national level, it is important to press ahead with the global networking of research, development and transfer centres. With the German federal government's support, UNEP has already laid the foundation stone for this process with the Global Network on Energy for Sustainable Development, which was launched at WSSD. In order to be able to perform this function satisfactorily, UNEP should be expanded and equipped with better financial resources.

The existing network of national energy agencies such as the German Energy Agency (DENA), which – as a support network for institutions with international and global responsibility – could drive forward the regional implementation of global targets, should also be supported.

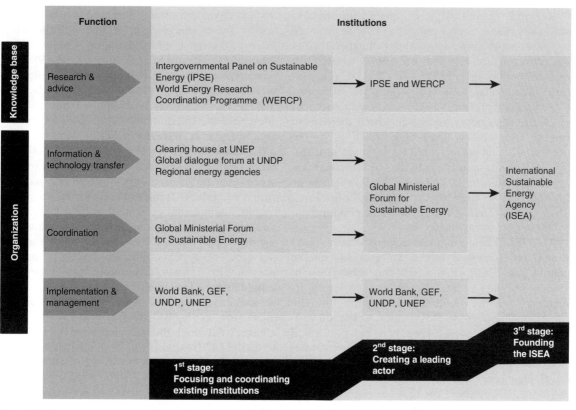

Figure 5.3-1
The path towards an International Sustainable Energy Agency.
Source: WBGU

Whereas other authors recommend generally limiting the institutionalization of a global energy strategy to establishing and reinforcing networks (Fritsche and Matthes, 2002), the Council regards a 'network solution' as inadequate in the energy policy field. Experience from other sectors indicates that countries would be inclined to absolve themselves of responsibility for the developing and financing of global sustainable energy policy by pointing out that they are involved in promoting networks. For individual aspects of global energy strategy, such as the above-mentioned exchange and dissemination of technologies, networks can be very useful. However, in the WBGU's view, they must be augmented by institutions which have the power to shape policy in a substantive way.

In the interests of sharing information, and disseminating 'best practices' and 'clean' technologies, establishing a central liaison office is a sensible approach. A clearing-house of this kind, with its own secretariat, could be integrated into an existing UN organization. UNEP, with its Division of Technology, Industry, and Economics, which is responsible, among other things, for promoting clean energy technologies, would be appropriate to take on this role.

However, UNEP could only deal effectively with this new task if its human and financial resources were substantially increased.

Finally, in order to promote international exchange and cooperation in the energy sector, a global dialogue forum should also be considered. Stakeholders from the public and private sectors would thus have the opportunity to exchange views on targets, mechanisms and new partnerships. It could be modelled on the forums held in advance of the WSSD, such as the UNDP Global Round Table on Energy for Sustainable Development and the Multi-Stakeholder Round Table on Energy for Sustainable Development, set up by the Tata Energy Research Institute (TERI) in India. In order to safeguard the global focus and the involvement of the developing countries, it would be advisable to establish the dialogue forum in a sub-organization of the UN. Ideally, this would be UNDP, as it could draw on its previous experience in organizing this type of forum. It could be launched at the 2004 International Conference for Renewable Energies announced by the German federal government. The forum could then be held every two years.

Regional, as well as global, energy forums also play a key role in energy policy. In order to promote regional cooperation and exchange of experience in the energy sector, it would be helpful to establish regional energy agencies akin to OLADE (Organización Latinoamericana de Energía). The European Commission is pressing ahead with the establishment of a European Energy Agency. In Africa, the African Energy Commission (AFREC), which is in the process of being set up, could form the core of this type of organization. The activities of these regional agencies should be in line with the World Energy Charter and be coordinated at international level. Their names should clearly reflect the commitment to sustainable energy policy.

The OECD's International Energy Agency (IEA) should be developed further towards sustainable energy policy. The IEA, which was established in 1974 in response to the oil crisis, aims to reduce its members' reliance on oil imports by promoting the use of fossil and renewable energies. As the industrialized countries' energy agency, it has performed valuable work in evaluating and processing information about national trends and instruments as well as developing best practices. Although established by the industrialized states, the IEA has already opened up to the developing, newly industrializing and transition countries in recent years, by establishing a Non-Member-Countries Department and holding dialogue forums. Furthermore, the focus of its work has also changed: Due to the desire to diversify their energy carriers in the interests of security of supply and also as a result of climate change, its members' interest in renewables has increased. The Council considers that this positive development must be pursued further.

Improving coordination between countries and institutions

The positive experiences with the Global Ministerial Environment Forum suggest that it would be useful to set up a Global Ministerial Forum for Sustainable Energy. Setting a new energy policy course can only be successful if it is underpinned by the necessary political leadership and support from the national governments. The Global Ministerial Forum for Sustainable Energy, which should meet at regular intervals and have its own small secretariat, would provide advice on the direction to be taken in the work of the various UN units and the World Bank in the energy sector and draft recommendations for the institutions. However, the World Bank, UNDP, UNEP and other relevant actors, depending on the expertise required, should continue to be responsible for coordinating projects in the field. The Forum should also be responsible for coordinating, monitor-

ing and developing the progressive institutionalization of global energy policy in line with the World Energy Charter. The World Energy Charter should safeguard the inclusion of environmental and development policy goals, e.g. by stating that the objectives set forth in the UNFCCC must be upheld in decision-making. The decision to establish the Forum could be adopted at the International Conference for Renewable Energies in Germany in 2004.

In addition, a group of like-minded states could act as pioneers in setting a course towards a sustainable energy policy. In theory, this could be achieved within the OECD framework. However, this would require fundamental agreement with the USA, Japan and Australia on the need to transform the energy sector towards renewables by utilizing bridging technologies. This appears to be an unlikely prospect at present. The EU would therefore be a more likely candidate for a leadership role. It is easier to achieve a common approach in highly integrated economic areas such as the EU than in other international alliances. Although the inclusion of an energy chapter in the EU Treaties is still viewed with restraint by Member States, the dynamics of the internal market are gradually leading to a shift in thinking. What's more, the EU is already playing a pro-active role in energy system transformation: The Commission's efforts to include environmental aspects to a greater extent in the Community's energy policy, its planned emissions trading system, relatively ambitious Kyoto commitments, and progressive policy approaches in the majority of its fifteen Member States make the EU a key player in energy system transformation.

The EU could also take on a leading role in a larger group of countries: At the WSSD, it launched a coalition of like-minded countries and regions committed to increasing their use of renewable energies through quantified targets. Numerous developing countries joined this initiative. Although it remains to be seen how the coalition will perform in practice, this process could certainly be used as a starting point.

Strengthening implementation and management functions

Management or implementation functions within the World Energy Charter framework are virtually impossible to perform without support and monitoring by a UN organization. In this context, the UN Secretary-General and UNDESA propose transforming UNESCO's Solar Programme into a 'World Sustainable Energy Programme'. However, as an organization, UNESCO deals with education, science and communications, and would thus seem ill-equipped for the tasks of formulating and implementing sustainable energy policies and strategies at national and regional level.

The World Bank and UNDP or UNEP would be more appropriate institutions to perform these tasks. Through their expertise and competencies in the policy development field, they are ideally suited to advising governments, implementing national programmes and undertaking capacity-building. As well as offering loans, the World Bank is increasingly providing policy advice, technical support and knowledge transfer. Furthermore, with its 180 members, it has considerable powers of enforcement. In the energy sector, however, it is exposed to various conflicts of interest. For example, its energy supply priorities are distinctly at odds with efforts to achieve an effective climate policy. UNDP admittedly has to grapple with its reputation as a poor performer, as well as with stagnating contributions and increasing competition from the World Bank, but it continues to be the central funding, coordination and steering body for the UN's development operations and is greatly trusted by the developing countries. UNEP could draw on its substantial environmental policy expertise, but its work is also hampered at present by under-funding and staff shortages. For the effective implementation of the tasks set out in a World Energy Charter, a clear and balanced division of competencies and coordinated cooperation between UNDP, UNEP and the World Bank should be defined. Instead of UNEP, the international environmental organization also demanded by the Council could take on appropriate implementation tasks in the long term (WBGU, 2001b).

Management functions at international level are also necessary for another key aspect of the desired energy system transformation: The phasing out of nuclear energy worldwide. The most appropriate organization to take on this responsibility is the International Atomic Energy Agency (IAEA). This organization, which currently has a remit to promote and monitor the civilian use of nuclear energy, is unlikely to undergo a swift or rapid transformation into an agency responsible for phasing out nuclear power. Nonetheless, the WBGU recommends working towards a change in the IAEA Statute, with the aim of removing all references to the specific objective of promoting the further expansion of nuclear energy. In the medium term, the IAEA should monitor and coordinate the winding down of the nuclear power industry worldwide. If the international community decides, en masse, to phase out nuclear power, the IAEA will continue to be indispensable to monitor fuel cycles, prevent proliferation of existing fissile material, and secure any nuclear plants which continue in operation during a transitional phase, along with final storage facilities.

PHASE 2: POOLING AND REINFORCING COMPETENCIES AT GLOBAL LEVEL

In a further step towards the institutionalization of global sustainable energy policy, the responsibilities and competencies of the Global Ministerial Environment Forum should be expanded. This requires an increase in the personnel and financial resources of its secretariat. As well as coordinating the relevant actors, the Forum should also promote and monitor the implementation of the World Energy Charter. For this purpose, it should review global developments in light of the Charter's objectives and with reference to new scientific findings, monitor the effectiveness of national activities in the energy sector, and make policy recommendations. As well as this steering and policy advice function, it could also increasingly take on tasks relating to the implementation of the World Energy Charter's objectives and technology transfer, e.g. supporting the developing countries in setting up research and development centres or providing training and development.

PHASE 3: SETTING UP AN INTERNATIONAL SUSTAINABLE ENERGY AGENCY

If it becomes apparent that even with the stronger institutional architecture resulting from Phase 2, the above-mentioned tasks cannot be managed satisfactorily at global level, the option of setting up an organization with a specific remit for sustainable global energy policy should be considered.

The establishment of an international organization responsible for renewable energies has been demanded, primarily by NGOs and producers of renewable energy conversion technology, for many years. However, two difficulties arise in this context:

- The WSSD showed that establishing a new international organization for renewable energies at the present time would be very difficult to achieve. It became apparent that the developing countries view this new institution very critically, regarding it as a vehicle for the promotion of exports from north to south, while the majority of industrialized countries are keen to avoid any additional financial and administrative burdens.
- The Council considers that a new organization to promote the use of renewables could play an important role in promoting sustainable forms of energy and the relevant industries. However, it would not be a suitable vehicle to push forward the global transformation of energy systems. For this purpose, a new institution, which could include the full spectrum of energy systems in the reform process, must be established.

The WBGU therefore recommends the establishment of a International Sustainable Energy Agency

(ISEA), whose remit would be based on the World Energy Charter.

This new agency could develop from the Ministerial Forum, mentioned above, which could then act as a steering committee. The advantage of this approach is that the ISEA would thus secure the necessary political leadership and support from the national governments. In setting up the ISEA, it should be clear that as well as environmental aspects, development policy issues are also a priority. This would encourage the developing countries to endorse the project.

Opposition to this way of institutionalizing global energy policy is likely to come from a number of industrialized countries, primarily the USA, Japan and Australia. Tactically speaking, it is therefore important for the interested industrialized countries (EU, etc.) and the developing countries to forge alliances from the outset. Whether or not countries support the proposal to set up an ISEA will depend largely on the perceived 'value added' offered by this new institution. The ISEA's great advantage is that it would offer the opportunity to draw together energy, environmental and development issues at global and institutional level for the first time. Based on the World Energy Charter, the new body could be successful in encouraging energy, environment and development ministers to commit themselves to a coordinated energy policy. In the long term, the global energy research undertaken by the Intergovernmental Panel on Sustainable Energy (IPSE) and the World Energy Research Coordination Programme (WERCP) could also be incorporated into the ISEA (Section 5.3.1). On environmentally relevant aspects of energy policy, the ISEA should work closely with the International Environmental Organization proposed by the Council (WBGU, 2001b).

5.3.3
Funding global energy system transformation

5.3.3.1
Principles for equitable and efficient funding of global energy policy

The starting point for the following hypotheses is the Council's recommendations on the financing of global sustainability policy ('Earth Funding'; WBGU, 2001b). In specific terms, the task is to develop a funding system for the transformation of global energy systems which responds effectively to two key challenges:

- Supplying the resources to cover the financial requirement;

- Creating transfer mechanisms to support economically weaker countries during the transformation process.

Devising an equitable and economically efficient funding system to develop a sustainable energy supply is an ambitious project, and a long-term implementation plan is essential. In the Council's view, it should be guided primarily by the subsidiarity principle. The subsidiarity principle relates, firstly, to the division of functions and competences in the public sector, and, secondly, restricts the state's roles to those tasks which, from a governance perspective, exceed the capacities of the private sector. Based on these twin functions, two significant conclusions can be drawn for the funding of global energy policy:

The private sector invests substantially in the energy industry. The states' task is to provide additional funds for those investments which, from a narrower micro-economic perspective, are not (yet) profitable but which contribute to developing a sustainable global energy system. In line with the subsidiarity principle, the international community should only bear the share of investment costs which delivers global benefits ('incremental costs'). In order to minimize the level of these incremental costs, existing disincentives (especially subsidies on non-sustainable and other market-distorting rules) must be dismantled and new incentives created for investment in energy system transformation. This can often take place without funds being deployed, and so improving the institutional framework at national and international level is seen as a major contribution to reducing the requirement for public funding and financial transfers.

In line with the subsidiarity principle, national sources of capital to fund the transformation should be deployed first. However, the economic capacity of states and hence their ability to finance the transformation of the global energy systems as a global public good vary significantly. Left to themselves, the poorest countries will be unable to restructure their energy systems according to sustainability criteria. International transfers are therefore required.

The financial resources which cannot be provided either by the market or the developing country itself must be supplied via official development assistance (ODA) or from other international sources. The Council considers that the process governing the delivery of these funds should be as equitable as possible. It identifies three principles as viable bases of an equitable funding system:
- The ability to pay principle,
- the benefit principle,
- the polluter pays principle.

In line with the ability to pay principle, charges are levied in accordance with taxpayers' individual contributive capacity. In a national context, this is generally assessed according to the individual's income and assets. There has also been a long debate about the extent to which consumption is an indicator of economic performance. At international level, it is national rather than individual economic performance which is the focus of interest. The ability to pay principle is currently the funding principle which is generally applied by international organizations. Per capita GDP is most commonly used as an indicator in this context.

'Benefit' is another recognized funding principle. This states that taxes should be based on the benefits received by people using the good financed with the tax. Individual contribution levels are based on the individual benefit received (benefit equivalence). However, as the individual benefit received from public goods is frequently difficult to measure or identify, the 'cost equivalence' principle is often used instead. Here, the individual's financial contribution is based on the costs arising from his or her use of the public good. The benefit principle results in a careful weighing up of the costs and benefits of services and encourages efficient service delivery. The benefit principle is rarely applied at international level, but it does flow into the financial formula used by a number of organizations (e.g. the WTO, Intergovernmental Organisation for International Carriage by Rail).

The polluter pays principle is of key importance to environmental policy and, in the WBGU's view, should play a central role in funding energy system transformation as well. The application of the polluter pays principle means that the person whose behaviour results in a requirement for the delivery of a publicly funded service must pay the ensuing costs. The lines between the polluter pays and the benefit principle are blurred, however.

The Council also bases its recommendations on the following criteria which must be considered when developing a funding system. They are: Target and system conformity, stability and reliability of revenue generation, political viability, and technical practicability. The Council attaches particular importance to political viability. Proposals for institutional reform should link in with the status quo, the prevailing discourses, and the interests of key actors so that endogenous forces can develop in support of the transformation. In other words, the WBGU is pursuing an incremental approach which focuses on initial pragmatic steps towards a funding system for global energy system transformation.

5.3.3.2
Provision of new and additional funding

REQUIREMENT FOR INTERNATIONAL FUNDING
Over the next 20 years, annual investment of US$$_{1998}$ 180–215 thousand million will need to flow into the developing and newly industrializing countries' energy sector in order to meet the growing energy demand (WEC, 2000; G8 Renewable Energy Task Force, 2001). A higher figure is required if highly efficient technologies and renewable energies are to be deployed (Ad Hoc Open-ended Intergovernmental Group of Experts on Energy and Sustainable Development, 2001). It is unclear which proportion of the required investment will, or should, come from the public purse or the private sector and from national governments or the international community. In the interests of efficiency, it is desirable for a substantial share of this investment to be funded by the private sector. Moreover, in view of the low GDP and extreme scarcity of capital in most developing countries, it is almost inevitable that a substantial amount will be provided by the international community.

MOBILIZING PRIVATE CAPITAL
The investment needs make it clear that private capital will be essential to fund energy system transformation. The last decade of the 20th century saw a dramatic rise in the amount of private direct investment flowing into the developing countries (Section 2.7.3). According to the World Bank, from 1990 to 1999, 733 energy projects were carried out with private-sector participation, with a total investment volume of US$186.7 thousand million, especially in Latin America and East Asia (Izaguirre, 2000). By integrating the private sector more fully into the work of the multilateral development banks, it has been possible to increase project volumes many times over. Furthermore, private-sector actors have an interest in long-term returns on their investment. They utilize their expertise and work pro-actively to improve economic efficiency. These are key factors in ensuring the success of projects to restructure the energy supply in the developing, newly industrializing and transition countries (G8 Renewable Energy Task Force, 2001; Enquete Commission, 2002).

Nonetheless, much more far-reaching efforts are required to secure adequate foreign private capital for direct investment in the energy sector, especially since only a handful of countries have benefited from significant levels of investment to date. The first step is to improve the framework conditions for a stable economic and monetary system at both national and international level, since avoiding economic and monetary crises is a key prerequisite in attracting direct investment. A priority at national level is to establish legal certainty, including the guarantee that contractual arrangements can actually be enforced. The liberalization of the energy markets contributes significantly to enhancing transparency and market access in the developing countries as well. Finally, the removal of subsidies on fossil energy carriers is also important in mobilizing private capital to develop a sustainable energy supply (Enquete Commission, 2002; Section 5.2.1.2).

Despite the importance of private capital, the extent to which private investment genuinely contributes to achieving energy system transformation targets must be critically examined. Until now, the main focus has been on large fossil-fuelled power stations. Renewables accounted for just 2 per cent of private direct investment in energy projects in developing countries during the 1990s (KfW, 2001). The German federal government should utilize the opportunities to restructure this trend:

- By reforming export guarantees in order to give targeted preference to project categories which meet sustainability criteria (Section 5.2.3.2);
- by increasing policy advice within the development cooperation framework, in order to boost partner countries' capacities to create conditions which are conducive to investment;
- by creating a 'public-private partnership for renewable energies' funding instrument, particularly to facilitate market access for small and medium enterprises which supply renewables in developing countries;
- by supporting the development of small grants schemes in developing countries in order to improve funding for smaller projects which focus on improving efficiency and renewables use (Section 5.3.3.3).

Additional private funds can also be generated through more intensive promotion of foundations. For example, the Ford and Rockefeller Foundations provide funding for energy policy projects which have social and environmental benefits (CENR, 2000). In Europe, there are fewer tax incentives to support the work of foundations than in the USA. There is scope to encourage the mobilization of private capital here (World Bank, 2002b).

The Council considers that an international financial requirement of several hundreds of thousand million US dollars per year during the first twenty years is plausible. Based on the current total foreign investment volume of US$205 thousand million annually in developing countries (, 2002) and a further US$7 thousand million or so, at the most, supplied for international development cooperation by the non-profit organizations (OECD, 2002), it is unlikely that private sources will be sufficient to fund energy system transformation. Indeed, private investors cannot be expected to fund in full the proportion of investment required to create external global benefits. An increase in direct public transfers to the energy sector is therefore essential.

INCREASING AND RESTRUCTURING THE RESOURCES FOR DEVELOPMENT COOPERATION

The regional priorities for private investment are currently East Asia and Latin America. By contrast, most African and South Asian countries have little appeal for private investors. These regions include the poorest developing countries with very low income levels, poor purchasing power and extreme capital poverty. Establishing a sustainable energy system here will be impossible without the support of development cooperation. In the Council's view, development policy can act as a catalyst for energy system transformation. Through the development policy framework, influence can be brought to bear on the key policy areas of relevance to energy system transformation towards sustainability: promoting efficiency and renewables, and improving access to modern energy services. However, the existing financial resources allocated to development cooperation fall a long way short of what is required to establish globally sustainable energy systems, especially in the poorest developing countries. The volume of ODA in 2000 totalled just US$54 thousand million. It is estimated that 3.1 per cent of bilateral ODA (which totalled around US$36 thousand million in 2000), i.e. just US$1.2 thousand million, flowed into the energy sector (OECD, 2002). For multilateral development funding, the figure rose to around 8 per cent due to the regional development banks' strong commitment to the energy sector (OECD, 2002). The World Bank Group, as the largest multilateral financing institution, provided just US$2.2 thousand million for environmentally relevant investment in the energy sector in 2001 (World Bank, 2002c). These figures graphically illustrate how much extra investment is required.

In its reports, the WBGU has repeatedly criticized the industrialized countries' failure, so far, to honour their pledge, announced as early as 1992 at UNCED, to boost their financial support for the developing countries significantly. In 2000, development cooperation amounted to just 0.22 per cent of the OECD states' GDP, falling well short of the internationally agreed target of 0.7 per cent. Furthermore, the proportion of development cooperation budgets allocated to supporting renewables is decreasing. To some extent, this decrease can be offset by the rise in private investment. Nonetheless, the need for programmatic support for the developing countries in restructuring their energy supplies towards more efficient and renewable technologies remains substantial (G8 Renewable Energy Task Force, 2001).

The Council regards the outcomes of the International Conference on Financing for Development (Monterrey 2002) and the USA's and EU's pledge, on that occasion, to increase their development budgets as an initial step. Yet even if these increases do indeed produce an additional US$12 thousand million for development cooperation in 2006, these commitments can only be a start. They would boost the EU's development budget to just 0.39 per cent of GDP – still a long way short of the 0.7 per cent target. At Monterrey, Germany committed itself to increasing its ODA to 0.33 per cent of GDP by 2006. The WBGU urges the German federal government to implement a far more substantial increase in its ODA funding beyond the 0.33 per cent announced for 2006 and recommends that by 2010, at least 0.5 per cent of GDP be allocated to ODA. Indeed, given the pressure of the problems faced, an increase to 1 per cent of GDP would be appropriate.

Moreover, the priorities governing the use of the funds should be changed. In this context, the Council welcomes the 'Sustainable Energy for Development' programme announced by the federal government at the WSSD, which aims to establish strategic energy partnerships. Over the next five years, a total of €1 thousand million will be provided for this purpose, comprising €500 million for renewables and €500 million to boost energy efficiency. However, in the WBGU's view, an annual figure of €100 million to promote the use of renewables is far too low a proportion of the total budget of the Federal Ministry for Economic Co-operation and Development (BMZ), which stands at around €3.8 thousand million (2001). Here, the Council recommends a significant increase in the share of funds allocated to sustainable energy projects within the ODA commitments, where the percentage has been far too low until now, amounting to 6.8 per cent (€282 million) in 1999 and 3.3 per cent (€105 million) in 2000. Giving higher priority to energy in the development cooperation context does not necessarily conflict with the Millennium Development Goals; on the contrary, funding sustainable energy policy can form a key element of a coherent poverty reduction strategy.

CREATING THE FINANCIAL SCOPE FOR ENERGY SYSTEM TRANSFORMATION IN DEVELOPING COUNTRIES THROUGH DEBT RELIEF

Debt relief for developing countries does much to establish the preconditions necessary for energy system transformation. The Council recommends the launch of new debt relief initiatives. The Heavily Indebted Poor Countries Initiative (HIPC Initiative) – the debt relief programme initiated by Germany at the G7 Summit in Cologne – plays a key role in shaping sustainable energy policy as it improves the framework conditions in the poorest developing countries. The total volume of debt relief for the poorest developing countries amounts to US$70 thousand million. The highly innovative aspect of this initiative is that debt relief is coupled to verifiable poverty reduction programmes, which include energy sector projects. The opportunity to reduce the debts of the poorest developing countries could be expanded substantially as part of this initiative.

INTRODUCING AN EMISSIONS-BASED USER CHARGE ON AVIATION

The advantage of directly funding global energy policy from official development assistance (ODA) is that the financial resources consist of allocations from the industrialized nations' budgets and are thus subject to regular parliamentary control. This encourages the efficient use of resources. However, a disadvantage is that the funding for ODA is provided voluntarily, which creates incentives to 'free-ride'. When budgets are tight, these resources are vulnerable to cuts. The Council therefore recommends that the funding of global energy policy be spread across many different mechanisms and programmes in order to ensure that funds flow as continuously as possible. In this context, innovative funding mechanisms must also be scrutinized in terms of their applicability.

Due to their ecological steering effects and funding implications, the Council is in favour of levying charges for the use of global public goods. It has explored this concept in detail in a special report and proposed a number of policy recommendations (WGBU, 2002). The 'user charge' concept is based on the benefit and polluter pays principles. It is therefore an improvement on many other international levy-based solutions. It establishes a direct link between the payment of the charge and the service being funded, thus highlighting the scarcity of environmental goods. This has a positive impact on the efficient use of environmental resources. Furthermore, the ringfenced use of the ensuing funds enhances political viability: Unlike national environmental levies, the Council recommends for international charge-based approaches that the accrued resources be ringfenced as precisely as possible since there is no democratically elected institution at international level which could decide on the use of the funds based on citizens' preferences. The user charge approach thus offers the opportunity to take initial pragmatic steps towards an international system of levies to underpin global sustainability policy.

With a view to transforming global energy systems, the Council urges the German federal government to lobby for the levying of an emissions-based charge on the use of the Earth's atmosphere by international aviation. Aviation is the fastest-growing source of greenhouse gases worldwide. Despite their substantial impact on climate, emissions produced by international aviation are still not subject to any reduction commitments. As long as this regulatory loophole remains, an emissions-based user charge should be levied. The revenue accrued will provide new additional funds for climate protection measures, the need for which arises as a result of aviation, among other things. The WBGU recommends that the financial resources thus accrued be distributed to the new climate funds (Special Climate Change Fund, Adaptation Fund, Least Developed Countries Fund) and the GEF climate window. However, it should be ensured that these new financial resources are not used to offset development budget cuts, but are deployed as additional funding.

TRADEABLE QUOTAS FOR RENEWABLES
International tradable quotas (e.g. flexible national or company permits, or Green Energy Certificates) would generate transboundary payment flows whose extent and payer-recipient structure depend, among other things, on the level of the individual quotas. This financing aspect should be taken into account from the outset when devising a global system of tradable quotas, for although the system is only likely to be implemented in the long term, it has an impact on the willingness of the poorer countries in particular to expand their renewables in the short to medium term. Developing countries are more likely to commit to minimum renewable energy quotas if there is an increasing probability that they will become net recipients of payment flows as soon as the quotas are transferred to a global system of tradable quotas.

PROVISION OF NEW FUNDING WITHIN THE
FRAMEWORK OF INTERNATIONAL CLIMATE
PROTECTION POLICY
The UNFCCC commits the developed countries included in Annex II to providing new and additional financial resources and to transferring technology to the developing countries so that the latter can make a contribution to climate protection. The developing countries' further integration into international climate protection will depend on whether the industrialized countries uphold this pledge to meet the agreed full incremental costs. This is a key issue for the Kyoto Protocol's second commitment period.

At the Sixth Session of the Conference of the Parties to the UNFCCC in 2001, the launch of three new funds was agreed, to be supported by voluntary contributions from the industrialized countries. At the

Conference, the EU and other states (Canada, New Zealand, Norway, Switzerland) made a joint political statement pledging to contribute US$410 million per year to these funds by 2005. The USA, Japan and Australia did not announce any contributions. Only the Adaptation Fund is funded by a regulated financing mechanism consisting of a 2 per cent tax on CDM projects. The Council welcomes the German federal government's statement that it will support the new climate funds. However, it recommends the further development of the funding structures and advocates measures to ensure that the funding commitment becomes legally binding in order to avoid any arbitrary ad hoc financing of these major funds. In this context, the Council also draws attention to its recommendation that a proportion of the revenue obtained from charging for use of the atmosphere by international aviation be allocated to these funds.

The climate regime's project-based mechanisms will also contribute to additional financial and technology transfer which can benefit energy system transformation. It is estimated that through the CDM, up to US$20 thousand million in incremental funding will be transferred to developing countries. This amounts to more than 50 per cent of total ODA in 2000. A CDM market volume amounting to US$10 thousand million could generate US$90–490 thousand million in additional investment in more sustainable technologies (Öko-Institut and DIW, 2001). Due to their smaller project volumes, renewable energy and energy efficiency projects have structural disadvantages compared with other – usually large-scale – CDM projects which aim to improve the energy efficiency of large power stations or create greenhouse gas sinks. In particular, higher transaction costs pose a major obstacle to the promotion of renewables use through the CDM. To overcome the blockades against the CDM and JI, the project-based Kyoto mechanisms should therefore be promoted by launching a fund based on the Dutch ERUPT and CERUPT programmes.

VISION: ESTABLISHING A REVENUE-STRONG
FUNDING AND TRANSFER MECHANISM
Far greater financial potential is offered by the further development of emissions trading, as envisaged in the Kyoto Protocol (Article 17). The plan until now has been to define countries' maximum permissible greenhouse gas emissions during the first commitment period from 2008 to 2012. Parties thus have a right to use the global good known as 'the atmosphere's capacity to absorb greenhouse gases' (Brockmann et al., 1999), while a ceiling is imposed on their greenhouse gas emissions at the same time. If the emissions fall below the maximum permissible limits, the surplus can be sold to other parties. Any

such trading is supplemental to domestic actions for the purpose of meeting quantified emission limitation and reduction commitments. However, the use of the flexible Kyoto mechanisms to fulfil reduction commitments has yet to be precisely defined in quantitative terms. Emissions trading during the first commitment period is limited to the countries listed in Annex B of the Kyoto Protocol and thus excludes developing and newly industrializing countries (Öko-Institut and DIW, 2001).

In order to achieve the objective, defined in the UNFCCC, of stabilizing greenhouse gases in the atmosphere at a non-hazardous level, the developing countries must be integrated more fully into the international climate protection regime in future and more ambitious reduction targets must also be achieved. The developing countries must therefore be incorporated into a global emissions trading regime during the next commitment period.

The Council has spoken out several times in favour of taking principles of equity into account when initially distributing emissions permits (e.g. WBGU, 2001c). The ethical standard underlying international climate policy should be the right to the same per capita emissions of greenhouse gases, which would also comply with the 'polluter pays' principle. This would enable maximum permissible greenhouse gas emissions to be defined on the basis of the same per capita emissions. The system could be designed to take account of the different energy needs depending on climatic zone. Any inappropriate incentives encouraging population growth could be avoided by a baseline rule, for example. Initial distribution based on this modified per capita approach would trigger a financial transfer amounting to several hundreds of thousand million euro from the industrialized countries to the developing countries while ensuring compliance with the Council's guard rails. This would exceed the current resources allocated to development cooperation many times over. In principle, this approach would establish a carbon-based system for the payment of global financial compensation. However, there is still a long way to go before this vision is achieved. Nonetheless, measures to develop climate protection policy towards this objective could start right away.

The economic impacts on countries with high per capita emissions depend both on the absolute level of emissions but also on the speed with which such an approach is implemented. A shift to the per capita principle during the next commitment period would impose too great a strain on the industrialized countries' economies. For this reason, the current approach should initially continue, with reduction quotas being defined on a differentiated basis, e.g. according to past reductions, energy policy starting

conditions, and economic costs. In the long term – over 20–30 years – however, the modified per capita approach should increasingly be used as the standard. This progressive implementation would not exceed the capacities of the national economies to adjust to more stringent reduction targets and would reduce the economic costs of environmentally effective climate protection to a tolerable level.

As a market economic alternative to a modified per capita approach in a global system of tradable emissions rights, the debate is currently focussing on a global CO_2 tax. This tax was being discussed as a possible global climate protection instrument even before UNCED in 1992 (Pearce, 1991; Cnossen and Vollebergh, 1992), although in the event, a quantitative approach was opted for instead, along with the greater flexibility afforded by the Kyoto mechanisms. The unresolved question of how to fund the ambitious development objectives adopted at the UN's Millennium Summit in 2000 has put the global CO_2 tax back on the international agenda.

The Council views the concept of an international CO_2 tax as an interesting approach. However, it is important to consider that the emissions trading regime envisaged in the Kyoto Protocol is likely to lead to a substantial financial transfer from north to south in the long term once the developing countries are integrated into the system and permits have been distributed accordingly. For this reason, an international CO_2 tax should be introduced simply as a temporary additional funding instrument until a global permits system based on the Kyoto Protocol and environmentally effective reduction commitments have been established. This would not only reinforce the Kyoto Protocol's steering incentives in the short to medium term, but would also fill existing funding gaps in international climate protection. If it were shaped appropriately, an international CO_2 tax would also be in line with the 'polluter pays' and, to some extent, the benefit principle. However, it is important to ascertain whether the proposal for an international CO_2 tax is likely to trigger further political blockades which would make it impossible to implement.

Based on this comparison of possible funding methodologies, the WBGU draws the following conclusion. In accordance with the precautionary principle, it recommends to the German federal government that the quantitative approach established in the Kyoto Protocol continue to be supported in combination with the flexible mechanisms. The progressive development of the Kyoto method towards a modified per capita approach could create a funding and transfer mechanism which would amount to the payment of global financial compensation and be unique in international political history.

5.3.3.3
Use of resources for energy system transformation by international financial institutions

STRENGTHENING THE GLOBAL ENVIRONMENT FACILITY

With the establishment of GEF, additional funds were made available for global environmental protection, while the adoption of double weighted majority voting created a mechanism which guarantees the industrialized and developing countries an equal amount of influence over the use of resources. Over the last ten years, GEF has established itself as an important international funding mechanism for global environmental protection. At the WSSD in Johannesburg, the third replenishment of GEF funds was agreed, amounting to US$3 thousand million to cover GEF's operations during the period 2003–2006. GEF's remit was also enhanced with the expansion of project funding to cover 'land degradation' and 'persistent organic pollutants'.

These positive developments indicate that it may also be appropriate to utilize GEF as a financing institution for global energy system transformation. In this context, however, the criticisms of GEF's work and the new challenges arising in the framework of global sustainability policy must be considered. The 'incremental costs' principle is a major bureaucratic obstacle to the funding of many projects whose implementation would be beneficial in terms of sustainability. The process of defining incremental costs is also extremely controversial (Horta et al., 2002). Projects which encourage the transfer of cheap technologies, build on the indigenous knowledge of the local population or support capacity-building in the developing countries through investment in public education are rarely covered by the 'incremental costs' concept and are generally ineligible for GEF funding. Global sustainability policy should build to a greater extent on the catalysing effects of local and regional approaches and pursue integrated concepts. Yet GEF is accused of failing to take adequate account of local conditions and, therefore, of not exploiting appropriate catalysing effects (Keohane, 1996).

Despite these criticisms, the Council recommends the expansion of GEF as a financing institution for global energy system transformation and as a catalyst for a more comprehensive financial framework. The alternative would be to establish a completely new financing institution, e.g. in the form of a Global Sustainable Energy Facility. However, this would require a lengthy negotiation process at international level which would currently have little prospect of success. In order to establish an effective international mechanism to fund the transformation of global energy systems as quickly as possible, GEF's structures should therefore be utilized. As part of this process, however, its working methods should be geared more strongly towards promoting global sustainable energy systems. The WBGU therefore recommends that the German federal government work proactively for the following reforms in GEF's organization, working methods and status in global sustainability policy:

- *Establishing a new window to promote globally sustainable energy systems:* Through the climate window, projects are already being funded which lower the barriers to promoting energy efficiency and renewables and reduce the costs of measures to cut greenhouse gas emissions in the long term. Furthermore, additional funds for adaptation to climate change are being made available. The WBGU recommends that by 2005, available funding for efficiency technologies and renewables be pooled in a new GEF window so that a new strategic direction can be adopted in GEF's funding policy in this area.

- *Modifying the criteria governing the granting of funding under the new 'window to promote globally sustainable energy systems':* The WBGU views the incremental costs principle as a key concept in the funding of global environmental protection. In order to achieve a global energy system transformation in line with the WBGU's transformation strategy, however, the funding criteria must be simplified so that development policy aspects can be taken into account more fully in resource utilization, e.g. by promoting rural development through renewable energies. This modification of the funding criteria should be based on a World Energy Charter. As many sustainable energy projects have a low volume of funding, the positive experiences with the GEF's successful Small Grant Programme (SGP) should be capitalized upon. This modification would establish a new strategic framework for the promotion of efficiency technologies and renewables which would ensure that projects are not just funded on an ad hoc basis without a broader programmatic context (UNDP et al., 2000).

- *Expanding cooperation with regional institutions, the private sector and local communities:* The WBGU welcomes the substantial increase in the number of implementing agencies. Alongside the World Bank, UNDP and UNEP, the regional development banks, UNIDO, the FAO, and the IFAD (International Fund for Agricultural Development) can now submit projects to GEF (Kutter, 2002). The WBGU also recommends the further expansion of cooperation with the private sector and the greater involvement of local populations. This substantially increases projects' prospects of success and triggers the necessary catalysing effects, such as the development of markets for 'green' products and the mobilization of additional private capital.

- *Further boosting GEF's financial resources and enhancing its status:* The WBGU welcomes the increase in GEF's resources for the third phase. However, it points out that GEF's mandate has also been expanded at the same time. In light of the substantial financial requirement to fund global energy system transformation, GEF's budget should therefore be further increased. At the same time, its status should be enhanced in order to prevent its efforts from being undermined by conflicting effects, particularly the programmes being implemented by other multilateral organizations.

INTEGRATING THE WORLD BANK AND IMF INTO ENERGY SYSTEM TRANSFORMATION TOWARDS SUSTAINABILITY

The World Bank, which not only functions as a provider of loans and credit but also gathers information about the solvency of developing countries and, through its sector investment programmes, directly influences the developing countries' national policies, defined new energy policy objectives in 2001 (Section 2.7.3). They include supporting energy sector reform, promoting competition, improving environmental protection in energy generation, and promoting solutions for the delivery of sustainable energy services to the poor. The World Bank is also coordinating the planned 'Global Village Energy Partnership', which is intended to guarantee better access to energy services to the poor. The German federal government has already announced its participation in this initiative.

However, the WBGU considers that the World Bank's activities will only move in the right direction if it abandons its commitment to the least-cost principle, i.e. to supporting profitable forms of energy, mainly on a micro-economic basis, without ensuring the internalization of external costs. It must also be ensured that long-term framework conditions are considered when implementing new technologies which are not yet market-viable. The World Bank should also see itself as a funding bank for sustainable energies and help to ensure that more effective financial incentives are created to increase the share of renewables in the developing countries. In general, the World Bank and the regional development banks should play a more active role in transforming energy systems worldwide than is currently the case. So far, the shift from the conceptual to the operational level

has not been adequately implemented. The WBGU therefore recommends the practical implementation of the World Bank's new funding strategies. The federal government should work actively to achieve this goal through its membership of the World Bank's Board of Governors.

UTILIZING THE EUROPEAN INVESTMENT BANK AS A FUNDING INSTRUMENT

The Cotonou Agreement was concluded between the EU and 77 ACP states in June 2000 and regulates, for a 20 year-period, the political, development and trade relations between the two groups of countries. The partnership is centred on the objective of reducing and eventually eradicating poverty in line with the objectives of sustainable development and the gradual integration of the ACP countries into the world economy (Article 1, paragraph 2). From the time of the Cotonou Agreement's entry into force, an Investment Facility is envisaged which consists of risk capital totalling up to €2.2 thousand million from the European Development Fund (EDF) and an additional €1.7 thousand million in loans from the own resources of the European Investment Bank (EIB) over a five-year period (2000–2005).

Within the framework of the Financial Protocol annexed to Lomé IV for the period 1995–2000, the EIB already administered risk capital amounting to €1.3 thousand million from the European Development Fund's resources, with a further €1.7 thousand million in loans from the EIB's own resources. Investment priorities for the EIB are infrastructure, especially energy and transport, and industrial development. For the energy sector, the Council recommends giving priority to promoting renewables in future. Until now, the EIB has focussed mainly on promoting fossil energy carriers.

STRENGTHENING THE REGIONAL DEVELOPMENT BANKS

In the energy industry, the regional development banks, which mainly support public-sector projects, have generally focussed on establishing and expanding the electricity grids and reforming the energy sector. When granting credit and loans, the banks pursue region-specific approaches which take account of the highly diverse problems faced, and are therefore important potential partners in overcoming energy poverty in Africa, Latin America, the Caribbean and Asia. However, the prerequisite is that the development banks' management capacities are progressively strengthened and developed.

The WBGU recommends that through its shareholdings in these banks and within the EU framework, Germany work to promote the funding of the developing countries' energy supplies via the regional development funds, which are administered by the regional development banks. In general, these countries do not have the resources to provide a secure energy supply to their poor populations. Moreover, many developing countries, faced with high levels of foreign debt, have very limited scope to manage the expected increases in the price of fossil fuels or to fund improvements in the efficiency of their energy supply. Due to their extremely high debt burden, too, few of them can afford to purchase plant and technologies in the field of renewables. Without extensive debt rescheduling and targeted support for the developing countries and help from the regional development funds, energy system transformation is highly unlikely to occur in these countries. In addition, Germany could lobby for the more intensive refocusing of the regional development banks' funding policies towards environmental protection. Germany has the opportunity to bring influence to bear since it holds shares in the regional development banks' capital stock. These amount to around 4.1 per cent in the African Development Bank (and, indeed, to 10 per cent in the group of non-regional members), around 5.8 per cent in the Caribbean Development Bank, 1.9 per cent in the Inter-American Development Bank, and around 4.5 per cent in the Asian Development Bank.

INTEGRATING ENERGY SUPPLY INTO THE PRSP PROCESS

The IMF and World Bank responded promptly to the decision, adopted at the G7 Summit in Cologne in 1999, to couple expedited debt relief with poverty reduction programmes (VENRO, 2001). At the end of 1999, they submitted preliminary papers outlining their poverty reduction policies for the poorest countries. Since then, around 70 countries, especially heavily indebted poor developing countries, have devised their own national poverty reduction strategies, known as 'Poverty Reduction Strategy Papers' (PRSP), which ought to be prepared by governments through a participatory process involving civil society and development partners. The BMZ supports countries engaging in this process. The PRSPs are also intended to act as steering mechanisms for the mid-term development of these countries as well as a basis for attracting international loans. A PRSP will in future be a prerequisite for debt relief through the HIPC Initiative. A more significant development, in the long term, is that the IDA (International Development Association) countries will only obtain new concessional lending from multilateral and bilateral donors on the basis of a PRSP. Currently, energy supply is not covered by the PRSP process. The WBGU recommends that energy supply feature prominently

in these strategies and that adequate participation by civil society actors be guaranteed.

5.3.4
Directing international climate protection policy towards energy system transformation

A key step in the further development of the Kyoto Protocol is to flesh out the objective contained in Article 2 of the UNFCCC, namely to 'prevent dangerous ... interference with the climate system'. This must be defined more precisely in order to establish more specific targets for the necessary emissions reductions. This should take place by 2005 so that the results can flow into the negotiations, scheduled to begin in 2005, on reduction targets for the second commitment period.

The exemplary path (Section 4.4) is based on a scenario in which the industrialized countries implement the Kyoto Protocol, in the form negotiated in 1997, by 2010. In the technical formulation of the Protocol, the industrialized countries were permitted to offset substantial carbon sinks against their reduction commitments. Furthermore, no quantitative reduction commitments exist for international maritime and air transport. Overall, instead of the original 5 per cent reduction compared with the base year of 1990, a slight increase in emissions until 2010 is therefore anticipated for the industrialized countries. There is still no effective protection of carbon stocks in the biosphere.

The guard rail analysis in Section 4.3 shows that the first decades of the 21st century are particularly critical as regards compliance with tolerable rates of warming. As an adequate reduction in the industrialized countries' emissions from 2012 (around 45 per cent from 2010 to 2020, based on WGBU, 1997) is viewed as unattainable, climate guard rails can only be obeyed if the developing countries curb their rapidly increasing emissions paths earlier than originally envisaged (WBGU, 1997b). Emissions control requirements should therefore come into force for developing countries by 2020, while the newly industrializing countries should adopt initial quantifiable targets even earlier than this.

However, this will be extremely difficult to implement because the industrialized countries are proving very sluggish in fulfilling their reduction commitments and the UNFCCC contains the principle that the industrialized countries must take the lead in reducing emissions. It can therefore be assumed that the developing countries will initially resist, justifiably, any form of quantified reduction targets.

The Council therefore recommends that the German federal government work pro-actively to ensure that the EU launches a 'developing countries initiative' with the following features:

- *Objective:* Closer cooperation is required to restore the confidence of the developing countries (historic alliance between the EU and the developing countries at COP1 in Berlin in 1995) and prepare the ground for future negotiations on emissions reductions. The developing countries will only be prepared to enter into commitments once they recognize that the Kyoto Protocol benefits them too.
- *Regional partnerships:* Individual EU Member States are seeking to establish contacts with specific developing countries or groups of countries with the aim of engaging in coordinated longer-term initiatives, e.g. through working groups which jointly set climate protection and other development targets. Technology transfer in particular should be promoted through joint projects (e.g. through the CDM). In principle, initiatives for such partnerships should also be launched by developing countries and lead to mutual benefits in the long term.
- *Best practice:* Germany and other Annex I countries should set an example to other countries in their fulfilment of the current provisions of the Kyoto Protocol. This applies especially to the 'demonstrable progress' achieved in implementing the Kyoto reduction commitments by 2005, the registration, development and management of CDM projects, the provision of resources for the three newly launched funds for developing countries, preventive measures to mitigate the effects of climate change, and capacity-building and technology transfer. This would enhance confidence in the process and create a basis for the developing countries' involvement.
- *Feasibility:* In order to persuade the developing countries, the short-, medium- and long-term emissions reduction capacities of the major developing countries and regions must first be ascertained. This is the only way to ensure that the developing countries do not reject all negotiations aimed at curbing emissions increases.
- *Closing gaps:* Greenhouse gas emissions produced by air and maritime transport are currently not subject to any quantitative reduction commitments. The WBGU therefore recommends that these gaps in international climate protection be closed, either by integrating these emissions into the Kyoto Protocol or by levying user charges (WBGU, 2002; Section 5.3.3.2).

In this context, the Council welcomes the EU's initiative, at the WSSD, to establish a coalition of like-minded states for the voluntary fulfilment of quantifiable targets to increase the use of renewables, which many developing countries have also joined. The task now is to breathe life into this initiative.

5.3.5
Coordinating international economic and trade policy with sustainable energy policy objectives

The process of turning energy systems towards sustainability must be flanked by measures to establish appropriate economic, legal and political framework conditions. From a global economic perspective, the first task is to create conditions which encourage domestic and foreign private investment in the energy sector and the international exchange of sustainable technologies and knowledge. Secondly, it must be ensured that international economic law does not obstruct the energy strategy but supports its practical implementation.

5.3.5.1
Conclusion of a Multilateral Energy Subsidization Agreement

The WBGU calls for the start of negotiations on a Multilateral Energy Subsidization Agreement (MESA) with the aim of progressively removing subsidies on fossil and nuclear energy carriers and adopting rules on the payment of subsidies for renewables and more efficient energy technologies. Experience at European and international level has shown that the removal of subsidies worldwide can be initiated most effectively through international agreements, because very few countries will be prepared to adopt a 'go-it-alone' approach to cutting subsidies due to their concerns about the ensuing loss of competitiveness on the international markets, especially for energy-intensive companies and the energy supply industry.

The international community should start negotiations as soon as possible on a MESA which should enter into force by 2008. The agreement should aim to achieve the following objectives:

- The removal of all subsidies on fossil and nuclear energy in industrialized and transition countries by 2015 and worldwide by 2030, and
- the removal of subsidies on fossil fuel extraction (coal mining, oil drilling, etc.) in developing countries by 2020.

In this context, the – now obsolete – 'traffic light' approach adopted for the WTO's Agreement on Subsidies and Countervailing Measures could be revived and modified. The key features of a MESA could be as follows: A blanket ban would apply to all obviously non-sustainable subsidies which are harmful to the environment, such as general energy price subsidies, which would be classified as 'red light' subsidies. Subsidies paid to producers and suppliers of fossil and nuclear energy should be classified as 'red light' subsidies except when they are targeted towards promoting higher environmental standards (energy efficiency, filters, 'greener' methods of oil, coal and gas extraction) or improving the safety of existing nuclear power plants. Timetables should be drawn up for the progressive dismantling, or at least the restructuring according to sustainability criteria, of existing 'red light' subsidies, while the introduction of new 'red light' subsidies would be prohibited. In addition, a list of 'green light' subsidies, which could not be the subject of legal challenges, should be drawn up in line with a World Energy Charter. These 'green light' subsidies could include, for example:

- Subsidies or tax credits for renewable energies to compensate for any remaining non-internalized costs of fossil/nuclear energies (including payments for feeding into the grid);
- Subsidies for research into renewables and efficiency improvements up to a specific proportion of total research expenditure;
- Subsidies for supplying electricity to poor households/rural regions in developing countries;
- Subsidies for the replacement of traditional biomass by sustainable fuels for heating and cooking;
- Subsidies for energy-efficient buildings and the use of solar technologies in construction;
- Social transfers to poor households (state support for heating and electricity);
- Cross-subsidies for electricity prices for households to the extent necessary to ensure compliance with the 'affordable individual minimum requirement' guard rail.

Energy subsidies which are not classified as 'red light' or 'green light' subsidies would be open to legal challenges on principle. These 'amber light' subsidies could be called into question by other parties to the Agreement, either on account of their incompatibility with the objectives of the World Energy Charter or because they violate principles already enshrined in the WTO (GATT and the Agreement on Subsidies and Countervailing Measures).

In the developing countries, the expansion of fossil energy use is unavoidable, also in the longer term, in the interests of social and economic progress. In order to take account of this specific starting position, one option is to grant them preferential treatment. This would mean, for example, that for a specified period, less stringent criteria would be applied to subsidies on fossil energies in these countries than proposed above in the context of 'red light' subsidies.

INSTITUTIONALIZING THE MESA

In institutional terms, the MESA could be established as a separate agreement or be incorporated into an existing institution. In the Council's view, the recommended option is the MESA's institutionalization within the framework of an International Sustainable Energy Agency (ISEA) (Section 5.3.2.3). However, to prevent any unnecessary delays in commencing the removal of subsidies worldwide, the Council is also considering the various options available within the existing institutional architecture. For example, the MESA could be integrated into an international energy, trade or environmental agency, such as the Energy Charter, WTO or UNEP. One factor currently militating against the MESA's institutionalization within the Energy Charter framework is the low number of states which are party to this Charter – which makes it little more than a regional agreement – and its lack of significance.

By contrast, the clear advantage of integrating the MESA into the WTO rules would be that it would thus cover virtually the entire international community. Furthermore, the WTO is a fairly effective organization which has acquired sufficient experience of subsidy agreements and is equipped with the necessary mechanisms, such as dispute settlement procedures. The WTO Secretariat regards the energy sector's full integration into GATT and GATS, and hence the application of the GATT Agreement on Subsidies and Countervailing Measures, as a win-win strategy for economic efficiency and environmental protection. However, the WBGU takes the view that in the short term at least, not all energy market segments should be fully subordinate to the WTO rules. Furthermore, the WTO Agreement on Subsidies and Countervailing Measures generally applies different criteria from those required to achieve the MESA's objectives. For this reason, integrating the energy sector into the WTO Agreement on Subsidies and Countervailing Measures is likely to produce little more than the partial dismantling of environmentally harmful subsidies, while the possibility that environmentally beneficial subsidies would be subject to legal challenges cannot be ruled out. Finally, the virtual universality of the WTO Agreement on Subsidies and Countervailing Measures is also an obstacle to a MESA's political viability. A more promising prospect would probably be to start work on an international subsidy agreement with a smaller group of countries, which would offer scope for the individual countries to make a variety of concessions. Membership could then be expanded from there. These factors, and the energy sector's specific characteristics, suggest that an Energy Subsidy Protocol within the WTO framework, similar to the protocols existing for individual services under the GATS regime, is the preferable option.

Alternatively, the OECD states could press ahead with an agreement, especially since the environment ministers of the OECD countries have already agreed to dismantle environmentally harmful subsidies. A further argument in favour of this option is the expertise acquired within the IEA. Accession should then also be opened to non-OECD states so that in the long term, the MESA can be transformed into a global convention.

Incorporating the MESA into UNEP or integrating it into a multilateral environmental agreement, such as the climate protection regime, is undoubtedly more likely to do justice to its overriding environmental policy focus, but also its development policy objectives, than incorporating it into the WTO, whose capacities must not be overtaxed with non-trade policy issues. As yet, however, UNEP has not acquired the necessary capacities and enforcement mechanisms. The climate protection regime is a very interesting alternative, although it has still to prove its enforceability and functionality in the coming years.

Based on these considerations, the WBGU recommends that the MESA concept be addressed both at the international climate protection negotiations and, in parallel, at the OECD. The long-term aim would be to integrate MESA into the International Sustainability Energy Agency (ISEA) recommended by the Council. More generally, however, there is still a substantial need for debate and research into the specific form to be adopted by a MESA as well as its institutional framework (Section 6.2).

5.3.5.2
Transformation measures within the GATT/WTO framework

At present, the energy sector is only partially covered by the World Trade Organization (WTO) rules. The more fully energy products are integrated into GATT, the more relevant the issue of compatibility between WTO agreements and energy system transformation measures will become. This applies even if far-reaching concessions on liberalization were to be achieved for the electricity and gas sector or energy-related services at the GATS negotiations. At present, there are few points of friction between energy policy transformation measures and the world trade regime.

For example, the subsidies on 'green' energy technologies, which the Council views as a useful energy policy tool, could conflict in principle with the WTO Agreement on Subsidies and Countervailing Measures. This might occur, for example, if direct subsidies on energy technologies result in real competitive disadvantages for related sectors in other countries. Although GATT explicitly permits measures necessary to protect the environment (GATT, Article XXb), such measures must distort trade as little as possible. Sector subsidies often do not satisfy this criterion. Nonetheless, the Council does not consider that this so far hypothetical dispute creates any urgent need for a reform of GATT. If such a dispute did occur and could not be averted in advance by a MESA (Section 5.3.5.1) or other energy agreement taking precedence over GATT/WTO, a decision can still be taken on whether possible countervailing duties by other countries are accepted by the subsidizing country or whether an exemption is sought under the Agreement on Subsidies and Countervailing Measures.

Further conflict potential between energy policy measures and GATT/WTO could arise from the main principles underlying the current world trade regime: The principle of national treatment – i.e. the equal treatment of domestic and foreign products – and the most-favoured-nation clause. In their practical implementation, the question of which specific goods must be treated equally, and hence the issue of 'like domestic products' (GATT, Article III 2), are of key importance. Unequal treatment of 'like' imported goods due to their indirect energy context or the method by which the energy required for their production was supplied is unlikely to be compatible with the world trade regime's core principles at the present time. There are fears, therefore, that national measures to promote sustainable energy systems will lead to competitive disadvantages for numerous domestic companies in the home and world markets. Firstly, this increases the costs of adjustment and decreases the acceptance of transformation efforts; secondly, the global environmental effectiveness of promoting sustainable energies will be reduced as other countries will increasingly specialize in the products which have become expensive on the domestic markets. Many countries are therefore planning to adopt energy tax exemptions for energy-intensive industries due to fears about competitive disadvantages. In general terms, this is compatible with GATT. However, the Council specifically points out that tax relief for energy-intensive sectors is certainly not an instrument which is designed to promote the establishment of sustainable energy systems.

An alternative is to introduce border adjustment taxes in order to compensate, to some extent, for the cost disadvantages faced by domestic companies resulting from such tax policy measures. The WTO rules allow all the product taxes applying in a given country to be imposed on imported goods, while goods for export can be exempted. In the interests of environmental efficiency and practicability, however, it is often necessary to impose a tax not on the end product but directly where the pollution occurs. The question which thus arises is whether border adjustment measures comply with the WTO rules even if the domestic final product is only indirectly subject to national environmental charges. These can include emissions charges, e.g. a CO_2 tax, or input taxes, such as a fuel tax. The WTO's Committee on Trade and Environment has already debated the issue of border adjustment taxes as compensation for national environmental levies, but has not reached any far-reaching conclusions as yet. In general, the GATT compatibility of compensating for emissions charges by taxing imports is a highly contentious issue. On exports, the majority view is that it is incompatible with GATT (Greiner et al., 2001). Border taxes to compensate for the imposition of national input taxes, on the other hand, are deemed to be compatible with GATT (Jenzen, 1998). Various decisions by the WTO's dispute settlement body point in this direction. The Council recommends to the German federal government that it clarify, within the WTO, the issue of the admissibility of border adjustment measures for CO_2, fuel and other environmental taxes and press for the development of pragmatic and – above all – transparent solutions to avoid unequal treatment of various forms of taxes and environmental protection instruments.

There is occasionally a debate about whether the Kyoto regime of tradable emissions rights is compatible with the WTO principles (Box 5.3-2); similarly, the issue of compatibility between the WTO rules and the trade in Green Energy Certificates, which is recommended by the Council, could also be discussed. In the Council's view, however, these certificates for green electricity or heating, which can be used to cover national renewable energy quotas, do not constitute a product or service as defined by GATT or GATS. If a global system of tradable Green Energy Certificates were to be implemented, the Council recommends that a separate trading regime be established here, as with the emissions trading set up under the Kyoto Protocol.

5.3.5.3
Preferential agreements in the energy sector

Through preferential agreements, it is possible to diverge from the most-favoured-nation principle within the WTO. The most important types of preferential agreements are customs unions with common foreign trade policies (e.g. the EU), free trade agreements (e.g. NAFTA), and trade preference systems

with particularly low tariffs or other trade benefits for imports from developing countries.

While national preference systems have, in practice, decreased in significance due to the general cut in the tariff level, the number and importance of regional trade associations are still increasing. In this context, in order to be recognized by the WTO, customs unions and free trade areas must fulfil three criteria relating to both products and services: Customs unions and free trade agreements must cover most of the trade in goods and services among their members, and individual sectors may not be excluded. In terms of the agreement's scope of application, all tariffs and quantitative restrictions must be progressively abolished within a 10-year period, and internal liberalization within the framework of a customs union or free trade area must not result in new and additional obstacles to market access for products from other WTO member countries (Cottier and Evtimov, 2000).

In the energy sector, preferential agreements are of particular interest as they allow liberalization of both goods and services which goes further than the agreements reached within the WTO framework. This offers the opportunity, especially for developing countries, to open up their energy sectors, initially within existing regional trade areas, without having to face direct competition with producers and service providers from industrialized countries. The Council recommends that the German federal government support such regionally restricted liberalization efforts among developing countries through appropriate capacity-building.

5.3.5.4
Technology transfer and the TRIPS Agreement

The international diffusion of 'green' technologies is a key element in turning energy systems towards sustainability. This includes the transfer of technology and knowledge from north to south. This transfer is influenced by the Agreement on Trade-Related Aspects of Intellectual Property Rights (TRIPS). TRIPS commits all WTO members to relatively high minimum standards of protection for intellectual property rights. While the industrialized countries were required to implement these standards by 1996, a general transitional period of five years applied to the developing countries until 2000 (with an additional five years, i.e. to 2005, for product patents), and for the Least Developed countries until 2010 and beyond. For the industrialized countries, the TRIPS commitments entail relatively minor changes to their laws applicable to intangible property, but many developing countries are required to undertake

Box 5.3-2

Compatibility of the Kyoto Protocol with WTO rules

The compatibility of the Kyoto Protocol's flexible mechanisms with basic commitments arising from the WTO rules is increasingly viewed as problematical. The WBGU considers, however, that through the appropriate development of the instruments established under the Kyoto Protocol and the evolution and interpretation of the WTO rules, the Kyoto Protocol approach and the WTO's core principles can both be safeguarded.

First of all, it should be noted that in the Council's view, the various tradable and transferable emissions reduction units are not products or services as defined by GATT or GATS. Emissions reduction units only become effective through legislative measures or via the institutions established under the Kyoto Protocol, and are therefore similar to a legal authorization which, on principle, is not covered by GATT or GATS. The situation is rather different, however, with regard to services supplied through emissions trading or as part of the certification process. These are generally covered by GATS. However, even if a WTO member commits itself to according equal treatment to suppliers of like services from another WTO member (Article II, GATS) and publishes all the relevant measures of general application in advance of their entry into force (Article III, GATS), the commitments will not be affected by the Kyoto Protocol.

Another point of conflict concerns the restrictions arising from the WTO Agreement on Subsidies and Countervailing Measures. Whereas GATS currently contains no specific subsidy restrictions, all financial transfers and tax exemptions fall within the scope of the Agreement on Subsidies and Countervailing Measures. Subsidies are therefore prohibited on principle if they are only available to one specific sector or company ('specificity') and if they cause severe adverse effects to the interests of other members (Article 5 of the Agreement on Subsidies and Countervailing Measures).

The issue which therefore arises in the context of emissions trading is whether all the various possible procedures for the initial distribution of emissions rights (free distribution, auction) are compatible with the Agreement on Subsidies and Countervailing Measures. The decision on the initial allocation has strong regulatory characteristics and is therefore more comparable to the setting of emissions standards or the imposition of a tax than the granting of subsidies. In the Council's view, there is therefore no incompatibility with the Agreement on Subsidies and Countervailing Measures.

A more difficult question to answer is whether government funding for Clean Development Mechanism (CDM) projects can be regarded as subsidies which are not compatible with GATT in individual cases. The first point to note in this context is that the financial transfers from industrialized countries or international institutions to develop these projects in developing countries must be regarded on principle as development assistance and not as a subsidy as defined by GATT or a future GATS subsidy agreement. However, problems can arise if the country hosting the CDM projects supports their development financially in order to create additional incentives for 'green' foreign investment. Although the Council considers that the GATS Agreement does not apply to emissions reductions within the framework of CDM projects, the products and services produced as part of CDM projects would profit from such subsidies and could thus result in a violation of the Agreement on Subsidies and Countervailing Measures. Compatibility with WTO rules thus depends, first and foremost, on the way in which the CDM regulations are developed at national level, and can be safeguarded by ensuring that such subsidies are not only available to specific sectors.

Overall, it is noted that few points of friction exist between the Kyoto Protocol and the WTO rules. To avoid conflicts, further efforts should be made, when developing the Kyoto mechanisms – especially at national level – to ensure that the necessary transparency continues to be guaranteed, that major decisions are taken on the basis of consensus wherever possible, and that potential conflicts can be addressed through a dispute settlement procedure within the Kyoto Protocol framework.

Sources: Werksmann, 2001; Petsonk, 1999; Wiser, 1999

---

major reforms. Among other things, TRIPS regulates the protection of technical inventions (patents) and thus has an impact on international technology transfer.

On the one hand, the implementation of TRIPS has impeded the transfer of patented technologies to the developing countries because the costs increase as a result of the licence fees; licensing negotiations also have to be conducted, for which many companies in developing countries lack both the resources and the expertise. There is also a risk that technology transfer will not take place if patent owners pursue a very restrictive licensing policy (Enquete Commission, 2002).

On the other hand, empirical studies show that countries with a high level of patent protection generally attract more foreign investors than other countries (Maskus, 2000), so that the introduction of Western patent standards is likely to encourage the transfer of environmental technologies, among other things. Furthermore, effective patent protection in developing countries encourages companies in both north and south to undertake research and development into energy technologies which are tailored specifically to the needs of these developing countries. The prerequisite, however, is that there is a demand – backed by real purchasing power – for the innovations, which are now patentable worldwide. Yet this is rarely the case, especially in the poorest developing countries.

Overall, then, the impact of TRIPS on the transfer of 'green' energy technologies is ambivalent. For this reason, the Council recommends a twin-pronged policy to enhance the positive effects and reduce the negative impacts. It includes providing training for institutions and companies in patent and licensing issues through the development cooperation framework. The WBGU also sees a need for research to determine which international mechanisms are suitable to increase the incentive potential of patent protection for innovations which are especially relevant to developing countries while contributing to innovation diffusion at the same time. This might include, for example, subsidizing the acquisition of patents and licences for 'green' energy technologies by companies in developing countries; in this context, the innovators must be made aware, in advance and at least in outline, of the criteria and scope of the subsidies. The proposal for an international patent fund (Enquete Commission, 2002) aims to achieve similar goals; this fund would acquire licences itself in order to grant licences, at subsidized prices and according to agreed criteria, to companies or institutions in developing countries. Such measures would accord with the developed countries' pledge to provide incentives to enterprises and institutions for the purpose of promoting and encouraging technology transfer to Least Developed Countries (TRIPS, Article 66(2)).

On the other hand, measures which interfere with the rights of the patent owner may conflict with the provisions of TRIPS. This applies especially to compulsory licences and the withdrawal of patents. TRIPS Article 27(2) grants Members the opportunity to exclude from patentability "inventions, the prevention within their territory of the commercial exploitation of which is necessary ... to avoid serious prejudice to the environment ...". Furthermore, compulsory licences are permissible if the patent owner refuses to authorize licences or applies anti-competitive licensing practices (TRIPS, Articles 31 and 40). Finally, WTO Members may adopt measures 'necessary to protect public health and nutrition, and to promote the public interest in sectors of vital importance to their socio-economic and technological development, provided that such measures are consistent with the provisions of this Agreement' (TRIPS, Article 8(1)). The flexibility of these exemptions is disputed, so that dispute settlement procedures will be required before the scope for restrictions on patent protection which are motivated by environmental policy considerations can be properly assessed.

The Council does not consider that the aim of transforming energy systems worldwide creates any need for a reform of TRIPS at the present time. However, it points out that anti-competitive practices by patent owners may constitute a barrier to the diffusion of sustainable (energy) technologies, and that Articles 31 and 40 of the TRIPS Agreement appear inadequate to respond to this problem. The Council recommends, among other things, that the German federal government continue to work pro-actively for the internationalization and, ultimately, the globalization of the core principles of competition law.

5.3.5.5
Liberalizing the world energy products market?

MOBILE PRIMARY ENERGY CARRIERS (ESPECIALLY OIL AND COAL)

While efficiency and therefore welfare gains are arguments in favour of the full integration of oil and coal into the WTO rules, specific interests of the exporting and importing countries conflict with such a move. The major oil exporters currently control prices by imposing restrictions on export quantities, among other things. This is only partially compatible with GATT. Importing countries naturally have an interest in the lower oil prices which would result from world market liberalization. On the other hand, they are pursuing policies to promote domestic fossil energies, to the detriment of imports, in order to reduce their import dependency.

GRID-BASED ENERGY (ESPECIALLY GAS AND ELECTRICITY)

The worldwide liberalization of the international trade in grid-based energies is slowed down, first and foremost, by problems with technical implementation. Firstly, there is no global main grid for electricity or gas, but only regional networks. Secondly, transport losses are an impediment to dynamic international trade, although these losses are likely to decrease in the foreseeable future with the introduction of more efficient technologies, such as high-voltage direct current power transmission or even hydrogen pipelines (Section 3.4). However, it is not only transport but also transit which is causing difficulties. Electricity transmission lines and especially gas pipelines are owned by private or national monopolies which charge monopoly prices for transport and give priority to their own or national interests, and the interests of importers and exporters, over free trade principles. Not surprisingly, the liberalization of the energy trade within the framework of the Energy Charter Treaty has failed until now, mainly due to the issue of the conditions under which Russia allows its pipelines to be used. The differences between countries in their level of energy market liberalization also impede the swift opening of these markets to foreign suppliers.

Finally, there is currently a risk that subjecting the electricity trade to GATT rules would tangibly restrict environmental policy scope at national level.

While there is broad agreement that coal, oil and gas are not 'like' forms of energy despite their mutual substitutability, electricity – from the WTO's perspective – would simply equal electricity. If electricity imports were treated unequally or were discriminated against compared with domestic energies, this would violate the core principles of GATT. This would make it more difficult to give political preference to 'green power', as opposed to other forms of electricity, by imposing environmental taxes or levies. There is controversy over the extent to which the general exceptions set out in Article XX b, g GATT (protection of the environment/exhaustible natural resources) would allow the unequal treatment of electricity from renewable energies and other sources.

Recent dispute settlement cases at the WTO indicate, however, that the Appellate Body, set up as part of the WTO's dispute settlement mechanism, does not view the unequal treatment of products as incompatible *per se* with GATT if, at least, individual production and process methods cause substantial transboundary pollution. Furthermore, environmentally motivated trade restrictions are more likely to be accepted if they are anchored strongly in a multilateral environmental protection agreement (WGBU, 2001a). If the further opening of the energy supply markets, e.g. between the parties to the Energy Charter Treaty, did indeed lead to dissent over the granting of trade policy preference to 'green' electricity, as opposed to other forms of electricity, there is a good chance that a dispute settlement panel would rule that such preference is compatible with the WTO rules, referring in this context to the Kyoto Protocol or a 'World Energy Charter'.

In any event, the full integration of electricity into GATT is not on the agenda at the current WTO negotiations, as very few WTO members classify electricity as a product (WTO, 1998). Instead, the negotiations are focussing, at the most, on the integration of electricity into the General Agreement on Trade in Services (GATS). The GATS negotiating system is relatively complicated, with each member identifying the sectors for which it wishes to grant market access and making detailed concessions. In this context, given the limited transportability of many services, the primary issue is freedom of establishment for foreign suppliers and their national treatment.

Will countries make concessions in the electricity sector? This will largely depend on whether the energy sector is classified as a separate area, thus paving the way for a GATS sub-agreement or protocol. So far, eight (groups of) countries have put forward negotiating proposals (USA, EU, Canada, Norway, Venezuela, Chile, Japan and Cuba). Although

there seems to be a consensus that energy-related 'non-core' services (WTO, 2001) – such as consulting, engineering and construction services, and maintenance – should be part of the negotiations, there are clear concerns about the inclusion of energy-related 'core services', i.e. transmission, distribution, and sales. It is a completely open question at present whether, in the foreseeable future, the GATS negotiations will deal with foreign suppliers' access to domestic resources (especially oil, coal and gas) or freedom of establishment for energy suppliers/producers. There is no doubt that in the long term, the use of renewables requires a more or less global network for electricity, and probably also for hydrogen, in order to function efficiently.

ENERGY-RELATED SERVICES

The opening of national markets to foreign trade in energy-related 'non-core' services offers national economic efficiency gains as the entry of foreign suppliers into the market strengthens competition, both on price and quality, in the domestic market and results in cost savings, e.g. in the construction and maintenance of grids or through operation and performance contracting. However, the static and dynamic efficiency gains are unlikely to develop fully at macroeconomic level unless liberalization of the electricity and gas markets leads to the separation of energy production and grid operation.

ENERGY EXTRACTION AND PRODUCTION

In the direct extraction and production of energy, the opening of markets to foreign trade also results in efficiency gains and helps to overcome the lack of capital and technology at domestic level which is a serious barrier to transformation, especially in the developing countries. On the other hand, these sectors are of such key strategic importance in political, economic and social terms that many governments are deciding not to open their markets to foreign suppliers for fear that this will result in a loss of national control over the energy sector (, 2001). This risk increases as the energy markets become more concentrated. In the least favourable scenario, when differences exist between countries in their level of energy market liberalization, opening the markets to foreign investors can lead to competitive advantages for some companies, notably those which dominate the market and/or are protected by the state. If these companies ease generally well-performing domestic suppliers out of the market, there is a risk that a global oligopoly with very few suppliers will emerge, and that this will be difficult to break apart due to the high start-up investment in the energy sector. Finally, negotiations on the opening up of grid-based energy supply would need to examine the extent to which freedom of establishment at local level, and non-discriminatory grid access, create any right to import electricity or gas. This right would cause similar environmental policy problems as integrating electricity as a 'product' into GATT.

RECOMMENDATIONS FOR ACTION ON LIBERALIZATION

The WBGU recommends that the German federal government bring influence to bear in the European Union to ensure that the EU lobbies for the further integration of unrestricted mobile primary energy carriers (oil, coal) into GATT. Better account can be taken of the issue of import independence by promoting renewable energies and efficiency increases than by protecting domestic coal or gas production. The Council makes the same recommendation in respect of the inclusion of all energy-related services into the GATS negotiations, as this can produce global welfare gains. The Council emphasizes, however, that this reinforces the need, at the same time, to press ahead speedily with energy system transformation. Otherwise, the risk that liberalization will result in rising external costs, leading on balance to welfare losses, cannot be ruled out.

The Council is very sceptical, on the other hand, about individual proposals to liberalize trade in electricity immediately and subject it to the GATT rules. If the energy supply sector is to be fully integrated into GATS, it is important to ensure, as a precautionary measure, that trade restrictions which are necessary for environmental policy reasons – e.g. for non-sustainable energies – cannot be subject to legal challenges. However, if renewable energy quotas and minimum health, environmental and socio-economic standards for energy generation are to apply worldwide, the Council can envisage the full international liberalization of the energy supply sector in the long term. This would promote the implementation of a global grid, which under certain conditions could result not only in economic and environmental efficiency gains but also export opportunities for developing countries.

The Council is in favour of swift liberalization of the trade in goods and services in the renewables and energy efficiency sector. Efforts by WTO members to exempt environmental protection products and technologies from tariffs must be pursued vigorously, especially in the area of 'green' energy technologies. As this is still a relatively new market, the opportunities associated with liberalization (market size advantage, knowledge transfer, technology diffusion) appear to outweigh any risks, especially in the developing countries. In order to reflect the significance of these sectors, especially in future, the Council considers that the conclusion of a separate protocol to the

GATT and GATS agreement is desirable. If possible, this should contain a commitment to providing access for renewable energy generation plants to the existing supply grids.

5.3.5.6
Rights and duties of direct investors

Private direct investment plays a central role in transforming energy systems, especially in developing, newly industrializing and transition countries (Section 5.3.3.2). Despite the continued need for improvements in this area – through bi- and multilateral agreements within the context of the Energy Charter Treaty, among other things (Waelde et al., 2000) – substantial progress has been made in recent years on the protection of direct investments. However, the lack of any obligation on direct investors to comply with social and environmental standards is increasingly regarded as a regulatory loophole (Esty, 1995; Subedi, 1998).

Implementing high-quality and effectively enforced standards for all investors in all countries is a worthwhile objective, but it is only likely to be achievable in the long term. The first task, then, is to persuade foreign investors, at least, to uphold specific environmental and social standards in the host country. In the Council's view, this can only be achieved through a mix of measures. Apart from the further development of international environmental law and the reform of export promotion, which is addressed above (Section 5.2.2.3), the following options should therefore be explored:

At international level, particular consideration should be given to the greater integration of environmental policy issues into the existing WTO rules and the Energy Charter Treaty. In the Council's view, the incorporation of commitments under environmental law into economic agreements, as envisaged in the Energy Charter Treaty, is desirable. This should aim to safeguard compliance with international minimum standards. As a further instrument, based on the NAFTA model, it would be desirable to prohibit any reduction of environmental standards undertaken for the purpose of attracting investors (Muchlinski, 1998).

In addition, stronger regulation of direct investors by their home countries should be considered. The Council is in favour of extending the geographical scope of national liability law in line with the US model, preferably through a European initiative. Under US law, foreign nationals can sue for compensation in the US courts if the damage they sustained is defined as unlawful under the foreign law and was caused by a subsidiary of an American corporation.

In the energy sector, the US legal system has shown that it has teeth, especially in relation to oil exploration and transport. The liability criterion applied to establish negligence by the subsidiary in question should be based on the current knowledge available to the parent company (Subedi, 1998).

The Council also supports the development of legally non-binding codes of conduct for the energy sector. In terms of content, a code of this kind could be established on the basis of Article 19 of the Energy Charter Treaty, for example. Although voluntary codes can contribute to supporting environmental and development policy objectives, their function is still limited due to the absence of any control and enforcement mechanisms (Dröge and Trabold, 2001). The OECD Guidelines for Multinational Enterprises, published in 2000, encourage companies to initiate a variety of measures, including the development of environmental management systems, granting access to key environmental information, compliance with statutory requirements in the host country, and adhering rigorously to the precautionary principle in corporate policy-making. The Guidelines attempt to reduce the shortcomings of voluntary codes through a specific implementation process whereby national liaison offices advise interested companies about the content of the Guidelines and the progress made on implementation. It remains to be seen whether the Guidelines will develop, in the longer term, into a generally recognized 'frame of reference for socially responsible corporate conduct' (OECD, 2000) for the energy sector as well.

More far-reaching proposals aim to develop a separate legally binding international convention on corporate responsibility (WEED, 2002a, b). Alongside measures to uphold and promote human rights and international labour standards, one proposal is that this convention should contain commitments on environmental protection. Issues under discussion include, firstly, the standards set out in the non-binding OECD Guidelines, and, secondly, more far-reaching commitments. In the Council's view, some of the proposals are worth further consideration. However, it emphasizes that first and foremost, there is still a substantial need for research into the political and legal obstacles, the practicability and effectiveness, in terms of achieving the desired objectives, and the economic and social side-effects and long-term impacts of implementing such a convention, especially as regards the extraterritorial application of environmental law. Other measures, in the Council's view, should be explored in more detail:
- Obligation to use the best technology available locally;
- Obligation to comply with the national environmental law of the host country, including standards arising from international environmental conventions ratified by the host country;
- Obligation to carry out environmental impact assessments in line with the World Bank's standards, including the development of environmental management plans to minimize damage;
- Supporting the transfer of environmentally relevant operating procedures, technologies and management know-how;
- Obligation to identify and analyse the key material flows and product life-cycle phases relevant to environmental impacts;
- Raising environmental awareness among staff through their active involvement in setting up and running environmental management systems.

5.3.6
Phasing out nuclear energy

The civilian use of nuclear energy has shown itself to be unsustainable: Reprocessing and final storage, but also proliferation and terrorism harbour a significant potential risk. The lack of economic efficiency in liberalized energy markets and the growing public criticism in many countries of the environmental and health risks associated with nuclear power have resulted in a dramatic fall in the number of nuclear power stations being connected to the grid each year (Section 3.2.2). The Council welcomes this development and advocates that no more new nuclear power stations be approved and built. The Council also recommends that efforts be made to commence negotiations, as swiftly as possible, on an international agreement on the phasing out of civilian nuclear energy use by 2050. The agreement should provide for the conversion of the International Atomic Energy Agency into a body responsible for winding down the industry, and should also dismantle the subsidies and special provisions for the nuclear industry, as these distort competition in the energy industry and tie up substantial financial resources in this non-sustainable industry which could otherwise be used to promote a sustainable energy sector (Section 5.3.2.3). In this context, the Council welcomes the German Bundestag's recommendation that the promotion of nuclear energy through the Euratom Treaty be allowed to expire.

The problematical dual use of nuclear technology is particularly evident in the context of proliferation. This concerns not only the reactors but the entire nuclear energy chain, from uranium extraction, conversion and use to interim and final storage. The IAEA monitors the non-proliferation commitments arising under the 1970 Treaty on the Non-Proliferation of Nuclear Weapons through 'safeguard agreements'. However, the discovery of Iraq's secret nuclear weapons programme after the 1991 Gulf War or North Korea's nuclear programmes have shown that these control systems are inadequate. Through its Additional Protocol, the Agency has now acquired the right to investigate undeclared nuclear materials and activities as well. However, just 22 states have ratified the Protocol, and only two of these states engage in nuclear activities (Froggart, 2002). The IAEA's powers are inadequate, for although it can record the proliferation of nuclear material, it cannot prevent it.

International rules to avert nuclear terrorism also fall short of the mark. The only agreement currently in place, namely the IAEA Convention on the Physical Protection of Nuclear Material, is limited to protecting international transports of nuclear material against theft or robbery. After 11 September 2001, the IAEA's General Assembly endorsed 12 new safeguards but rejected any form of mandatory reporting or international control. The Council therefore recommends that more stringent IAEA security standards be adopted for all plutonium storage facilities by 2005 and that the IAEA's powers to initiate the controls and measures necessary to guard against terrorism and proliferation be expanded.

A problem relating to the normal operation of civilian nuclear power plants is that safety levels vary, and there is no binding international standard. The WBGU recommends that safety standards be harmonized internationally at a high level by 2010. Furthermore, the insurance obligations relating to nuclear power plants should be borne solely by the operators, and tax advantages should be dismantled. Possible starting points here are the European Commission's two draft proposals for directives, dated November 2002, which concern the safety of nuclear installations and the management of radioactive waste. The long-term solution for the storage of nuclear waste continues to be one of the major challenges facing the nuclear energy industry. Currently, there are just three potential countries with sites for final storage, i.e. Finland, the USA and Russia. It is almost impossible to predict, at this stage, whether these sites will ever be able to accept nuclear waste. The Council therefore recommends that from 2010, the operation of nuclear power plants be permitted only when the operators can certify that they have reserved space for their nuclear waste in an existing final storage facility. This should be regulated by the IAEA.

The reprocessing of spent fuel elements at Sellafield and La Hague represents the largest release of anthropogenic radioactivity worldwide (WISE, 2001). This violates the WGBU guard rail, which states that risks must be kept within the normal range

(Section 4.3). In the Council's view, the German federal government should therefore press the European Commission for the adoption of moratoria on the operating licences of the reprocessing plants at Sellafield and La Hague by 2010: Licensing of the plants under Article 6 of Directive 96/29/Euratom should be suspended as long as their operation violates international agreed limit values. In this context, the Convention for the Protection of the Marine Environment of the North-east Atlantic (OSPAR Convention) can serve as a strategic starting point. This came into force in 1998 and aims to achieve concentrations in the environment close to zero for harmful substances.

5.3.7
Development cooperation: Shaping energy system transformation through global governance

In the WBGU's view, the group of developing countries is too diverse to enable energy system transformation here to be promoted through an overarching strategy or a single policy approach. In policy formulation, the group of Least Developed Countries (LDCs) in particular must be differentiated from the other developing countries. In the poorest, generally heavily indebted developing countries, liberalization approaches for the energy market often have little bearing on reality. Financial, personnel and technical support at all levels is essential here.

Key recommendations for action in the field of development cooperation are set out in Sections 5.2 und 5.3 above. In this context, special emphasis should be placed on the strategic partnerships between industrialized and developing countries which were launched at the WSSD (Box 5.3-3). In the WBGU's view, they should be supported and developed further.

However, without conducive framework conditions at national and international level or coherent sectoral policies, these initiatives and projects have little prospect of success. The WBGU therefore considers that the primary task must be to change the structures of international political processes in the long term to ensure that they support globally sustainable development. In this context, changes in the policies of key international institutions have recently been observed. For example, new guidelines have been introduced for the OECD countries' development policies, while the World Bank has recently prioritized poverty reduction. Other examples include the decisions adopted at major negotiating processes, especially the Millennium Summit, the WTO Conference in Doha, the UNFCCC, the HIPC Initiative, the International Conference on Financing

Box 5.3-3

Strategic partnerships for global energy system transformation launched at the WSSD

Existing or emerging initiatives to promote global energy system transformation can serve as a framework for global energy policy. The Council recommends that the following international initiatives, which were launched at the WSSD in 2002, be used as catalysts to promote global transformation:

EU ENERGY INITIATIVE: ENERGY FOR POVERTY ERADICATION AND SUSTAINABLE DEVELOPMENT
This EU initiative aims to play a catalyzing role in achieving the WSSD objectives and the Millennium Development Goals and serves as a platform for the coordination and coherent development of energy projects with developing countries at EU level. This strategic energy partnership is intended to promote partnerships for energy access, integrating civil society and the private sector in this process. Phase 1 of the initiative is anticipated to last until 2004.

Phase 2, on implementing the agreed action, will begin subsequently.

GLOBAL VILLAGE ENERGY PARTNERSHIP
The Global Village Energy Partnership aims to ensure access to modern energy services by the poor. To this end, investment funds, framework conditions conducive to the establishment of rural energy systems, and networking by key players are supported. The initiative is sponsored, among others, by UNDP, numerous governments (including Germany), the World Bank and the private sector and is currently in its preparatory phase.

GLOBAL NETWORK ON ENERGY FOR SUSTAINABLE DEVELOPMENT
This Network, which was initiated by UNEP, aims to promote research, development and distribution of sustainable energy systems in developing countries and establish a global network of energy 'centres of excellence' linking governments and the private sector. It involves numerous governments (including Germany), energy institutions, UN agencies, the World Bank and the private sector (Shell Foundation, World Energy Council, UN Foundation). The network is currently being established.

for Development in Monterrey, and the replenishment of GEF until 2006. The outcomes of these recent international political processes are crucial for the successful implementation of sustainable energy policies in the developing countries. They are also starting points for the development of a global governance policy.

Transforming energy systems in the developing countries makes intervention within the framework of a global structural policy a necessity. To this end, the key points of leverage must first be identified. The WBGU notes that:

- A universal energy supply cannot be achieved without adequate income opportunities, even if electricity and energy prices are low, i.e. subsidized. Access for the LDCs to the industrialized countries' markets, as already announced by the EU ('Everything But Arms'), and the swift dismantling of farm subsidies in the EU, USA and other OECD states are therefore necessary. Farm export subsidies are particularly problematical as they threaten the survival of farming communities in the developing countries.
- As a result of the developing countries' often high levels of foreign debt, these countries often have little scope to implement a sustainable energy strategy. Energy system transformation in the southern countries is therefore highly unlikely to occur without a comprehensive debt relief package. The WBGU recommends that the German federal government take the initiative here within the G7/G8 framework.

- In the WBGU's view, overcoming energy poverty should also be included in the 'social basic services' funding priority and in the 20:20 Initiative.
- In the development cooperation undertaken by the OECD countries, the principles of coherence, convergence and complementarity should be upheld to a greater extent. To this end, the new development policy guidelines adopted by the Development Assistance Committee (DAC) in 2001 for its member countries must be implemented in practice (OECD, 2002). A key task in this context is to integrate the principles of sustainable development. The document emphasizes that the development process must be driven by the needs and priorities of the developing countries themselves. It also underlines the need to integrate various sectoral policies and calls for coherent policy development on the donors' side as well. The new DAC Guidelines also contain recommendations on how development cooperation could be developed further. A priority, in this context, is the development of a long-term energy strategy.

A sustainable energy strategy should fit into these existing structures and programmes. Coherence must be a particular priority when several donors are engaged in activities in one country. Here, the World Bank's Comprehensive Development Framework, which is currently being established, could serve as a useful tool. This instrument offers a conceptual framework which draws together all elements of development and focuses them on the recipient country's development strategy.

5.3.8
Launching 'best practice' pilot projects with a global impact

Progress with the introduction and expansion of renewable energy carriers is still very sluggish. This is due, among other things, to the high start-up costs, inadequate knowledge of what is technically feasible and – especially in the developing countries – poor infrastructure, as well as uncertain or low profit expectations. The WBGU therefore recommends utilizing a small number of large-scale 'best practice' pilot projects as a strategic lever for global energy system transformation. These 'best practice' pilots should send out a signal worldwide that improving energy efficiency or establishing a renewable energy supply is already feasible under current technological, political and socio-economic conditions in many areas and can generate long-term profits. Successful 'best practice' projects would act as a positive incentive for private investors and also increase the political viability of energy system transformation. They should all be underpinned by research programmes. The WBGU proposes the following 'best practice' projects:

SAHARA POWER FOR EUROPE
Based on the WBGU's quantifications, it is estimated that western Europe's energy consumption in 2050 will amount to around 100EJ per year. Two-thirds of this should be supplied from renewables. This amount of energy is roughly eight times the European Union's current total energy consumption. It is sensible, therefore, to integrate North Africa's solar and wind energy into the European energy supply in the medium term as well. The WBGU recommends establishing a strategic energy partnership between the EU and North Africa. For Europe, this would not only be a cost-effective way of securing a climate-relevant volume of renewable energy; it would also be a major step towards more intensive economic and foreign policy cooperation with North Africa. For North Africa, this partnership would offer an opportunity to link climate protection with industrial and social development. The energy partnership could be a driving force for development in the region. This strategy has three key elements:

1. The construction of major power plants in North Africa to generate electricity from renewable sources;
2. The provision of transmission capacities to the European grid;
3. The establishment of a European liaison office for the North African project partners and European investors.

Using 'best practice' pilot projects, the feasibility of supplying energy from renewable sources using existing technologies could be demonstrated, obstacles identified and overcome, and the necessary structures developed in advance of private-sector involvement. In order to establish and utilize a learning curve, the projects should contain a strong research element. As well as establishing a coordinating body at European level, the WBGU recommends that the following pilot projects be carried out:

• Planning and tendering for a large-scale photovoltaic and a solar thermal power station in cooperation with one or more North African countries;
• Planning and tendering for a large wind farm in cooperation with one or more North African countries;
• Planning and tendering for a transmission line from North Africa to Europe;
• The output of solar and wind power generation should be adapted to the minimum feasible transmission capacity of the power transmission line. If necessary, the power stations should be divided into sub-units. Provision must also be made for the use of the generated electricity at local level.

As far as is compatible with competition law, the EU should ensure that the project package is sufficiently economically attractive by signing time-limited electricity purchase agreements at guaranteed prices in order to secure private-sector support for the project's implementation. It is also important to create the political and diplomatic framework for the strategic partnership. This could take place through the Economic Partnership Agreements currently being negotiated between the EU and the ACP countries.

DISTRIBUTED ENERGY SUPPLY THROUGH CLIMATE-NEUTRAL LIQUEFIED GAS
In developing countries, the traditional use of biomass often poses a major problem (health impairment caused by fumes, over-exploitation of local timber stocks; Section 3.2.4.2), which could be reduced by progressively replacing the traditional three-stone hearth with liquefied gas cookers. However, the large-scale use of liquefied gas from fossil sources cannot be viewed as sustainable in the long term in the interests of climate protection. There is, however, the option of producing this energy carrier from biomass: Through gasification or anaerobic digestion/conversion, synthesis gas (CO/H_2) can be produced from biomass which can then be converted into longer chain hydrocarbons. In this way, biogenic liquefied gas could be manufactured. The chemical processes involved can be supported through the use of solar thermal energy. A starting point for this project could be the EU Energy Initiative: Energy for

Poverty Eradication and Sustainable Development. The WBGU recommends:
- Initiating the substitution of traditional three-stone hearths with liquefied gas cookers through German development cooperation;
- Through research cooperation with developing countries, creating facilities for the environmentally compatible synthesis of liquefied gas, adapted to local conditions.

ENERGY-EFFICIENT BUILDINGS IN THE LOW-COST SECTOR, PILOTED BY SOUTH AFRICAN TOWNSHIPS

Since 1994, more than one million new dwellings have been constructed in South African townships in order to improve the living conditions of disadvantaged population groups. However, the aspect of sustainable construction was largely overlooked during this process. For example, tin roofs without any insulation have generally been constructed, resulting in unbearable indoor temperatures in summer and winter alike. Open coal fires cause pollution which is up to eight times higher than the international standards, resulting in health costs amounting to €244 million per year (Holm, 2000). The WBGU recommends that through German development cooperation and in conjunction with South African partners, pilot projects on energy-efficient construction in the low-cost sector be carried out. Due to the multiplier effects, it is specifically recommended that these projects be undertaken near well-frequented sites (e.g. railway stations). This type of project could be carried out as part of the 'Global Village Energy Partnership', which is a WSSD initiative.

IMPROVING THE POWER QUALITY IN WEAK ELECTRIC GRIDS IN RURAL AFRICAN REGIONS

When bringing electricity to rural regions in developing countries, a frequent problem is that due to low user density, large distances have to be bridged in weak electricity grids. This reduces electricity quality (voltage, frequency and reliability of the grid), especially for users in more remote areas. The technologies developed in Europe to integrate the different renewable energy sources into the grid could be utilized profitably and cost-effectively to improve this situation, but they are still largely unfamiliar to the local grid operators. The WBGU recommends that within the framework of technical and financial cooperation, a selected rural region be connected to the grid in cooperation with a larger African energy supplier and using appropriate new technologies. Cooperation with the local grid operator is essential in order to produce a multiplier effect. Here ,too, the starting point could be the EU Energy Initiative 'Energy for Poverty Eradication and Sustainable Development'.

'ONE MILLION HUTS' PROGRAMME

As part of the process of supplying electricity to rural regions in developing countries, distributed concepts such as individual photovoltaic systems and microgrids are essential – alongside intelligent grid expansion – as a response to low population density. Until now, this type of practical project has taken place on too small a scale to develop the necessary momentum, and social and technical conditions have generally been overlooked. The WBGU therefore recommends launching a 'One Million Huts' programme on the required scale and with the necessary duration, which must also include a new dimension of technical and socio-economic support. For the project to have a sustained impact, it must draw on the expertise of leading companies from industrialized countries, launch regional training programmes and develop local financing structures and supply industries. The Global Village Energy Partnership Initiative offers an appropriate framework for project implementation.

Research for energy system transformation

6

The task of energy system transformation has a magnitude comparable to that of a new industrial revolution. It will continue to pose major technological and societal challenges for many decades. Therefore, for the process to succeed, a substantial research effort is essential to support transformation – both in advance and throughout the process. In Section 5.3.1 of the present report, the German Advisory Council on Global Change (WBGU) has already made concrete recommendations on ways to further develop research structures in institutional and financial terms. The present chapter focuses exclusively on questions of research content.

The Council's aim here is not to present a comprehensive research strategy or analysis of research activities surrounding the theme of energy. The purpose of this chapter is rather to identify those research themes which, in the course of work on this report, have emerged as key preconditions to implementing the energy system transformation that this report outlines.

No attempt is made to specify which research actors at which levels are best suited to tackle the research questions presented here. The suggestions made here do not target specifically the German or European research landscape, but are directed to all countries and actors with an interest in energy system transformation. These suggestions can represent important components of various research programmes at various levels. Consequently, the individual points are not placed in relation to the great diversity of already ongoing and highly committed German or European research programmes.

The Council always examines energy systems from the perspective of environment and development issues. Research on systems analysis, presented in Section 6.1 below, therefore plays a key role. When implementing and applying system transformation recommendations, economic, political and societal tasks arise that need to be prepared and supported through research. These include the market penetration of new technologies, the comparative analysis of socio-economic instruments, the management of technology transfer or the transition to sustainable lifestyles (Section 6.2). Finally, research on and development of new technologies are key to the success of energy system transformation (Section 6.3).

6.1
Systems analysis

KNOWLEDGE BASE FOR GUARD RAILS
'Guard rails' demarcate the viable action space, and thus play a key role in the policy advice provided by the WBGU (Section 4.3). It remains difficult, however, to determine these socio-economic and ecological boundaries of the tolerable domain of human activities. Neither is the necessary knowledge available in satisfactory quality on, for instance, the precise position of 'catastrophe domains' that must be avoided unconditionally, nor are modelling approaches sufficiently mature to give adequate consideration to all important factors impinging upon predictions. As a first step, the knowledge base for setting normative guard rails thus needs to be improved. This can be done by means of modern and partly unconventional modelling approaches. Systems analysis is thus part of an iterative process, continuously improving the basis for predictions about future development trajectories.

MODELLING
Models help to explore the entire action space and thus to develop consistent predictions and scenarios constrained by normative guard rails. There is a substantial need for research to further develop the methodologies of existing models, for instance with regard to the coupling, regionalization, sectoralization and integration of climate, land-use and macro-economic energy system models (integrated assessment models). There is also a need to develop novel modelling techniques (qualitative, semi-quantitative and hybrid models) that do greater justice to the inherent uncertainties. It would be useful in this context to endogenize various processes (such as technological progress), which should no longer be examined in isolation from macro-economic development.

Important specific questions include the effects upon long-term investment behaviour of emissions reduction targets pre-announced in binding form (Section 4.5.1), the potential risks of path dependency raised by sequestration strategies (Section 3.6), the impacts of land uses upon greenhouse gas emissions, and the economic effects of control instruments such as certificate trading.

Developing theoretical concepts of sustainability

There is still a lack of fundamental theoretical concepts of sustainability, which would need to be linked with Earth System analysis and the guard rail concept. These should help to answer the question, for instance, which of the possible development paths can be termed acceptable under the conditions of a certain sustainable development paradigm (pessimization, standardization, etc.; Schellnhuber, 1998). How must control or management strategies – which can range from e.g. access to electricity over changing lifestyles through to strict non-intervention – be designed in order that the guard rails can be obeyed? How to develop early warning systems that register in time a slip into non-sustainable domains? Which synergies among anthropogenically influenced, but not yet sufficiently understood, Earth System dynamics can be harnessed? May there even be a 'global conscience' that, as an emergent feature, commands over a collective perceptual faculty to the benefit of humanity and the Earth System? The answers to these questions will ultimately determine the implementation and further development of international regimes (such as the UNFCCC).

Scenario development

The findings presented in this report show that it is also recommendable to develop CO_2 stabilization scenarios for low equilibrium concentrations (<450ppm). To this end, the regional and sectoral resolution of the corresponding models needs improvement. Building upon regional quantifications, it may be possible to develop stabilization scenarios at low global warming levels, which are able to model the specific properties of sectoral and regional units and thus provide an improved basis on which to identify concrete options for action.

Climate sensitivity

It is evident that climate research can make a key contribution to energy system transformation, for it delivers the knowledge and the substantiation for the climate protection guard rail (Section 4.3.1.2). If CO_2 and climate should be found to be coupled more closely than has previously been thought, or if unknown amplifying ecosystem processes are found,

then this would call for not only an accelerated energy system transformation, but also a transformation of land uses (Section 4.6). Sensitivity to anthropogenic perturbations therefore continues to be one of the crucial factors of the climate system that is not yet sufficiently quantifiable (Sections 4.3.1.2 and 4.5.2.1). As a consequence, the IPCC refrained in its 2001 Third Assessment Report from stating a mean sensitivity of the climate system that would follow from a doubling of the CO_2 concentration. Considerable research efforts need to be undertaken in order to be able to better assess climate sensitivity.

CO_2 sources and sinks

Since the 1950s, the USA has been operating important measuring stations (e.g. on Mauna Loa in Hawaii, at the South Pole, and other island stations), which form the backbone for assessments of the carbon cycle to date. There is an urgent need for Germany and Europe to make stronger contributions in the future to efforts to verify trends in CO_2 sources and sinks: A worldwide observation system should be established on the continents and linked to the existing marine observation system. With such a system it would become possible to check the hypothesis that, at present, a major part of anthropogenic emissions is stored in the boreal forests of Siberia. Responsible ecological management to strengthen this natural sink could provide a valuable additional option for the energy system transformation process.

Impacts of high CO_2 concentrations and of climate change upon terrestrial ecosystems

The forecasts of the impacts of climatic changes upon terrestrial ecosystems in Europe still have a wide range and are thus greatly in need of improvement. A first step is to optimize the database, which is the precondition for more accurate model computations and predictions. Moreover, the models themselves need further development – this is a point also stressed by the EU's 6th Framework Programme. Consideration further needs to be given to the responses of silvicultural and agricultural plant species and varieties.

Impacts upon soils

Europe's soils store large quantities of carbon accumulated in approx. 10,000 years of vegetation development following the ice age. If substantial parts of this carbon were released through land-use changes, then even complete implementation of the measures for energy system transformation recommended in this report (Chapter 5) would not suffice to obey the climate guard rail (Section 4.3.1.2). There is a need for basic research to improve the understanding of accumulation and degradation processes in soils. The findings of such research could also create the pre-conditions for harnessing further control options, and for the accounting of carbon changes in soils.

6.2
Social sciences

A key task of social science research for energy system transformation must be to elaborate options for action at the political and economic levels, and to assist in selecting those options that are indispensable and most suitable for system transformation. Potential barriers standing in the way of system transformation need to be identified and analysed, and ways to remove them found.

EFFECTS OF LIBERALIZATION AND GLOBALIZATION IN THE ENERGY SECTOR

In the course of globalization and liberalization, complex new conditions have emerged in the energy sector. To identify their implications for sustainable development adequately, it will be essential to establish a joint, international research effort. In particular, research is needed to develop recommendations that combine sector liberalization with compliance with ecological and social guard rails (Section 4.3). Research on the benefits and drawbacks of various deregulation and regulatory instruments and on prevailing market barriers remains one of the key tasks of socio-economic research relating to the energy sector. It needs to be clarified which market structures promote energy system transformation goals, and what influence liberalization and globalization exert upon structures, notably upon supplier structures of energy production and supply markets.

There is a need for intensified research on the effects of private foreign direct investment (FDI) in liberalized energy markets. It needs to be clarified under which conditions FDI will tend to rather promote or hamper the development of sustainable energy systems. Furthermore, there continues to be a need for research on the opportunities for and limits to orienting the behaviour of transnational corporations abroad to sustainability requirements, for instance through quality labels, voluntary codes of conduct, international soft law, extraterritorial application of liability and environmental law, or international environmental and social standards.

Analysis of small and medium-sized enterprises (SMEs), which include many providers of wind and solar energy technology, and of their importance to energy system transformation should become a focus of research in economics and political sciences. A recent study has shown for Switzerland that most state assistance provided to SMEs ultimately fails to address their specific needs and difficulties (Iten et al., 2001). It therefore needs to be clarified in particular to what extent SMEs may contribute to a worldwide diffusion of renewable energy use, and how SME foreign direct investment in the energy sector could be facilitated.

TRANSFORMING ENERGY SYSTEMS IN DEVELOPING COUNTRIES

There is a need for supplementary study on minimum energy requirements, and on the socio-economic linkages between poverty, lack of energy and development barriers. Comprehensive primary data on energy use is lacking, above all in newly industrializing and developing countries. The World Energy Outlook 2002 provided for the first time a country-specific analysis of electricity and biomass consumption (IEA, 2002c). However, comparable data on the use of other energy carriers as well as analyses of the potential to deploy the various renewable energy carriers are still largely absent. Data collection needs to distinguish more clearly between the conditions governing access to energy in urban areas and those in rural areas. Given that urban-rural divergence is expected to intensify, research must provide specific answers concerning the ways to ensure supply in these two types of area. It needs to be kept in mind that different barriers to energy access prevail in the two settlement types. When researching the barriers preventing the poor from gaining access to sufficient and affordable energy services and to forms of energy acceptable in environmental and health terms, careful consideration needs to be given not only to the opportunities and risks of privatization and liberalization. The ambivalent role of major private sector and state power supply companies also needs to be explored.

Evaluation of the potential to reduce and control greenhouse gas emissions in developing countries carries particular importance for sustainable climate protection in view of anticipated economic growth and improved private household access to energy services in these countries. Here there is a need for further research, in particular regarding the design and monitoring of the flexible Kyoto Protocol mechanisms.

The Council has called in this report for a long-term phase-out of traditional biomass use; this gives rise to a considerable need for social science research on the suitable path towards this goal. The number of people using health-endangering traditional biomass for cooking and heating in developing countries will not drop over the next 30 years if no targeted measures are taken (IEA, 2002c; UNEP, 2002). Even if all people were to gain access to power or gas, past cooking and heating habits would not be abandoned immediately; traditional biomass use would persist, at least partly. It remains unclear which incentive systems and which logistic effort would be required to overcome these behavioural patterns and deliver the transition to healthy and culturally as well as financially appropriate energy carriers. Here the Council sees a need for interdisciplinary case studies across a range of environmental, economic and cultural settings, integrating economics and cultural studies, engineering sciences and health studies.

The same techniques are handled differently by different cultures. Technologies therefore not only need to be developed, but also transferred, adapted to local circumstances and integrated within existing social systems. To overcome barriers, quantitative and qualitative improvements are needed in development cooperation in training with regard to energy systems as well as with regard to savings and investment. Furthermore, research on the acceptance of technical and financial systems needs to be intensified, together with representatives of the countries concerned and of indigenous as well as local communities.

HEALTH IMPACTS OF ENERGY SYSTEMS

A programme of empirical research needs to be launched that is capable of identifying the adverse health effects of different energy systems (in extraction, transportation and use), building upon, for instance, the Disability Adjusted Life Years (DALYs; Section 4.3.2.7) methodology developed by the WHO. The goal of this research would be to quantify the connection between energy use and disease burden.

FINANCING REQUIREMENTS FOR SYSTEM TRANSFORMATION

Global energy system transformation requires the redirection and mobilization of substantial investment funding, particularly in the initial phases of the process. Aspects of particular interest include the differentiated quantification of the short to medium term, regionally specific investment requirement, and of the necessary capital transfer from industrialized to developing, newly industrializing and transition countries (IEA, 2003). Options for financial arrangements at an international level also need further research, e.g. the question of the extent to which the Kyoto mechanisms and an adaptation fund may contribute to capital transfer.

INSTITUTIONAL DIMENSIONS

To reduce the economic uncertainties attaching to climate protection efforts, new ideas expanding upon certificate trading are under debate. There is a need for further research on their prospects for implemen-

tation. They include e.g. the concept of a 'safety valve' combining quantitative restrictions (through certificates) and levies (acting in a manner similar to price caps for certificates). Proposals seeking to supplement or substitute internationally agreed quantitative approaches by means of a CO_2 levy also deserve further study.

Implementation of the Kyoto mechanisms in energy projects is currently a focus of policy analysis and at the same time a research task for numerous institutions such as the World Bank, GEF, OECD or IEA. These efforts must be continued and, where possible, integrated within the overall context of a global research strategy (Section 7.6). Research efforts should also address the further development of the Kyoto Protocol after 2012, for instance the issue of how flexible mechanisms can be developed in order to involve newly industrializing and developing countries in reduction efforts more closely than in the first commitment period.

Implementation of the Multilateral Energy Subsidization Agreement (MESA) proposed by the Council will require in-process research, in particular on the Agreement's concrete design, institutionalization and enforcement mechanisms.

ENERGY USE

The debate on energy system transformation needs to give greater attention to the high savings potential on the demand side – particularly given continuing population growth. The development of sustainable patterns of consumption, as called for at the Rio de Janeiro Conference on Environment and Development (UNCED) in 1992, can be achieved mainly by means of efficiency measures on the demand side. This will need to be supported by changed attitudes among consumers, and a transformation of the 'western-industrialized lifestyle'. This goes beyond energy efficiency, and concerns the prospects for reducing energy demand (sufficiency); major barriers yet stand in the way of any such development. To develop socially equitable options for removing these barriers, research needs to tackle the issue of 'lifestyles', which has not yet been studied systematically on a global scale. There is a great need for research in this field. Furthermore, in view of the inadequate implementation of international resolutions on consumer-focussed sustainability policy, it is currently unclear which policy solutions should be further pursued to tackle consumption patterns at global level. Further research is needed here in order to better shape and implement guidelines for sustainable consumption (UNEP, 2002).

There is still a major lack of understanding of the links between information availability, energy requirement and economic performance. For instance, it is still unclear whether the 'New Economy' has led to a reduction of the private sector's energy requirement. An empirical study conducted in the USA identified such a trend in the late 1990s, but was unable to verify causality unequivocally (Sanstad, 2002). This debate ties in with questions surrounding the dematerialization of the economy, and developments towards a service society. Both are variable quantities with contradictory impacts upon energy demand. It will be essential to have more knowledge about these connections to make targeted recommendations on how to shape the transition towards sustainable energy systems.

INSTRUMENTS FOR THE DIRECT PROMOTION OF RENEWABLES

An extensive range of potential mechanisms by which to promote the use of renewables is available. Although there is broad debate in industrialized countries on the individual mechanisms, a need for further research remains on the long-term effects of these instruments, and on their transferability to other country groups or to the global level. Ways to link instruments within integrated packages of policies (e.g. levies and quotas for the various different sectors) need further study. It also needs to be examined in this context what the relationships among the various instruments are, i.e. whether, for instance, it is possible or purposeful to combine different instruments within one sector or within one energy or technology field, or whether the application of one instrument excludes other measures (e.g. CO_2 or energy tax and Green Energy Certificates). A question of particular relevance to the transformation process is the extent to which a universally valid 'promotion roadmap' can be drawn up for the deployment of instruments, and whether it is possible to plan shifts between instruments over the course of time (e.g. initial financing through price instruments such as state subsidies or fixed rates for power sold to the grid, followed by a long-term selective transition to quantitative instruments such as quotas and Green Energy Certificates). Promotion in the field of electricity supply must give attention to the economic-technical aspect of how to tie renewables into integrated mains-borne services (suitability for distributed feed-in with high investment and operating costs, irregularly fluctuating power output; Section 3.4.3). These aspects still require major research efforts, for instance on how to shape distributed power plants or distributed feed-in structures and the question of their long-term effects, including socio-economic effects.

GEOPOLITICAL RESEARCH REQUIREMENTS

In Germany, geopolitical research lags far behind international standards. In other western countries there are research centres and specialized journals on geopolicy, but not in Germany. Peace and conflict research has albeit focussed increasingly since the 1990s upon intra- and inter-state resource conflicts and the threats that these pose to peace. However, as long as the major dependency upon energy imports could be resolved relatively easily through markets, as long as no scarcity problems were perceptible, and as long as the USA ensured in both political and military terms the resource security of Western Europe and Japan in addition to its own, security policy research and research promotion lacked a vital interest in the geopolitics of resource security. This tended to be treated as a marginal theme, on the fringes of the debate, emerging after the end of the cold war, on 'new threats' and the concept of 'extended security'.

There is a major need for research, but a lack of research resources in Germany. This is especially apparent in comparison to France and Great Britain, both of which have a global policy tradition and – still – ambition. While in these two countries about 2,500–3,000 (and in the USA even around 10,000) researchers work on international affairs, in Germany they number just 250–300.

Key research issues from a German and European perspective are:

- Are there geopolitical tendencies indicating that competition over energy resources is once again making war a means of policy? How could the United Nations be placed in a position to fulfil its mandate to maintain peace, as established in its founding charter, despite heightening international conflicts over resources and growing unilateralism on the part of the USA?
- What opportunities does Germany have, together with the other EU member states, to safeguard its resource security through peaceful means, i.e. through market relations and scientific as well as technical cooperation?
- How can the EU influence political, social and economic developments in the CIS states at its periphery (the Caucasus and Central Asia) in order to promote peaceful development in this region rich in energy resources which, while instable, is of great importance to the global economy?
- How can Germany and the EU provide targeted support to renewables and energy efficiency in order to reduce dependency upon fossil fuel imports, and thus also mitigate potentially violent competition over energy extraction regions?

6.3
Technology research and development

The exemplary transformation path (Section 4.4) developed by the Council relies equally upon the strong expansion of renewable energy use and major efficiency improvements. Such a transformation of the global energy system will only succeed if research and development is pursued or intensified across a broad range of highly diverse fields of technology.

The sources that can only be expanded to a limited degree (e.g. wind, hydropower; Section 3.8) are partly already available today at competitive prices, so that research now needs to focus above all upon further improving efficiency, tapping new fields of application and reducing environmental and social impacts.

In contrast, those sources that can be expanded almost without limit, such as solar-electric energy conversion, are still relatively expensive today in micro-economic terms (Section 3.2.6). Nonetheless, the appraisals of sustainably utilizable potentials show that, over the long term, solar-electric energy conversion must become the main pillar of global energy supply. For cost-reducing learning processes to take place swiftly in this field, it is essential to continue pursuing the related research and development activities vigorously, besides ensuring a committed and sustained rate of expansion. An excellent basis has already been established in this respect in Germany and Europe, thanks to state R&D programmes as well as industrial activities. Learning must be accelerated so that solar energy is available at sufficiently low cost at the point in time when the expansion of other renewable forms of energy meets the limits of its sustainably utilizable potential (Section 3.8).

At the same time, the integration into global energy supply structures of renewable energy from mostly fluctuating sources requires the further development of broad-scale, networked energy distribution structures. In this connection, suitable energy storage systems will need to be developed over the long term (Section 3.4).

6.3.1
Technologies for supplying energy from renewable sources

PHOTOVOLTAIC ELECTRICITY GENERATION (SOLAR CELLS)

Along with solar thermal power generation, photovoltaics (PV) is one of the two key technologies for solar-electric energy conversion (Section 3.2.6). The Council welcomes the current intense research and

development effort in this area and stresses that these efforts should be continued in a dedicated fashion, since in the long term they represent an important element of the exemplary transformation path proposed by the Council. Several promising approaches for cost reduction and increased efficiency are currently being pursued. Since a sound evaluation of the different approaches in terms of long-term developments is currently not possible, support for a diverse range of technologies should be maintained. The emphasis for the medium term should be on environmentally benign silicon technology. Current research activities relating to manufacturing processes of thinner wafers (150μm) and ultra-thin wafers (target: 50μm) should be intensified. Crystalline silicon thin film technologies on foreign substrates continue to require high research and development effort. Furthermore, previous efforts regarding thin film technologies based on other environmentally justifiable materials should be re-intensified speedily.

For the development of power plant applications in sunny regions, research activities concerning PV power plants with optical concentration should be intensified, and the development of appropriate stacked solar cells, e.g. based on III-V semiconductors, should continue. For the long-term development of photovoltaics, the continuation of the research activities regarding organic and dye-sensitized solar cells is essential. Such concepts may form the basis for completely new photovoltaic conversion techniques that are not based on semiconductors. Ultimately, application-oriented basic research should be undertaken in order to assess concepts that are currently rather speculative but may offer high potential (e.g. quantum well structures, thermo-photonics, multi-band cells, auger cells, cells with extraction of hot load carriers, self-organizing organic photovoltaic structures, application of molecular antenna structures for energy conversion).

For all solar cell technologies, appropriate module encapsulation technology should be developed. Particular emphasis should be placed on fully automatic production, low material use and reusability of the photovoltaic elements and materials. Another important criterion for photovoltaics is security of raw material supply.

In addition to the development of actual solar cells and modules, photovoltaic systems technology should be considered more strongly in the allocation of research projects. For concentrator power plants, this also includes optical components. In order to achieve further strong cost degression in system technology, highly integrated power electronics and digital control technology as well as new network monitoring techniques are required. Significant progress is also required in terms of integration into buildings, so that in future solar technology will become an integral component of the building envelope, rather than merely being an add-on.

SOLAR THERMAL POWER PLANTS

For the long term, solar thermal power plants with optical concentration are the second important cornerstone of a solar power supply in the exemplary transformation path, along with photovoltaics (Sections 3.2.6 and 4.4). For large plants, developments concentrate on tower and trough power plants. Both approaches are promising, and potential for improvements should continue to be opened up through research and development. Current activities inadequately reflect the importance of this technology in the Council's exemplary transformation path, and they should therefore be strengthened significantly. For power plants based on optical linear concentration, new optical concentration concepts (e.g. Fresnel concentrators) should be investigated, not least from a cost reduction point of view. Moreover, particular emphasis should be placed on materials research for optical and thermal components (mirrors, selective optical absorbers, etc.) and process technology, e.g. for direct water evaporation. Power towers, which can achieve higher temperatures and therefore higher efficiency compared with linear concentration plants, should be developed further.

For cost reduction and for operation during the hours of darkness, the further development of hybrid power plants combining solar and fossil technologies and of large heat stores for high temperatures is important. In view of the large proportion of global primary energy derived from solar-electric energy conversion, as envisaged from the middle of this century within the Council's exemplary transformation path, advanced concepts for the smooth integration of such power plants into energy supply systems should be developed.

SOLAR THERMAL ENERGY CONVERSION (SOLAR COLLECTORS)

Solar heat can be used for space heating, water heating, cooling and process heat applications (e.g. in the food processing industry) (Section 3.2.6). Its sustainable potential is largely determined by the local demand for heat and not by the supply. If the aims formulated in the Council's exemplary transformation path are to be achieved, research should also focus on solar technologies for the cooling of buildings. For water heating and space heating applications, the further development of heat stores with high energy density and low heat losses should have priority. For tapping new areas of application for solar heat, process heat collectors in the temperature

range 100–200°C should be developed further. The application of solar process heat for water desalination and purification and for food cooling is particularly relevant for global sustainability. For these varied applications, system technology, including associated control techniques, should not be neglected.

WIND POWER PLANTS

After solar energy, the utilization of wind power is the second most important renewable energy source in the Council's exemplary transformation path (Section 3.2.5). For the further rapid tapping of the sustainable wind energy potential, particular emphasis should be placed on the development of the offshore sector, where specific questions relating to installation at sea and the development of systems with larger capacity need to be addressed. The WBGU welcomes the current environmental research associated with offshore wind power (BMU, 2002c), since such research is an important prerequisite for the sustainable large-scale utilization of wind energy. Further improvements in rotor blade quality and the application of reusable materials should be amongst the aims, with emphasis on improved stability and self-cleaning surfaces. The assured duration of plant operation for wind energy conversion plants should be increased. System management strategies for integration into the grid, associated grid control and the control behaviour and early detection of faults should be developed further. With regard to future export markets and in addition to innovative areas of application (e.g. water desalination), the integration of wind power plants into weak grids under different climatic conditions requires further research.

HYDROPOWER

Due to increased requirements in terms of environmental and social acceptability, the WBGU views the sustainable potential of hydropower cautiously, with only 15EJ per year being considered in the exemplary transformation path for the year 2100 (Section 3.2.3). An important prerequisite for a sustainable expansion of this order of magnitude is a significant improvement of the scientific database over the next 5–15 years. Currently, there is a lack of ecological and socio-economic regional data for the sustainability analysis of the socio-economic, landscape, ecological and health impacts of large hydropower projects. Regional studies are also a prerequisite for comparing alternative design options as required by international guidelines. This database cannot be developed short-term and simultaneously with the preparation of project-specific environmental impact studies. Further important research questions include the differences in social and environmental consequences between large and small hydropower installations in developing countries and associated research for the practical implementation of the proposals of the World Commission on Dams.

GEOTHERMAL ENERGY CONVERSION

In the long term, geothermal energy conversion should be able to contribute to a demand-oriented and location-independent energy supply system, providing a complement to supply from other renewable energy sources that is independent of weather and seasons. Notwithstanding the possible benefits of geothermal heat utilization, the Council has identified a large number of unresolved questions regarding the technological implementation and various sustainability aspects, so that in the exemplary transformation path the realistic sustainable potential for 2100 is estimated cautiously as 30EJ per year (Section 3.2.7). Utilization of the potential initially requires cost reductions for deep drilling operations. New, yet to be developed stimulation techniques of hot rocks in the ground will be able to increase the productivity of hot deep water and therefore the yield of geothermal systems. The development of suitable and cost-effective district heat networks is an important prerequisite for the application of geothermal energy for heating purposes. However, further research effort is needed for the efficient conversion of heat from deep water (including low-temperature water) into electricity, so that power plants suitable for covering base loads can be developed.

In order to avoid environmental damage through brine or gases at the earth's surface, the compatible reintroduction of the extracted water into the ground should be examined further. Moreover, ecological management of the large quantities of waste heat that are generated due to comparatively lower process efficiencies is required for geothermal power plants.

UTILIZATION OF BIOMASS FOR ENERGY GENERATION

The exemplary transformation path (Section 4.4) envisages an expansion of modern biomass utilization for energy generation to approximately 100EJ per year by 2040, i.e. five times the current value (Section 3.2.4). In view of this ambitious target, research activities should initially concentrate on optimum land use, given the competing requirements of food production, energy generation and carbon storage. The structures for transporting the raw biomass to the associated conversion plants should be examined and improved. For combustion technologies, further research should focus on cost reduction and emission reduction. Since the Council's exemplary transformation path points towards a hydrogen economy, further research should focus on technolo-

gies for the efficient gasification of biomass, the production of fuels from biomass and for associated distribution and utilization structures. The production of hydrogen from biomass should be further developed, using fermentation and reforming techniques on the one hand, and also through direct production of synthesis gas. Since many of these technologies can also be realized modularly, they are suitable for both centralized and distributed applications. The latter may be particularly appropriate for application in developing countries. Appropriate production methods for biogenic liquefied gas from bio-energy should be developed.

INNOVATIVE CONVERSION TECHNOLOGIES

In future, unforeseen technological developments are expected to lead to better tapping of renewable energy sources or to innovative conversion technologies. Therefore, application-oriented, customized basic research should be increased significantly. This would include scientific studies with particularly uncertain outcomes and 'speculative' research (see also examples in the photovoltaic power generation section). Some examples are listed here.

- *Photochemistry:* membrane structures similar to those in photosynthesis, hydrogen generation via photo-electrochemical techniques;
- *Solar chemistry:* synthesis of storable energy carriers, synthesis techniques that simultaneously use thermal, optical and electrical energy;
- *Biotechnology:* microbial hydrogen generation.

6.3.2
System technology for sustainable energy supply

The special features of fluctuating energy sources make research and development for system technologies a basic prerequisite for the transformation of energy systems (Section 3.4) since renewable energy sources have to be integrated in global energy supply structures without disturbing them. In addition to adaptation of the structures of global electricity supply, the further development of the technological basis for a hydrogen economy (Section 3.4.4) is a key part of such a transformation in accordance with the Council's exemplary path.

ELECTRICITY TRANSPORT AND STORAGE

The generation of electricity from renewable sources is not only possible in mid latitudes, but particularly effective in arid, sunny areas. Hence, transporting large amounts of electricity across long distances at high capacity with little loss is a key technology towards matching the supply of and demand for power on a continental scale.

High-voltage direct current transport and – in the long term – high-temperature superconducting thus should be stepped up. In the long run, strong intercontinental, bi-directional power grids up to the level of a 'global link' are to be developed for virtual electricity storage or to compensate for fluctuations (in connection with dispatchable power plants) (Section 3.4.3). The design and management of power grids is to be improved to accommodate for the large-scale inclusion of fluctuating renewable energy sources. Alternative concepts for the storage of electricity and other types of energy should be further researched (such as compressed air, centrifugal masses, superconducting magnets). Electrochemical storage for distributed applications should be further developed for use in automobiles, off-grid solar power systems, etc.

DISTRIBUTED GENERATION IN ELECTRICITY GRIDS
The use of renewable energy sources in the exemplary transformation path (Section 4.4) can be divided into distributed, off-grid applications (such as Solar Home Systems), centrally connected power plants (such as geothermal electricity generation) and distributed generation within grids (such as fuel-cell combined heat and power plants; Section 3.4). In particular, distributed generation in power grids poses a great challenge to future grid control strategies. Communications technologies for distributed power generators and the overriding grid should be further developed to accommodate for these trends (such as controls, log-in and log-out of generators as in the Internet, etc.). In addition, strategies for optimal electricity and (distributed) heat use along with customized bidirectional grid architectures and safety systems are on the research agenda. The further development of power generators adapted to this purpose (fuel cell systems, microturbines, etc.) is indispensable as a basis. Power electronics will also act as an important interface between distributed generators and the grids, handling many additional functions (improvements in voltage quality, etc.). Finally, the development of heat accumulators with great storage density and negligible heat loss should be stepped up as this is especially promising with regard to decoupling the generation of electricity and heating needs in homes. Intelligent control technology can minimize losses here.

HYDROGEN GENERATION, TRANSPORT AND STORAGE
Hydrogen is an essential element both as a secondary energy source and as an energy storage medium in a sustainable energy system such as the one designed in the Council's exemplary path (Sections 3.4.4 and 4.4). Many of the technologies needed are not, however, ready for the market so that broad research and development remains necessary. In producing hydrogen, research should take account both of the various electrolysis methods using electricity and different kinds of thermochemical methods based on hydrocarbons (such as biomass). Hydrogen storage systems for distributed applications (such as cars) and the large-scale, central storage of hydrogen must be further developed in combination with gas transport, with globally relevant leaks in hydrogen systems kept to a minimum (Section 3.4.4.5). Technologies for the use of hydrogen in engines and turbines – especially the important field of fuel cell technology – require further technological progress.

ENERGY METEOROLOGY
Information about the potential of solar and wind energy fluxes are of great global interest for the reliable, large-scale use of renewable, fluctuating energy sources, especially in developing countries. Forecasts of local energy fluxes (such as wind speeds near the ground in the range of minutes to days) via remote sensing (earth observation satellites) should be further developed.

APPROPRIATE SYSTEM TECHNOLOGY FOR DISTRIBUTED, OFF-GRID APPLICATIONS
When technically useful distributed energy from renewable energy sources is available (Section 3.4.2), related energy service technologies will have to be (further) developed. Some examples are off-grid water purification technologies and communication technologies (Internet access, etc.). In addition, electric systems technology, power electronics, and control technology will have to be optimized for use with special applications in everything from individual solar power supply (Solar Home Systems) over microgrids (for villages) through to integration in larger networks.

6.3.3
Development of techniques for more efficient energy use

More efficient energy utilization along the complete chain of the energy system (from the conversion of primary energy, for example in power plants, to the supply of energy services through technologies such as domestic appliances, thermal building insulation or lighting systems) is a significant aspect of the transformation of energy systems (Sections 3.5 and 4.4). The analysis of potentials and barriers to their implementation (UNDP et al., 2000) clearly shows that further research is required, not only in terms of technology and development in the narrower sense, but also in terms of accompanying and supplementary socio-economic research with a view to breaking down barriers and to creating appropriate incentive structures and energy policy frameworks. Questions of the social acceptance of modified user behaviour, the application of more efficient technologies, the more intense use of consumer goods (e.g. car sharing) and the development of settlements and transport structures that aim to reduce overall energy consumption require intensified research (Sections 3.5, 6.2).

COMBINED HEAT AND POWER (CHP)

The main individual technology for efficiency improvements on the supply side is combined heat and power (Section 3.3). In particular, research efforts for the expansion of decentralized applications should be intensified (engines, gas and micro gas turbines, fuel cells, micro-cogeneration units, Stirling engines).

SOLAR AND ENERGY-EFFICIENT BUILDINGS

The utilization of solar energy in the building sector (Section 3.5.2) leads to a reduction in primary energy use and is therefore frequently included in energy efficiency measures. Since a large proportion of total energy requirements is consumed in the building sector, this is an essential element of the transformation path (Section 4.4). It includes the following areas: solar-optimized windows with optical switching properties, solar-active opaque façade elements (e.g. translucent insulation), development of new thermal insulation systems (e.g. vacuum insulation), the development of flat heat stores with high energy density for surface implementation in walls and ceilings, and the development of central, compact heat stores with low heat losses. In tandem with solar building technology, appropriate building services have to be developed, including new air conditioning technologies for the low-energy building of the future, very small heating and refrigeration units, heat pumps, components for distributed electricity and heat production, etc. Daylighting systems for the internal illumination of buildings should also be developed further, e.g. light guidance and distribution systems with integrated switching properties. The integration of solar energy technologies into the building envelope should also be optimized in terms of aesthetic and cost considerations. Another important aspect is the development of urban planning procedures that enable optimum solar energy utilization in buildings.

EFFICIENT ENERGY UTILIZATION IN INDUSTRY

Research efforts are currently directed more towards energy converter technologies, with the improvement of energy efficiency at the user energy level being rather neglected. Government support for research projects should address these issues. Significant potential exists for the replacement of thermal production processes by physical-chemical or biotechnological processes, the recovery and storage of kinetic energy, increased material efficiency, the recycling of energy-intensive materials and their substitution by less energy-intensive materials. Appropriate research in these areas is therefore recommended. For more efficient industrial electricity utilization, research efforts should concentrate on industries with very high electricity demand (e.g. aluminium plants and metal smelting plants).

EFFICIENCY IMPROVEMENTS IN THE TRANSPORT SECTOR

Whilst in the long term the application of new technologies will be an important component of energy system transformation, particularly in road transport, in the short and medium term research efforts for higher efficiency in road and rail transport should continue and indeed be intensified. Examples are increased efficiency of combustion processes and weight reductions through new materials. Other aspects are the development of multi-modal infrastructures and the application of information technology (e.g. telematics). Finally, research regarding modern concepts of regional, urban and transport planning with the aim of reducing transport demand and energy use should be supported.

7.1
From vision to implementation: Using the opportunities of the next 10–20 years

Building upon the analysis of long-term energy scenarios in Chapter 4 and the options for action set out in Chapter 5, the present chapter proposes key policy objectives and activities, with time frames for their implementation (Fig. 7-1). These objectives and activities aim to prevent ecological and socio-economic guard rails being overstepped, and to return any non-sustainable state beyond the guard rails to a state within them (Fig. 7-2). The objectives and activities recommended here by the German Advisory Council on Global Change (WBGU) point to the direction that needs to be taken to permit a global transformation of energy systems towards sustainability. In view of the uncertainties that attach to all assessments of future developments, it will remain essential to continuously review the objectives, take into consideration new scientific findings and technological advances, and adjust the objectives and activities accordingly. The goals and measures set out here constitute key elements of the World Energy Charter proposed by the Council (Section 5.3.2). There is a particular need for action in the coming 10–20 years. That period presents the main window of opportunity to transform global energy systems. The intended effects will only occur with a certain time lag. This lag makes swift action all the more urgent. The German federal government should make use of its international weight, taking all steps to vigorously advance the transformation of energy systems within the context of global governance.

7.2
Protecting natural life-support systems

One of the two overarching objectives of the WBGU transformation strategy is to protect natural life-support systems; the other is to eradicate energy poverty (Section 7.3).

7.2.1
Reducing greenhouse gas emissions drastically

To keep global warming within tolerable limits, global carbon dioxide emissions need to be reduced by at least 30 per cent from 1990 levels by the year 2050 (Chapter 4). This concerns above all the CO_2 emissions from fossil fuels used to produce heat and power and the emissions of the transport sector; taken together, these account for some 85 per cent of all CO_2 emissions worldwide today. For industrialized countries, this means a reduction by some 80 per cent, while the emissions of developing and newly industrializing countries are allowed to rise by at most 30 per cent. Developing and newly industrializing country emissions should peak earlier, between 2020 and 2030 (Chapter 4). Without a fundamental reconfiguration of energy systems, emissions must be expected to double or even quadruple in developing and newly industrializing countries over that period. This is why in these countries, too, a rapid redirection of energy production and utilization towards an alternative technology path is essential. The focus of such activities needs to be placed on promoting renewables *and* enhancing efficiency. Further supplementary measures need to be taken in the agricultural and forestry sectors. In particular, the stocks of carbon stored in vegetation and soils must be protected.

In view of the considerable uncertainties, e.g. regarding climate sensitivity, these emissions reduction goals are minimum requirements. To this end, the WBGU recommends
- adopting a pioneer role in the implementation of Article 3.2 of the Kyoto Protocol (demonstrable progress) at the national and European level in the period up to 2005. This would strengthen confidence in the process and would create a basis for integrating the developing and newly industrializing countries.
- updating, by 2008, the Kyoto Protocol's reduction goals for industrialized countries (agreed in a step-by-step process in further commitment periods)

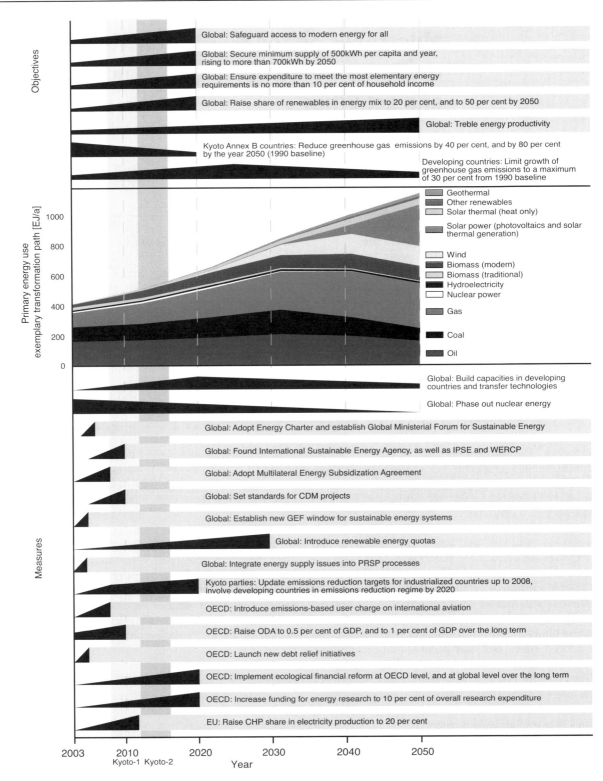

Figure 7-1
The transformation roadmap of the German Advisory Council on Global Change (WBGU). *CDM* Clean Development Mechanism, *CHP* Combined Heat and Power, *GDP* Gross Domstic Product, *GEF* Global Environment Facility, *IPSE* Intergovernmental Panel on Sustainable Energy, *ODA* Official Development Assistance, *OECD* Organisation for Economic Co-operation and Development, *PRSP* Poverty Reduction Strategy Papers, *WERCP* World Energy Research Coordination Programme.
Source: WBGU

Figure 7-2
Connection between guard rails, measures and future system development.
The figure shows possible states of a system in terms of its sustainability, plotted over time. The current state of a system relative to the guard rail can be in the green area (the 'sustainable area' according to best available knowledge) or in the red area (the 'non-sustainable area'). If a system is in the non-sustainable area, it must be steered by appropriate measures in such a way that it moves 'through' the guard rail into the sustainable area. The guard rail is thus permeable from the non-sustainable side. If a system is in the sustainable area, there are no further requirements upon it at first. The system can develop in the free interplay of forces. Only if

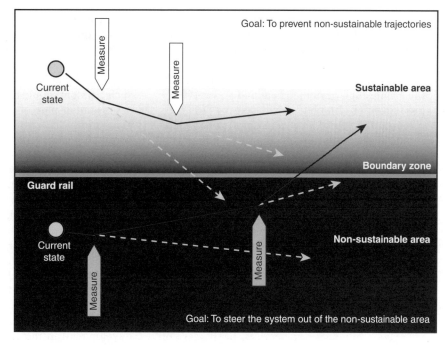

the system, moving within the sustainable area, is on course for collision with a guard rail, must measures be taken to prevent it crossing the rail. The guard rail is thus impermeable from the sustainable side. As guard rails can shift due to future advances in knowledge, compliance with present guard rails is only a necessary criterion of sustainability, but not a sufficient one.
Source: WBGU

and introducing emissions control requirements for developing countries by 2020 at the latest. Newly industrialized countries should accept first quantified requirements at an even earlier date. Furthermore, by 2005, Article 2 UNFCCC (prevention of greenhouse gas concentrations that lead to 'dangerous' climate change) should be concretized.

- taking up in international climate protection policy the issue of effective preservation of biospheric carbon stocks.
- tapping emissions reductions potentials in developing and newly industrializing countries by means of intensified cooperation with industrialized countries. This can be done through voluntary partnerships, through promoting the Clean Development Mechanism and through technology transfer above and beyond the CDM.
- swiftly integrating into the Kyoto Protocol quantified reduction commitments for aviation and shipping emissions.
- giving particular attention to emissions from agriculture and land-use changes. The importance of these is expected to grow in the future. Specific emissions reduction potentials should therefore be researched.
- completely removing all subsidies for fossil energy carriers and nuclear power in industrialized and transition countries by 2020, and worldwide by

2030. Moreover, subsidies for fossil energy carriers should be removed in developing countries by 2020 (Section 7.7.2)

7.2.2
Improving energy productivity

In order to minimize resource consumption, global energy productivity (the ratio of gross domestic product to energy input) needs to be improved by 1.4 per cent every year initially, and then by at least 1.6 per cent as soon as possible. Energy productivity needs to be doubled by the year 2030 from 1990 levels. This increase involves higher efficiencies in the conversion of primary to final energy, demand-side efficiency improvements as well as structural changes in national economies. There are various suitable forums and starting points for the political implementation of these goals: EU directives, a Global Ministerial Forum for Sustainable Energy yet to be established or, if necessary, an International Sustainable Energy Agency (ISEA), also yet to be established (Section 7.5). Moreover, minimum efficiencies of more than 60 per cent should be aimed at by 2050 for large fossil-fuel power plants. To this end, the WBGU recommends

- establishing international standards prescribing minimum efficiencies for fossil-fuelled power plants in a stepwise process from 2005 onwards, based on the European Union (EU) directive concerning integrated pollution prevention and control (IPPC Directive).

- generating, by 2012, 20 per cent of electricity in the EU through combined heat and power (CHP) production (EU target: 18 per cent by 2012). To promote this, the German federal government should argue within the ongoing negotiations on an EU CHP Directive for a challenging definition of 'quality CHP' and for the swift setting of binding national CHP quotas.
- initiating ecological financial reforms as a key tool by which to create incentives for more efficiency. This includes measures to internalize external costs (e.g. CO_2 taxation, certificate trading) and the removal of subsidies for fossil and nuclear energy.
- improving the information provided to end users, in order to promote energy efficiency, e.g. by means of mandatory labelling for all energy-intensive goods, buildings and services. In the case of goods traded internationally, cross-national harmonization of efficiency standards and labels is recommendable.
- exploiting the major efficiency potentials in the use of energy for heating and cooling through instruments of regulatory law targeting the thermal insulation and performance of buildings.

7.2.3
Expanding renewables substantially

In order to safeguard the protection of the natural environment despite growing worldwide demand for energy services, and in order to reduce to an acceptable level the risks associated with energy production, the proportion of renewable energies in the global energy mix should be raised from its current level of 12.7 per cent to 20 per cent by 2020 (of which at most 2 percentage points come from traditional biomass), with the long-term goal of reaching more than 50 per cent by 2050 (of which at most 0.5 percentage points come from traditional biomass). Ecological financial reforms will make fossil and nuclear sources more expensive and will thus reduce their share in the global energy mix. Consequently, the proportion of renewables will rise. However, this rise will remain well below the envisaged increase to 20 per cent and, respectively, 50 per cent. The WBGU therefore urges that renewables be expanded actively. For nature conservation reasons, hydropower should not be used to its full potential. In particular, the WBGU recommends

- that countries agree upon national renewable energy quotas. In order to minimize costs within such a scheme, a worldwide system of internationally tradable renewable energy credits should be aimed at by 2030. Its flexibility notwithstanding, such a system should commit each country to meet a substantial part of its quota through domestic generation.
- continuing and broadening market penetration strategies (e.g. subsidy schemes over limited periods, fixed rates for power sold to the grid, renewable energy quota schemes). Until significant market volume has been achieved, fixed rates for power sold to the grid with tapering payments over time are a particularly expedient option. When a sufficiently large market volume of individual energy sources has been reached, assistance should be transformed into a system of tradable renewable energy credits or green energy certificates.
- further intensifying investment in research and development in the energy sector, building upon the existing basis. Investment needs to grow at least ten-fold from present levels by 2020. The focus needs to be shifted to renewables and energy efficiency (Chapter 6).
- upgrading energy systems to permit the large-scale deployment of fluctuating renewable sources. This includes in particular enhancing grid control, implementing appropriate control strategies for distributed generators, upgrading grids to permit strong penetration by distributed generators as well as expanding grids to form international energy transport structures ('global link'). This should be followed later by the establishment of an infrastructure for hydrogen storage and distribution, using natural gas as a bridging technology.
- building and strengthening human-resource and institutional capacities in developing countries (e.g. through partnerships between German and developing-country institutions) and intensifying technology transfer.
- setting, within export credit systems, progressive minimum requirements for the permissible carbon intensity of energy production projects from 2005 onwards.
- providing vigorous support to disseminate and further develop the technologies involved in solar and energy-efficient construction.

7.2.4
Phasing out nuclear power

The use of nuclear energy has proven non-sustainable in the past, above all because of the risks associated with final storage, reprocessing, proliferation and terrorism (Section 3.2.2). For this reason, no new nuclear power plants should be given planning permission, nor be built. The goal should be to phase out the use of nuclear power worldwide by 2050. To this end, the WBGU recommends

- seeking to launch international negotiations on the phase-out of nuclear power. This process could begin with an amendment to the statutes of the International Atomic Energy Agency (IAEA).
- establishing by 2005 new, stricter IAEA safety standards for all sites at which plutonium is stored, as well as expanded monitoring and action-taking competencies of the IAEA in the field of safeguards relating to terrorism and proliferation.
- permitting nuclear power plant operation from 2010 onwards only if proof is furnished that there is a disposal path for nuclear fuel rods; a process needs to be initiated within the IAEA in this context.
- reviewing by 2010 EU moratoriums for the reprocessing plants in Sellafield and La Hague. This could take as a starting point the Convention for the Protection of the Marine Environment of the North-east Atlantic (OSPAR).
- harmonizing safety standards at an international level by 2010, with significantly increased levels of mandatory insurance cover for nuclear power plants; the goal would be for mandatory insurance schemes to operate entirely without state cover. Moreover, tax concessions should be removed. Points of departure here include the planned EU directive on safety standards for nuclear power plants, as well as IAEA's new anti-terrorism programme.

7.3
Eradicating energy poverty worldwide

The second overarching goal of the WBGU transformation strategy is to safeguard and expand access to modern forms of energy in developing countries and thus to eradicate energy poverty worldwide. This is a fundamental contribution to poverty reduction – attainment of the Millennium Development Goals is determined critically by questions of energy supply. To overcome energy poverty, in the age of globalization developments must be put on track not only through measures within the affected countries, but also by creating an appropriate international setting.

7.3.1
Aiming towards minimum levels of supply worldwide

IMPROVING ACCESS TO MODERN FORMS OF ENERGY
2,400 million people still lack access to modern forms of energy. The poor in the Least Developed Countries are those most affected. Particular challenges are presented by the switch from health-endangering biomass use for cooking and heating to modern energy carriers, and the provision of energy services that depend upon access to electricity. The WBGU recommends, as an international goal, that everyone should have access to the following per-capita annual quantities of modern energy forms in order to meet the most elementary energy needs: At least 500kWh from 2020 onwards, 700kWh from 2050 and 1,000kWh by 2100.

All measures undertaken to transform energy systems should take care to reduce regional and socio-economic disparities. Disadvantaged groups need to be supported particularly, and special cultural or gender-specific aspects observed.

The WBGU considers it only just acceptable if poor households must spend at most one tenth of their income to meet their most elementary energy service needs (500kWh per person and year). In some cases, this may require cross-subsidies or social transfers (state support for electricity and heating). It should be ensured by 2050 at the latest that no household is forced to spend more than 10 per cent of its income to meet its most elementary energy requirements.

IMPLEMENTING THE MILLENNIUM DEVELOPMENT GOALS
Access to modern energy is a key contribution to attainment of the development goals adopted in the United Nations Millennium Declaration. Particular importance attaches to reducing indoor air pollution in view of the major health hazards that this represents. This is joined by ambient air pollution in urban areas. In order to prevent respiratory diseases, health-endangering forms of traditional biomass use need to be phased out (Sections 3.2.4, 4.3.2.7).

7.3.2
Focussing international cooperation on sustainable development

IMPLEMENTING NEW WORLD BANK POLICY IN ASSISTANCE DELIVERY PRACTICE

The WBGU takes the view that the World Bank, which supports countries in efforts to expand their energy systems, should also regard itself as a bank delivering assistance for sustainable energy in order to facilitate leapfrogging. In efforts to promote the reconfiguration of energy systems, the World Bank has not yet moved sufficiently from the conceptual to the operational level. An urgent need thus remains to redirect its assistance delivery procedures, which until now have predominantly financed fossil fuels according to the least-cost principle. This has meant that micro-economically profitable forms of energy have been supported without ensuring the internalization of negative externalities. In a transitional phase, bridging technologies such as modern gas-fired power plants, together with their associated infrastructure, deserve support. However, for the least-cost principle to be accorded less weight, it would be necessary for the financial resources available for multilateral development financing to be increased substantially. The Council recommends that

- the new assistance delivery approach of the World Bank is implemented swiftly in practice. The German federal government should use its membership on the Board of Governors of the World Bank to work towards this.

INTEGRATING SUSTAINABLE ENERGY SUPPLY WITHIN POVERTY REDUCTION STRATEGIES

Sustainable energy supply needs to be integrated sufficiently within the poverty reduction strategies of multilateral organizations such as the IMF and World Bank. These began in late 1999 to focus their policies vis-à-vis Least Developed Countries upon poverty reduction. Poverty Reduction Strategy Papers (PRSPs) serve to steer the medium-term development of countries and provide a basis for eliciting international support (Section 5.3.3.3). Eradicating energy poverty is not among the issues being negotiated within the current PRSP process. The WBGU recommends

- integrating sustainable energy supply within PRSPs in order to raise the profile of energy-related issues in development cooperation. This would ensure that, within a development cooperation context, energy policy is linked even more consistently with poverty reduction policy.

STRENGTHENING THE ROLE OF REGIONAL DEVELOPMENT BANKS

The role of regional development banks should be strengthened. These have good regional connections and intimate knowledge of local problems. They can therefore be important partners in overcoming energy poverty in low-income countries. As a prerequisite to this, however, the management capacities of the development banks first need to be strengthened and expanded in a step-wise process. The WBGU recommends that

- Germany, in connection with its involvement in these banks and within the EU context, works towards ensuring that the regional development funds administered by the development banks promote energy supply in low-income countries.
- the EU makes targeted use of the European Development Fund to promote renewables in the ACP (African, Caribbean, Pacific) states. These countries generally lack the resources to ensure the supply of their population with modern forms of energy.

7.3.3
Strengthening the capabilities of developing countries

PROMOTING ECONOMIC AND SOCIAL DEVELOPMENT IN LOW-INCOME COUNTRIES

To turn energy systems towards sustainability, a minimum degree of economic development is a precondition. Many countries fall far short of the per-capita income required for this (Section 4.3.2.5). The WBGU therefore recommends not only intensifying development cooperation in the field of basic services and for sustainable energy supply, but also intensifying cooperation with low-income countries in particular, in both quantitative and qualitative terms. Furthermore, within the context of the WTO 'Development Round', improved access for goods from all low-income countries to the markets of industrialized and newly industrializing countries should be urged.

LAUNCHING NEW DEBT RELIEF INITIATIVES

In general, heavily indebted developing countries have little scope to cope with price fluctuations on world energy markets. Their ability to finance improvements to the efficiency of their energy supply systems and to advance the deployment of renewable energy technologies is similarly limited. To embark on transformation, wide-ranging debt relief is needed. The WBGU recommends that

- the German federal government argues for new debt relief initiatives within the G7/G8 context.

7.3.4
Combining regulatory and private-sector elements

It is essential to embark upon activities on both the supply and demand side in order to improve access to advanced low-emission energy forms and to renewable energy sources, and to improve the efficiency of energy use in developing, newly industrializing and transition countries.

SUPPLY SIDE: COMBINING LIBERALIZATION AND PRIVATIZATION WITH REGULATORY INTERVENTIONS
On the supply side, privatization and liberalization need to be combined with regulatory interventions undertaken by the state. The mix of these three spheres will need to vary depending upon the specific circumstances of a region. Liberalization and privatization require attractive framework conditions for private-sector investors and the tapping of international sources of capital. Stronger state intervention requires the setting of standards, and also an expansion of public-private partnerships, possibly supported by bilateral and multilateral development cooperation activities.

DEMAND SIDE: INCREASING THE PURCHASING POWER OF THE POOR
On the demand side, the aim must be to increase purchasing power in relation to energy, particularly of the poor. This can be done by target-group specific subsidies, or by expanding micro-finance systems. To increase not only purchasing power but also the willingness to use energy more sustainably, measures taken on the demand side need to give consideration to culture-specific and gender-specific framework conditions.

7.4
Mobilizing financial resources for the global transformation of energy systems

To finance the global transformation of energy systems towards sustainability, there is an urgent need to mobilize additional financial resources, as well as to create new transfer mechanisms or strengthen existing ones in order to support economically weaker countries in this transformation process. The WBGU welcomes the programme on 'Sustainable energy for development' geared to establishing strategic partnerships announced in 2002 at the World Summit on Sustainable Development. Over the next five years, a total of €1,000 million will be budgeted for this programme by the German government: 500 million for renewables and 500 million for energy efficiency.

MOBILIZING PRIVATE-SECTOR CAPITAL
It is desirable for efficiency reasons that a considerable part of the requisite investment is provided by the private sector. To mobilize private-sector capital for the global transformation of energy systems, the WBGU recommends
- intensifying policy advice within the context of development cooperation activities, in order to place partner countries in a position to create framework conditions conducive to investment.
- facilitating access to developing country markets for small and medium-sized suppliers of renewable energy technologies within the context of public-private partnerships.
- establishing by 2010 a German and, if possible, EU standard for the CDM. This standard should permit exclusively, with exceptions to be substantiated in each case, projects that promote renewables (excluding large hydropower), improve the energy efficiency of existing facilities or engage in demand-side management.

BOOSTING DEVELOPMENT COOPERATION FUNDING
At 0.27 per cent of GDP in 2001, German official development assistance (ODA) funding is far removed from the internationally agreed target – reaffirmed at the 2002 UN Conference on Financing for Development (UNFfD) – of 0.7 per cent. It is also at the lower end of the European range. Even an increase of these contributions to some 1 per cent of GDP would be commensurate to the severity of the problems prevailing. Germany committed itself at UNFfD to raise ODA funding to a level of 0.33 per cent of GDP by 2006.
- The WBGU recommends, as a matter of urgency, raising ODA funding beyond the level of 0.33 per cent announced for 2006, and proposes allocating, as a first step, at least 0.5 per cent of GDP for ODA by 2010.

HARNESSING INNOVATIVE FINANCING TOOLS
To implement the global transformation of energy systems, it will be essential to tap new sources of finance. In particular, the potential of raising charges for the use of global commons deserves examination. The WBGU recommends
- raising from 2008 onwards an emissions-based user charge on international aviation, to the extent that this sector is not yet subject by then to international emissions reduction commitments.
- that the initial allocation of emission certificates, with a time horizon of 20–30 years, is oriented to a modified per-capita approach. While complying with the WBGU climate guard rails, this could (depending upon system design) trigger an estimated transfer of financial resources in the order

of several hundred thousand million euro from industrialized to developing countries.

STRENGTHENING THE GLOBAL ENVIRONMENT
FACILITY AS AN INTERNATIONAL FINANCING
INSTITUTION

The Global Environment Facility (GEF) is a key catalyst for global environmental protection measures. It has proven its usefulness and should be further strengthened. The GEF only meets the incremental costs of projects of global benefit. The WBGU recommends

- concentrating by 2005 the financial assistance provided for efficiency technologies and renewable resources in a newly created GEF window ('window for sustainable energy systems'). In order to be able to give greater consideration to development policy aspects in the deployment of funds, a simplification of the incremental costs approach should be considered. With a view to the high levels of funding required to promote the global transformation of energy systems, GEF resources (currently US$3,000 million for the third phase from 2002 to 2006) need to be expanded considerably.

7.5
Using model projects for strategic leverage, and forging energy partnerships

SENDING OUT SIGNALS THROUGH MODEL PROJECTS

The WBGU argues in favour of using model projects to introduce new renewables on a large scale to deliver strategic leverage for a global transformation of energy systems towards sustainability. Such model projects could have global knock-on effects. They would showcase how technology leaps can be implemented in energy projects. The WBGU recommends initiating the following model projects (Section 5.3.8):

- *Energy partnership between the European Union and North Africa*: The EU should establish a strategic energy partnership with North Africa. Such a strategy would involve: building large-scale power plants for renewable electricity production in North Africa; creating capacities for transmission to the European interconnected power grid; and setting up a European focal point for North African project partners and European investors.
- *Distributed energy supply through liquefied gas:* In developing countries, traditional biomass use causes substantial health impacts. These could be averted by the step-wise substitution of three-stone hearths through liquefied gas cookers. It would be beneficial to base the production of liq-

uefied gas – or of a similar energy carrier – on modern biomass.

- *Energy-efficient buildings in the low-cost sector, piloted by South African townships:* Within the context of development cooperation activities, projects demonstrating energy-efficient and low-cost building techniques should be implemented in cooperation with South African partners. To produce a multiplier effect, such projects should be located in the vicinity of frequently visited places.
- *Improving the power quality in weak electric grids in rural African regions:* Within the context of development cooperation activities, a sufficiently densely populated rural region should be electrified, in cooperation with a major African energy supplier. In order to improve power quality in grids technologies should be deployed that are currently being developed in connection with distributed generation strategies.
- *1 million huts programme for developing countries:* To promote rural electrification in developing countries, not only the expansion of grids into sufficiently densely populated areas is important, but also distributed approaches and micro grids. This programme needs a certain volume and duration, and should also involve a new dimension of in-process socio-economic support.

FORGING STRATEGIC PARTNERSHIPS TO TURN
ENERGY SYSTEMS TOWARDS SUSTAINABILITY

Policy initiatives – existing or emergent – promoting a global transformation of energy systems towards sustainability provide a framework for action. In particular, the International Conference for Renewable Energies in Bonn 2004, announced by German Chancellor Gerhard Schröder at the World Summit on Sustainable Development (WSSD), is an important contribution to advancing this theme at the international level. The WBGU recommends that, in this context, the following policy processes in particular are used as catalysts to promote the transformation process:

- The initiatives adopted at the WSSD (Box 5.3-1), e.g.:
  - the Energy Initiative for Poverty Eradication and Sustainable Development, a strategic energy partnership between the EU and developing countries,
  - the Global Village Energy Partnership, involving, among others, the United Nations Development Programme (UNDP), the World Bank and the private sector,
  - the Global Network on Energy for Sustainable Development, involving, among others, the United Nations Environment Programme

(UNEP), energy institutions, the World Bank and the private sector.

- The economic partnership agreement currently being negotiated between the EU and the ACP states.

Following the new OECD Development Assistance Committee guidelines adopted in 2001, the principles of coherence, convergence and complementarity need to be observed. It follows that the energy strategies to be developed within the context of the above initiatives should be suitable for integration into the numerous already existing structures and programmes of the partner countries. In particular, the Council takes the view that overcoming energy poverty should become a constituent part of the 'basic social services' promotion focus within German development cooperation, and of the 20:20 initiative agreed at the 1995 World Summit for Social Development.

7.6
Advancing research and development

Turning energy systems towards sustainability is a major technological and social challenge on a scale comparable to that of a new industrial revolution. For it to succeed, a major research and development effort is a prerequisite. This concerns renewable energy sources, infrastructure, end-use efficiency technologies as well as the provision of knowledge on the conservation and expansion of natural carbon stocks and sinks. The social sciences also need to contribute, by researching the individual and institutional barriers to this transformation process, and developing and assessing strategies to overcome these barriers.

To develop the necessary diversity of options, it will be essential to promote a broad range of research activities (Chapter 6). This challenge is not currently being tackled. Expenditure for research and development in the energy sector has been declining for many years: At present, across the OECD only some 0.5 per cent of turnover in the energy sector is devoted to research and development activities, and the percentage is dropping. Without research and development, it will be impossible to achieve e.g. the high growth rates for renewable energy sources that are envisaged in the exemplary transformation path. This applies to all spheres: from private companies through to state support, from renewable energy innovation through to fossil bridging technologies. Only if there is sustained, high investment in research and development can there be a prospect of renewable-energy technologies and efficiency-enhancing measures coming into wide-

spread use over the medium and long term at low cost. The WBGU recommends

- increasing direct state expenditure for research and development in the energy sector in industrialized countries at least ten-fold by 2020 from its current level of about US$1,300 million annually (average across the OECD for the 1990–1995 period), above all through re-allocation of resources from other areas. This will be essential to tackle the tasks set out here. This level of resource allocation corresponds roughly to the average expenditure undertaken in the EU throughout the 1980s for research on energy production by means of nuclear fission alone. The focus needs to be shifted rapidly away from fossil and nuclear energy towards renewables and efficiency.
- establishing within the UN system a World Energy Research Coordination Programme (WERCP) to draw together the various strands of national-level energy research activities, in analogy to the World Climate Research Programme (WCRP).

7.7
Drawing together and strengthening global energy policy institutions

7.7.1
Negotiating a World Energy Charter and establishing coordinating bodies

Integrating the two objectives – to conserve natural life-support systems and eradicate energy poverty – requires a coordinated approach at the global level. To do this, international institutions and actors need to be drawn together. The WBGU recommends strengthening and expanding the institutional architecture of global energy policy in a stepwise process, building upon existing organizations:

- As a first step, a World Energy Charter should be agreed or negotiated at the International Conference for Renewable Energies in 2004. This should contain the key elements of global energy policy (targets, time frames and key measures) and provide a joint basis for action by the relevant actors at global level.
- Moreover, this conference should decide upon – or better still establish – a Global Ministerial Forum for Sustainable Energy responsible for coordinating and determining the strategic direction of the relevant actors and programmes.
- In parallel, a Multilateral Energy Subsidization Agreement (MESA) should be negotiated by 2008. This agreement could provide for the step-

wise removal of subsidies for fossil and nuclear energy, and could establish rules for subsidizing renewable energy and energy efficiency technologies.

- In support of these activities, a group of like-minded, advanced states should adopt a pioneering role on the path towards sustainable energy policies. The European Union would be a suitable candidate for such a leadership role.
- Building upon the above, the institutional foundations of sustainable energy policy could be further strengthened by concentrating competencies at global level. To this end, the role of the Ministerial Forum could be further expanded.
- Using the experience gained until that date, by about 2010 the establishment of an International Sustainable Energy Agency (ISEA) should be examined.

7.7.2
Enhancing policy advice at the international level

It is important that the political implementation of a global transformation of energy systems towards sustainability receives continuous support through independent scientific input, as is currently the case in climate protection policy. To this end, the WBGU recommends

- establishing an Intergovernmental Panel on Sustainable Energy (IPSE) charged with assessing global energy trends and identifying options for action.

7.8
Conclusion: Political action is needed now

As this report has shown, to protect humankind's natural life-support systems and eradicate energy poverty in developing countries it will be essential to transform energy systems worldwide. A global transition to renewable energy sources would have the added benefit of yielding a peace dividend: For one thing, energy poverty in low-income countries would be reduced. For another, the geostrategic importance of oil reserves would decline significantly over the long term.

This transformation of energy systems will be feasible without severe interventions in the societal and economic systems of industrialized and transition countries if policy-makers grasp the opportunity to shape this process over the next two decades. The costs of inaction would be much higher over the long term than the costs of initiating this transformation.

Every delay will make it more difficult to change course.

The direction of transformation is clear: The efficiency with which fossil fuels are used must be increased, and massive support must be provided to launch systems using renewables. It will be particularly important in this endeavour to swiftly reduce path dependency on fossil fuels. The long-term objective is to break the ground for a solar age. Both distributed solutions and the establishment of international energy transport structures should receive support in equal measure.

In the view of the WBGU, the transformation is feasible. It is also financeable if, in addition to intensified use of existing mechanisms (e.g. GEF, ODA, World Bank and regional development bank loans) and enhanced incentives for private-sector investors (e.g. through public-private partnerships), innovative financing avenues (such as user charges for global commons) are pursued. This is the message of the present report which highlights the key opportunities to steer energy systems towards sustainability, guided by a transformation roadmap.

The transformation of energy systems towards sustainability will not be achieved through any single strategy defined today. For worldwide transformation to succeed, it will need to be shaped in a stepwise and dynamic manner, for no one can predict today with sufficient certainty the technological, economic and social developments over the next 50–100 years. Long-term energy policy is thus also a searching process. It is the task of policy-makers to shape this difficult searching process vigorously and bring it to fruition.

References

Ad Hoc Open-ended Intergovernmental Group of Experts on Energy and Sustainable Development (2001) Report of the Ad Hoc Open-ended Intergovernmental Group of Experts on Energy and Sustainable Development. Second Session 26 February-2 March 2001 (E/CN.17//2001/15). United Nations website, http://www.un.org/esa/sustdev/enrexpert.htm

AEBIOM (Association Européenne Pour la Biomasse) (1999) Statement of AEBIOM on the Green Paper From the Commission on Renewable Sources of Energy. AEBIOM website, http://www.ecop.ucl.ac.be/aebiom/publications/Paper1.html

Agence France Press (2002) *Central Asia: World Bank Chief in Talks Over Pipeline*. Agence France Press, Paris.

Andronova, N G and Schlesinger, M E (2001) Objective estimation of the probability density function for climate sensitivity. *Journal of Geophysical Research* **106**: pp22605–611.

Angerer, G (1995) Auf dem Weg zu einer ökologischen Stoffwirtschaft, Teil I: Die Rolle des Recyclings. *GAIA* **4**(2): pp77–84.

Azar, C, Lindgren, K and Persson, T (2001) Carbon Sequestration from Fossil Fuels and Biomass – Long-term Potentials. Paper Presented at the Second Nordic Minisymposium on Carbon Dioxide Capture and Storage. Goeteborg, October 26, 2001. Entek website, http://www.entek.chalmers.se/~anly/symp/symp2001.html

Bachu, S (2000) Sequestration of CO_2 in geological media: criteria and approach for site selection in response to climate change. *Energy Conversion and Management* **41**: pp953–70.

Bächler, G, Böge, V, Klötzli, S, Libiszewski, S and Spillmann, K R (1996) *Kriegsursache Umweltzerstörung: Ökologische Konflikte in der Dritten Welt und Wege ihrer friedlichen Bearbeitung. Band 1*. Rüegger, Chur, Zürich.

Baklid, A, Korbol, R and Owren, G (1996) *Sleipner west CO_2 disposal, CO_2 injection into a shallow water underground aquifer*. Society of Petroleum Engineers. Paper 36600.

Bates, J, Brand, C, Davison, P and Hill, N (2001) *Economic Evaluation of Sectoral Emission Reduction Objectives for Climate Change. Economic Evaluation of Emissions Reductions in the Transport Sector of the EU. Bottom-up Analysis*. AEA Technology Environment, Abingdon.

Batley, S L, Colbourne, D, Fleming, P D and Urwin, P (2000) Citizen versus consumer: challenges in the UK green power market. *Energy Policy* **29**: pp479–87.

Bauer, M and Quintanilla, J (2000) Conflicting energy, environment, economy policies in Mexico. *Energy Policy* **28**, pp321–6.

Beauchamp, E G (1997) Nitrous oxide emission from agricultural soils. *Canadian Journal of Soil Science* **77**: pp113–23.

Behera, D, Sood, P and Singhi, S (1998) Respiratory symptoms in Indian children exposed to different cooking fuels. *JAPI* **46**(2): pp182–4.

Bell, R G (2002) Are market-based instruments the right choice for countries in transition? *Resources for the Future* (146): pp12–4.

BfN (Bundesamt für Naturschutz) (ed) (2002) Daten zum Naturschutz. BfN, Bonn.

BGR (Bundesanstalt für Geowissenschaften und Rohstoffe) (ed) (1998) Reserven, Ressourcen und Verfügbarkeit von Energierohstoffen. BGR, Hannover.

BGR (Bundesanstalt für Geowissenschaften und Rohstoffe) (2000) Geht die Kohlenwasserstoff-Ära zu Ende? BGR website, http://www.bgr.de/b123/kw_aera/kw_aera.htm

Bharati, K, Mohanty, S R, Rao, V R and Adhya, T K (2001) Influence of flooded and non-flooded conditions on methane flux from two soils planted to rice. *Chemosphere – Global Change Science* **3**: pp25–32.

BMBF (Bundesministerium für Bildung und Forschung) (1995) *Wasserstoff als Energieträger – Ergebnisse der Forschung der letzten 20 Jahre und Ausblick auf die Zukunft. Bericht zum Statusseminar*. BMBF Projektträger BEO, Berlin.

BMU (Bundesministerium für Umwelt, Naturschutz und Reaktorsicherheit) (ed) (2000) *Nationales Klimaschutzprogramm. Fünfter Bericht der Interministeriellen Arbeitsgruppe "CO_2-Reduktion"*. BMU, Berlin.

BMU (Bundesministerium für Umwelt, Naturschutz und Reaktorsicherheit) (ed) (2002a) *Entwicklung der Erneuerbaren Energien – Aktueller Sachstand*. BMU, Bonn.

BMU (Bundesministerium für Umwelt, Naturschutz und Reaktorsicherheit) (ed) (2002b) *Erneuerbare Energien und nachhaltige Entwicklung*. BMU, Berlin.

BMU (Bundesministerium für Umwelt, Naturschutz und Reaktorsicherheit) (ed) (2002c) *Ökologische Begleitforschung zur Offshore-Windenergienutzung. Fachtagung des Bundesministeriums für Umwelt, Naturschutz und Reaktorsicherheit und des Projektträgers Jülich. Bremerhaven 28. und 29. Mai 2002*. BMU, Berlin.

BMZ (Bundesministerium für wirtschaftliche Zusammenarbeit und Entwicklung) (1999) *Erneuerbare Energie für nachhaltige Entwicklung und Klimaschutz*. BMZ, Bonn and Berlin.

Boman, U R and Turnbull, J H (1997) Integrated biomass energy systems and emissions of carbon dioxide. *Biomass and Bioenergy* **13**: pp333–43.

Bond, G and Carter, L (1995) Financing energy projects. Experience of the International Finance Corporation. *Energy Policy* **23**: pp967–75.

Bosch, H-S and Bradshaw, A M (2001) Kernfusion als Energiequelle der Zukunft. *Physikalische Blätter* (November): pp55–60.

Boyd, P W, Watson, A J, Law, C S, Abraham, E R, Trull, T, Murdoch, R, Bakker, D C E, Bowie, A R, Büßeler, K O, Chang, H, Charette, M, Croot, P, Downing, K, Frew, R, Gall, M, Hadfield, M, Hall, J, Harvey, M, Jameson, G, LaRoche, J, Liddicoat, M, Ling, R, Maldonado, M T, McKay, R M, Nodder, S, Pickmere, S, Pridmore, R, Rintoul, S, Safi, K, Sutton, P, Strzepek, R, Tanneberger, K, Turner, S, Waite, T and Zeldis, A (2000) A mesoscale phytoplankton bloom in the polar Southern Ocean stimulated by iron fertilization. *Nature* **407**: pp695–702.

Börjesson, P, Gustavson, L, Christersson, L and Linder, S (1997) Future production and utilisation of biomass in Sweden: Potentials and CO_2 mitigation. *Biomass and Bioenergy* **13**: pp399–412.

Brauch, H G (1998) *Klimapolitik der Schwellenstaaten Südkorea, Mexiko und Brasilien*. Afes-Press Report, Mosbach.

Brockmann, K L, Stronzik, M and Bergmann, H (1999) *Emissionsrechtehandel – Eine neue Perspektive für die deutsche Klimapolitik nach Kioto*. Physica, Heidelberg.

Brown, C E (2002) *World Energy Resources*. Springer, Berlin, Heidelberg and New York.

Bruce, N, Perez-Padilla, R and Albalak, R (2000) Indoor air pollution in developing countries: a major environmental and public health challenge. *Bulletin of the WHO* **78**(9): pp1078ff.

Bruckner, T, Petschel-Held, G, Toth, F L, Füßel, H M, Helm, C, Leimbach, M and Schellnhuber, H J (1999) Climate change decision report and the tolerable windows approach. *Environmental Modelling and Assessment* **4**: 217–34.

BTM Consult (2001) *World Market Update 2000*. BTM Consult, Ringkøbing.

Bundesverband Windenergie (2001) *Von A bis Z – Fakten zur Windenergie*. Bundesverband Windenergie, Osnabrück.

Bunn, G (2002) *Reducing the Threat of Nuclear Theft and Sabotage*. IAEA Discussion Paper, IAEA-SM-367/4/08. International Atomic Energy Agency (IAEA), Vienna.

Burger, A and Hanhoff, I (2002) Unterwegs in die nächste Dimension. Kassensturz – Strukturwandel durch Ökologische Finanzreform. *Politische Ökologie* (77/78): pp15–8.

CEFIR (Center for Economic and Financial Research and Club 2015) (2001) Russia in the WTO: Myths and Reality. CEFIR website, http://www.hhs.se/site/Publications/WTO-report0907.pdf

CENR (Committee on Energy and Natural Resources for Development) (2000) Financial Mechanisms and Economic Instruments to Speed up the Investment in Sustainable Energy Development. United Nations Economic and Social Council website, http://www.un.org/esa/sustdev/energy/cenr/e_c14_2000_1.pdf

CESCR (committee on Economic Social and Cultural Rights) (1991) The Right to Adequate Housing (Art. 11 (1) CESCR General Comment 4). UNHCHR website, http://www.unhchr.ch

Chen, H, Zhou, H and Ma, S (2002) Brightness Program and its Progress. Mimeo. Beijing.

Chisholm, S W, Falkowski, P G and Cullen, J J (2001) Discrediting ocean fertilization. *Science* **249**: pp309–10.

Ciais, P, Rasool, I, Steffen, W, Raupach, M, Donney, S, Hood, M, Heinze, C, Sabine, C, Hibbard, K and Cihlar, J (2003) 'The carbon conservation challenge'. In Integrated Global Observing Strategy (IGCO) (ed) *Carbon Theme Report 2003*. IGCO, Paris.

Clark, A (2001) 'Implications of power sector reform in South Africa on poor people's access to energy: lessons for Africa'. In United Nations Environment Programme (UNEP) (ed) *African High-level Regional Meeting on Energy and Sustainable Development*. UNEP, Nairobi: pp127–35.

Cnossen, S and Vollebergh, H R J (1992) Toward a global excise on carbon. *National Tax Journal* **35**(1): pp23–36.

Coles, S L (2001) 'Coral bleaching: what do we know and what can we do?' In Salm, R V and Coles, S L. (eds) *Coral Bleaching and Marine Protected Areas. Proceedings of the Workshop on Mitigating Coral Bleaching Impact Through MPA Design, Honolulu, Hawaii, 29-31 May 2001*. Asia Pacific Coastal Marine Program Report 0102. The Nature Conservancy, Honolulu: pp25–35.

COM (2000) *EU Policies and Measures to Reduce Greenhouse Gas Emissions: Towards a European Climate Change Programme (ECCP)*. Communication from the Commission, Brussels, COM(2000)88. European Commission, Brussels.

Cook, J H, Beyea, J and Keeler, K H (1991) Potential impacts of biomass production in the United States on biological diversity. *Annual Review of Energy and the Environment* **16**: pp401–31.

Cook, I, Marbach, G, Di Pace, L, Girard, C and Taylor, N P (2001) Safety and Environmental Impact of Fusion. European Fusion Development Agreement (EFDA) Report EFDA-S-RE-1. EUR (01) CCE-FU\FTC 8\5. EFDA, Brussels.

Copeland, B and Taylor, M (2000) *Free trade and global warming: a trade theory view of the Kyoto protocol. National Bureau of Economic Research (NBER) Working Paper Series 7657*. Cambridge University Press, Cambridge, Ma.

Cottier, T and Evtimov, E (2000) Präferenzielle Abkommen der EG: Möglichkeiten und Grenzen im Rahmen der WTO. *ZEuS* **4**: pp477–505.

CRED (The Centre for Research on the Epidemiology of Disasters) (2003) Disaster Profiles. CRED website, http://www.cred.be/emdat/disindex.htm

CSE (Centre for Science and Environment) (2001) *Poles Apart. Global Environmental Negotiations No. 2*. CSE, New Delhi.

Davidson, O and Sokona, Y (2001) 'Energy and sustainable development: key issues for Africa'. In United Nations Environment Programme (UNEP) (ed) *African High-level Regional Meeting on Energy and Sustainable Development*. UNEP, Nairobi: pp1–19.

DeLallo, M R, Buchanan, T L, White, J S, Holt, N A and Wolk, R H (2000) Evaluation of innovative fossil cycles incorporating CO_2 removal. Paper presented on the Gasification Technologies Conference, San Francisco California. Gasification Technologies Council website, www.gasification.org/98GTC/Gtc00360.pdf

Deutsche Bank Research (2001) Hoffnungsträger Erneuerbare Energien. *DBR: Aktuelle Themen* (195): pp1–10.

de Vries, B, Bollen, J, Bouwan, L, den Elzen, M, Janssen, M and Kreileman, E (2000) Greenhouse gas emissions in an equity-, environment- and service-oriented world: an IMAGE-based scenario for the 21st century. *Technological Forecasting & Social Change* **63**(2-3): pp137–74.

DFID (Department for International Development) (ed) (2002) *Energy for the Poor. Underpinning the Millenium Development Goals*. DFID, London.

Dixon, J A, Talbot, L M and Le Moigne, G J M (1989) *Dams and the Environment: Considerations in World Bank Projects. World Bank Technical Paper 110*. World Bank, Washington, DC.

DLR (Deutsche Gesellschaft für Luft- und Raumfahrt) and DIW (Deutsches Institut für Wirtschaftsforschung) (1990) *Bedingungen und Folgen von Aufbaustrategien für eine solare Wasserstoffwirtschaft. Untersuchung für die Enquete-Kommission Technikfolgenabschätzung und -Bewertung des Dt. Bundestages*. DLR and DIW, Bonn.

Drange, H, Alendal, G and Johannessen, O M (2001) Ocean release of fossil fuel CO_2: a case study. *Geophysical Research Letters* **28**(1): pp2637–40.

Dreier, T and Wagner, U (2000) Perspektiven einer Wasserstoffwirtschaft. *Brennstoff Wärme Kraft (BWK)* **12**: p52.

Dröge, S and Trabold, H (2001) Umweltbezogene Verhaltenskodizes für ausländische Direktinvestitionen: Möglichkeiten und Grenzen. DIW-Arbeitspapier. Deutsches Institut für Wirtschaftsforschung (DIW) website, http://www.diw.de

Duijm, B (1998) Das Europäische System der Zentralbanken – ein Modell für ein Europäisches Kartellamt? *Zeitschrift für Wirtschaftspolitik* **47**(2): pp123–41.

Dunkerley, J (1995) Financing the energy sector in developing countries. Context and overview. *Energy Policy* (23): pp929–39.

Dwinnell, J H (1949) *Principles of Aerodynamics*. McGraw-Hill, New York.

EBRD (European Bank for Reconstruction and Development) (ed) (2001) *Transition Report 2001. Energy in Transition*. EBRD, London.

ECMT (European Conference of Ministers of Transport) (ed) (2001) *Sustainable Transport Policies*. ECMT, Paris.

Edenhofer, O, Kriegler, E and Bauer, N (2002) *Szenarien zum Umbau des Energiesystems*. Expertise for the WBGU Report "World in Transition: Towards Sustainable Energy Systems". WBGU website, http://www.wbgu.de/wbgu_jg2003_ex02.pdf

EEA (European Environment Agency) (ed) (2001) *TERM 2001. Indicators Tracking Transport and Environment Integration in the European Union*. EEA, Copenhagen.

EIA (Energy Information Administration) (2000) The Changing Structure of the Electric Power Industry 2000: An Update. EIA website, http://www.eia.doe.gov/bookshelf/electric.html

EIA (Energy Information Administration) (2001) International Energy Outlook 2001. EIA website, Internet: http://www.eia.doe.gov/oiaf/ieo/index.html

EIA (Energy Information Administration) (2002) Country Brief: Brazil. EIA website, http://www.eia.doe.gov/emeu/cabs/brazil.html

Energy Charter Secretariat (2000) Work Programme 2001. Energy Charter Secretariat website, http://www.encharter.org/cDetail.jsp?contendID=6&topicID=13

Energy Star Australia (2002) Promoting Energy Star. Australian Government website, http://www.energystar.gov.au/promo.html

Enquete Commission "Protecting the Earth's Atmosphere" (ed) (1995) *Climate and Mobility*. Economica, Bonn.

Enquete Commission "Sustainable Energy Supplies in View of Globalization and Liberalization" (ed) (2001) *First Report of the Enquete Commission. Interim Report. Bundestags-Drucksache 14/7509*. German Bundestag, Berlin.

Enquete Commission "Sustainable Energy Supplies in View of Globalization and Liberalization" (ed) (2002) *Final Report of the Enquete Commission. Bundestags-Drucksache 14/9400*. German Bundestag, Berlin.

Erdmann, G (2001) Energiekonzepte für das 21. Jahrhundert. *Internationale Politik* **56**(1): pp3–10.

ESMAP (Energy Sector Management Assistance Programme) (1998) *Malawi Rural Energy and Institutional Development*. ESMAP, Washington, DC.

ESMAP (Energy Sector Management Assistance Programme) (1999) *Uganda Rural Electrification Strategy Study. Final Report*. ESMAP, Washington, DC.

ESMAP (Energy Sector Management Assistance Programme) (2000) Subsidies and Sustainable Rural Energy Services: Can We Create Incentives Without Distorting Markets? ESMAP website, http://www.worldbank.org/html/fpd/esmap/ pdfs/010-00-subs_services.pdf

ESMAP (Energy Sector Management Assistance Programme) (2001) Best Practice Manual: Promoting Decentralized Electrification Investment. ESMAP website, http://www.worldbank.org/html/fpd/esmap/publication/bestpractice.html

Espey, S (2001) *Internationaler Vergleich energiepolitischer Instrumente zur Förderung von regenerativen Energien in ausgewählten Industrieländern*. Books on Demand, Norderstedt.

Esty, D C (1995) Private Sector Foreign Investment and the Environment. *RECIEL* **4**: pp99–105.

ETH-Rat (1998) 2000 Watt-Gesellschaft – Modell Schweiz Strategie Nachhaltigkeit im ETH-Bereich. ETH Wirtschaftsplattform website, http://www.novatlantis.ch/projects/2000W/brochure/resources/pdf/ge_brochure.pdf

ETSU (Renewable and Energy Efficiency Organisation) (1998) A Structured State-of-the-Art Survey and Review. FANTASIE Contract No. ST-96-AM-1109, Deliverable 8. ETSU website, http://www.etsu.com/fantasie/public.html

EU Commission (1998) *Energie für die Zukunft: Erneuerbare Energieträger. Weißbuch für eine Gemeinschaftsstrategie und Aktionsplan*. European Commission, Brussels.

EU Commission (1999a) *Elektrizität aus erneuerbaren Energie-trägern und der Elektrizitätsbinnenmarkt.* Working Document of the Commission, 14.4.1999; SEK (1999) 470 endg. European Commission, Brussels.

EU Commission (1999b) Energy in Europe. Economic Foundations for Energy Policy. Special Issue: December 1999. European Commission website, http://www.shared-analysis. fhg.de/Publ-fr.htm

EU Commission (Ed) (2000a) *Grünbuch – Hin zu einer Strategie für Energieversorgungssicherheit.* European Commission, Brussels.

EU Commission (2000b) Nukleare Sicherheit in den Neuen Unabhängigen Staaten und den mittel- und osteuropäischen Ländern. European Commission website, http://www.europa. eu.int/scaplus/leg/de/lvb/127036.htm

EU Commission (2001a) Energy & Transport in Figures. European Commission website, http://www.europa.eu.int/ comm/ energy_transport/etif/index.html

EU Commission (2001b) Förderung der Stromerzeugung aus erneuerbaren Energiequellen (Förderung des EE-Stroms). European Commission website, http://www.europa. eu.int/ comm/energy/library/renouvelables-de.pdf

EU Commission (2001c) White Paper. European Transport Policy for 2010: Time to Decide. European Commission website,http://europa.eu.int/comm/energy_transport/en/ lb_en.html

EU Commission (ed) (2001d) *Directive of the European Parliament and of the Council of 27 September 2001 on the Promotion of Electricity Produced from Renewable Energy Sources in the Internal Electricity Market. Official Journal of the European Communities 2001/77/EC, L 283/33-40.* European Commission, Brussels.

EU Commission (2002) *Mitteilung der Kommission an den Rat und das Europäische Parlament: Halbzeitbewertung der Gemeinsamen Agrarpolitik.* European Commission, Brussels.

EU Commission (ed) (2003) *Directive 2002/91/EC of the European Parliament and of the Council of 16 December 2002 on the energy performance of buildings. Official Journal of the European Communities L 1/65, 4.1.2003.* European Commission, Brussels.

EU Parliament (2001) *Possible Toxic Effects From the Nuclear Reprocessing Plants at Sellafield (UK) and Cap de La Hague (France). Report of STOA, PE. Nr. 303.110.* European Parliament, Luxembourg.

EUREC Agency (ed) (2002) *The Future for Renewable Energy.* James & James, London.

Eurostat (2002) 6% of energy consumed in the EU comes from renewables. UN World Summit on Sustainable Development Johannesburg, 26/08–04/09 2002. Statistical Office of the European Communities website, http://europa.eu.int/ comm/ eurostat/Public/datashop/print-product/DE?catalogue= Eurostat&product=8-23082002-DE-AP-DE&mode=download

FAO (Food and Agriculture Organisation) (2002) FAOSTAT Statistical Database. FAO website, http://apps.fao.org/ default.htm

FAO (Food and Agriculture Organisation) Land and Plant Nutrition Management Service (2002) TERRASTAT Land-resource Potential and Constraints Statistics at Country and Regional Level. FAO website, http://www.fao.org/ag/agl/agll/ terrastat/

Fearnside, P M (1995) Hydroelectric Dams in the Brazilian Amazon as Sources of "Greenhouse" Gases. *Environmental Conservation* **22**(1): pp7–19.

Fearnside, P M (1997) Greenhouse-gas emissions from Amazonian hydroelectric reservoirs: the example of Brazil's Tucurui Dam as compared to fossil fuel alternatives. *Environmental Conservation* **24**(1): pp64–75.

Feldmann, G C and Gradwohl, J (1996) *Oil Pollution.* Smithsonian Institution, Cambridge, Ma.

fesa (Förderverein Energie- und Solaragentur Freiburg) (2002) *Land- und forstwirtschaftliche Biomasse zur Energiegewinnung.* fesa, Freiburg.

Fischedick, M, Nitsch, J, Lechtenböhmer, S, Hanke, T, Barthel, C, Jungbluth, C, Assmann, D, vor der Brüggen, T, Trieb, F, Nast, M, Langniß, O and Brischke, L-A (2002) *Langfristszenarien für eine nachhaltige Energieversorgung in Deutschland. Forschungsvorhaben für das Umweltbundesamt.* Umweltbundesamt (UBA), Berlin.

Fischer, G and Schrattenholzer, L (2001) Global bioenergy potentials through 2050. *Biomass and Bioenergy* **20**: pp151–9.

Fischer, G, van Velthuizen, H, Shah, M and Nachtergäle, F O (2001) *Global Agro-ecological Assessment for Agriculture in the 21st Century.* IIASA and FAO, Laxenburg and Rome.

Fischer, G, van Velthuizen, H, Shah, M and Nachtergäle, F O (2002) *Global Agro-ecological Assessment for Agriculture in the 21st Century: Methodology and Results.* IIASA and FAO, Laxenburg and Rome.

Fleig, J (ed) (2000) *Zukunftsfähige Kreislaufwirtschaft. Mit Nutzenverkauf, Langlebigkeit und Aufarbeitung ökonomisch und ökologisch wirtschaften.* Schäffer-Poeschel, Stuttgart.

FNR (Fachagentur Nachwachsende Rohstoffe) (ed) (2000) *Leitfaden Bioenergie – Planung, Betrieb und Wirtschaftlichkeit von Bioenergieanlagen.* FNR, Gülzow.

Forest, C E, Stone, P H, Sokolov, A P, Allen, M R and Webster, M D (2002) Quantifying in climate system properties with the use of recent climate observations. *Science* **295**: pp113–7.

Freibauer, A (2002) *Biogenic Greenhouse Gas Emissions from Agriculture in Europe – Quantification and Mitigation. Agrarwissenschaften II (Agrarökonomie, Agrartechnik und Tierproduktion).* Institut für Landwirtschaftliche Betriebslehre (410b) der Universität Hohenheim, Hohenheim.

Freibauer, A, Rounsevell, M D A, Smith, P and Verhagen, J (2002) *Background Paper on Carbon Sequestration in Agricultural Soils under Article 3.4 of the Kyoto Protocol. European Climate Change Programme (ECCP), Working Group Agriculture, Subgroup Soils.* European Commission, Brussels.

Freund, R (2002) Einspar-Contracting bei Gemeindegebäuden. *Energiewirtschaftliche Tagesfragen* **52**(7): pp472–5.

Fritsche, U R and Matthes, F C (2002) *Changing Course. A Contribution to a Global Energy Strategy (GES). External Review Draft.* Heinrich Böll Foundation, Berlin.

Froggart, A (2002) *World Nuclear Industry Status Report. Study on behalf of Greenpeace International.* Greenpeace International, Amsterdam.

G8 Renewable Energy Task Force (2001) Final Report. G8 website, http://www.renewabletaskforce.org/report.asp

Gattuso, J P, Allemand, D and Frankignolle, M (1999) Photosynthesis and calcification at cellular, organismal, and community levels in coral reefs: a review on interactions and control by carbonate chemistry. *Am. Zool.* **39**: pp160–83.

GEF (Global Environment Facility) (2000) GEF Programmes. GEF website, http://www.gefweb.org/PROGLIST.pdf

Gerling, P and May, F (2001) Stellungnahme vor der Enquete-Kommission Nachhaltige Energieversorgung des Deutschen Bundestages. German Bundestag website, http://www.bundestag.de/gremien/ener/ener_stell_gerl.pdf

Ghosh, S (2002) Electricity consumption and economic growth in India. *Energy Policy* **30**: pp125–9.

Göttlicher, G (1999) *Energetik der Kohlendioxidrückhaltung in Kraftwerken. Fortschritt-Berichte VDI zur Energietechnik.* VDI-Verlag, Düsseldorf.

Goldemberg, J (1996) A note on the energy intensity of developing countries. *Energy Policy* **24**: pp759–61.

Goldemberg, J (2001) Energy and Human Well Being. HDRO-Human Development Office. UNDP website, http://www.undp.org/hdro/occ.htm#2

Golob, T F and Regan, A C (2001) Impacts of information technology on personal travel and commercial vehicle operations: research challenges and opportunities. *Transportation Research, Part C, Emerging Technologies* **9**: pp87–121.

Goor, F, Dubuisson, X and Jossart, J-M (2000) Adéquation, impact environnemental et bilan d'énergie de quelques cultures énergétiques en Belgique. *Cahiers Agricultures* **9**: pp59–64.

Goulder, L H (1995) Environmental taxation and the double dividend: a readers guide. *International Tax and Public Finance* **2**: pp157–83.

Goulder, L and Mathai, K (2000) Optimal CO_2 abatement in the presence of induced technological change. *Journal of Environmental Economics and Management* **39**: pp1–38.

Graham, R L, Downing, M and Walsh, M E (1996) A framework to assess regional environmental impacts of dedicated energy crop production. *Environmental Management* **20**: pp475–85.

Grassi, G (1999) Modern bioenergy in the European Union. *Renewable Energy* **16**: pp985–90.

Green, M J B and Paine, J (1997) *State of the World's Protected Areas at the End of the Twentieth Century. Paper presented at IUCN World Commission on Protected Areas Symposium on Protected Areas in the 21st Century: From Islands to Networks. Albany, Australia, 24-29th November 1997.* World Conservation Monitoring Centre (WCMC), Cambridge, UK.

Green, M A, Emery, K, King, D L, Igari, S and Warta, W (2002) Progress in Photovoltaics. *Research and Applications* **10**: pp55–61.

Greenpeace (ed) (2002) Große Tankerkatastrophen 1967-2002. Greenpeace website, http://www.greenpeace.org/deutschland/fakten/klima/erdoel-gefahr-fuer-die-umwelt/große-tankerkatastrophen-1967-2001

Greiner, S, Großmann, H, Koopmann, G, Matthies, K, Michaelowa, A and Steger, S (2001) *WTO-/GATT-Rahmenbedingungen und Reformbedarf für die Energiepolitik sowie die Rolle der Entwicklungspolitik im Kontext einer außenhandels- und klimapolitischen Orientierung. Gutachten im Auftrag der Enquete-Kommission "Nachhaltige Energieversorgung" des Deutschen Bundestages.* Hamburg Institute of International Economics (HWWA), Hamburg.

Grimston, M C, Karakoussis, V, Fouquet, R, van der Vorst, R, Pearson, P and Leach, M (2001) The European and global potential of carbon dioxide sequestration in tackling climate change. *Climate Policy* **1**: pp155–71.

Gröger, K (2000) Structural Fiscal Measures and Electronic Tolling. The German Experience. ECMT Joint Conference on Smart CO_2 Reductions. Non-product Measures for Reducing Emissions from Vehicles, TURIN 2-3 March 2000. OECD website, http://www1.oecd.org/cem/Topics/env/CO2turin/CO2groeger.pdf

Groscurth, H, Beeck, H and Zisler, S (2000) Erneuerbare Energien im liberalisierten Markt. *Elektrizitätswirtschaft* **99**(24): pp26–32.

Grupp, M, Bergler, H, Biermann, E, Klingshirn, A, de Lange, E and Wentzel, M (2002) *Solar Cooker Field Test in South Africa – Phase 2 – Memo: Greenhouse Gas (GHG) Emissions by Different Cooking Fuels and the Potential GHG Impact of Solar Stoves – Draft 2.* GTZ and DME, Eschborn.

Gsänger, S (2001) Germany's EEG goes global. *New Energy* 2: pp36–8.

GTZ (Gesellschaft für Technische Zusammenarbeit) (1998) GTZ Microfinance-Multinationals: Microfinance-Netzwerke und Geberinitiativen im Überblick. GTZ website, http://www.gtz.de/themen/ebene3.asp?Thema=112&ProjectId=212&Reihenfolge=5&spr=1

Gupta, J, Vlasblom, J and Kroeze, C (2001) *An Asian Dilemma: Modernising the Electricity Sector in China and India in the Context of Rapid Economic Growth and the Concern for Climate Change. Report number E-01/04.* Dutch National Research Programme on Global Air Pollution and Climate Change (NOP), Amsterdam and Wageningen.

Hall, D O and House, J I (1995) Biomass energy in Western Europe to 2050. *Land Use Policy* **12**: pp37–48.

Halsnaes, K, Markandya, A, Sathaye, J, Boyd, R, Hunt, A and Taylor, T (2001) Transport and the Global Environment. Accounting for GHG Reductions in Policy Analysis. UNEP Collaborating Centre on Energy and Environment website, http://www.uccee.org/OverlaysTransport/TransportGlobalOverlays.pdf

Hamacher, T and Bradshaw, A M (2001) *Fusion as a Future Power Source: Recent Achievements and Prospects. Proceedings of the 18th World Energy Congress.* World Energy Congress, Buenos Aires.

Hanegraaf, M C, Biewinga, E E and van der Bijl, G (1998) Assessing the ecological and economic sustainability of energy crops. *Biomass and Bioenergy* **15**(4-5): pp345–55.

Hebling, C, Glunz, S W, Schumacher, J O and Knobloch, J (1997) *High-Efficiency (19.2%) Silicon Thin-Film Solar Cells with Interdigitated Emitter and Base Front-Contacts. 14th European Photovoltaic Solar Energy Conference, Barcelona.* Stephens, Bedford.

Hein, M, Meusel, M, Baur, C, Dimroth, F, Lange, G, Siefer, G, Tibbits, T N D, Bett, A W, Andreev, V M and Rumyantsev, V D (2001) *Characterisation of a 25% high-efficiency fresnel lens module with GaInP/GaInAs dual-junction concentrator solar cells. Proceedings of the 17th EU-PVSEC Munich* (submitted).

Hendricks, C A and Turkenburg, W C (1997) Towards Meeting CO_2 Emission Targets: The Role of Carbon Dioxide Removal. *IPTS Report* **16**: pp13–21.

Hendriks, C A, Wildenborg, A F B, Blok, K, Floris, F and van Wees, J D (2001) 'Costs of Carbon Dioxide Removal by Underground Storage'. In Williams, D, Durie, B, MacMullan, P, Paulson, C and Smith, A (eds) *Proceedings of the 5th International Conference on Greenhouse Gas Control Technologies*. CSIRO Publishing, Collingwood: pp967–72.

Herzog, H (2001) What future for carbon capture and sequestration? *Environmental Science and Technology* **35**: pp148A–153A.

Hoegh-Guldberg, O (1999) Climate change, coral bleaching and the future of the world's coral reefs. *Marine and Freshwater Research* (50): pp839–66.

Hoffert, M I, Caldeira, K and Jain, A K (1998) Energy implications of future stabilization of atmospheric CO_2 content. *Nature* **395**: pp881–4.

Holloway, S (1997) Safety of the underground disposal of carbon dioxide. *Energy Conversion and Management* **38** (Supplement): pp241–5.

Holm, D (2000) *Interventions and Cost Savings in Environmentally Sound Low-Cost Energy Efficient Housing Study for USAID.* The United Nations Energy for International Development (USAID), Washington, DC.

Horlacher, H-B (2002) *Globale Potenziale der Wasserkraft.* Expertise for the WBGU Report "World in Transition: Towards Sustainable Energy Systems". WBGU website, http://www.wbgu.de/wbgu_jg2003_ex03.pdf

Horta, K, Round, R and Young, Z (2002) The Global Environmental Facility. The First Ten Years – Growing Pains or Inherent Flaws. Halifax Initiative website, http://www.halifaxinitiative.org/hi.php/Publications/333

Hu, P S (1997) Estimates of 1996 U.S. Military Expenditures on Defending Oil Supplies from the Middle East: Literature Review. Revised August. Oak Ridge National Laboratory for the U.S. Energy Department website, http://www-cta.ornl.gov/Publications/military.pdf

Hulscher, W (1997) Renewable Energy in Thailand. FAO Regional Wood Energy Development Programme in Asia (RWEDP) website, http://www.rwedp.org/acrobat/p_carbon.pdf

Huynen, M and Martens, P (2002) Future Health. The Health Dimension in Global Scenarios. ICIS Maastricht University, Maastricht.

IABG (Industrieanlagen-Betriebsgesellschaft mbH) (ed) (2000a) FANTASIE "Synthesis of Impact Assessment and Potentials". Project Funded by the European Commission Under the Transport RTD Programme of the 4th Framework Programme. IABG website, http://www.etsu.com/fantasie/fantasie.htm

IABG (Industrieanlagen-Betriebsgesellschaft mbH) (ed) (2000b) FANTASIE "Forecasting and Assessment of New Technologies and Transport Systems and their Impacts on the Environment". Final Report. Project funded by the European Commission under the Transport RTD Programme of the 4th Framework Programme. IABG website, http://www.etsu.com/fantasie/fantasie.htm

IAEA (International Atomic Energy Agency) (1998) *Technical Basis for the ITER Final Design Report, Cost Review and Safety Analysis (FDR). ITER EDA Documentation Series No. 16.* IAEA, Vienna.

IAEA (International Atomic Energy Agency) (2001) *Final Report of the ITER EDA, ITER EDA Documentation Series No. 21.* IAEA, Vienna.

ICOLD (International Commission on Large Dams) (1998) *World Register of Dams.* ICOLD, Paris.

ICRP (International Commission on Radiological Protection) (ed) (1991) *Recommendations of the International Commission on Radiological Protection. Annals of the ICRP.* Pergamon Press, Oxford and New York.

IEA (International Energy Agency) (ed) (1997) *Transport, Energy and Climate Change.* IEA, Paris.

IEA (International Energy Agency) (ed) (1998) Technical Basis for the ITER Final Design Report. Cost Review and Safety Analysis (FDR). IEA, Vienna.

IEA (International Energy Agency) (ed) (1999) *World Energy Outlook 1999. Looking at Energy Subsidies: Getting the Prices Right.* OECD and IEA, Paris.

IEA (International Energy Agency) (ed) (2000) *Energy Technology and Climate Change – A Call to Action.* IEA, Paris.

IEA (International Energy Agency) (ed) (2001a) *Competition in Electricity Markets.* IEA, Paris.

IEA (International Energy Agency) (ed) (2001b) *World Energy Outlook 2000.* OECD and IEA, Paris.

IEA (International Energy Agency) (ed) (2001c) *Things That Go Blip In the Night – Standby Power and How to Limit it.* IEA, Paris.

IEA (International Energy Agency) (ed) (2002a) *Russia Energy Survey 2002.* OECD and IEA, Paris.

IEA (International Energy Agency) (ed) (2002b) *Energy Policies of IEA Countries: The United States 2002 Review.* IEA, Paris.

IEA (International Energy Agency) (ed) (2002c) *World Energy Outlook 2002.* IEA, Paris.

IEA (International Energy Agency) (ed) (2003) *World Energy Outlook 2003. Insights: Global Energy Investment Outlook.* IEA, Paris.

IfE/TU Munich (2003) Lehrstuhl für Energiewirtschaft und Kraftwerkstechnik der Technischen Universität München. *Notes on a lecture.*

IFRC-RCS (International Federation of Red Cross) and RCS (Red Crescent Societies) (2001) *World Disasters Report.* IFRC-RCS, Geneva.

IPCC (Intergovernmental Panel on Climate Change) (ed) (1996) *Climate Change 1995. Impacts, Adaptations and Mitigation of Climate Change: Scientific Technical Analyses. Contribution of Working Group II to the Second Assessment Report of the IPCC.* Cambridge University Press, Cambridge.

IPCC (Intergovernmental Panel on Climate Change) (ed) (2000a) *Land Use, Land-Use Change, and Forestry*. Cambridge University Press, Cambridge.

IPCC (Intergovernmental Panel on Climate Change) (ed) (2000b) *Emissions Scenarios. Special Report of Working Group III of the IPCC*. IPCC, Geneva.

IPCC (Intergovernmental Panel on Climate Change) (ed) (2001a) *Climate Change 2001: The Scientific Basis. IPCC Third Assessment Report. Summary for Policymakers*. Cambridge University Press, Cambridge.

IPCC (Intergovernmental Panel on Climate Change) (ed) (2001b) *Climate Change 2001: Impacts, Adaptation and Vulnerability. Third Assessment Report. Summary for Policymakers. Draft*. Cambridge University Press, Cambridge.

IPCC (Intergovernmental Panel on Climate Change) (ed) (2001c) *Climate Change 2001: Mitigation. Contribution of Working Group III to the Third Assessment Report of the IPCC*. Cambridge University Press, Cambridge.

IPCC (Intergovernmental Panel on Climate Change) (ed) (2001d) *Climate Change 2001: Synthesis Report. Contribution of Working Groups I, II and III to the Third Assessment Report of the Intergovernmental Panel on Climate Change*. Cambridge University Press, Cambridge.

IPS (Institute for Policy Studies) (1999) OPIC, Ex-Im and Climate Change: Business as Usual? IPS website, http://www.seen.org/oeordon1.html

IPTS (Institute for Prospective Technological Studies) (2001) Expert Panel on Sustainability, Environment and Natural Resources. Enlargement Futures Project. Final Report. IPTS website, http://www.jrc.es/projects/enlargement/

ISIS (Institute for Science and International Security) (2000) Special Section: Nuclear Terrorism. ISIS website, http://www.isis-online.org

Iten, R, Oettli, B, Jochem, E and Mannsbart, W (2001) *Förderung des Exports im Bereich der Energietechnologien. Studie im Auftrag des Bundesamtes für Energie*. INFRAS und Fraunhofer Institut Systemtechnik und Innovationsforschung, Zurich and Karlsruhe.

Izaguirre, A K (2000) Private Participation in Energy. Private Sector Note No. 208. World Bank website, http://rru.worldbank.org/Viewpoint/HTMLNotes/250/250Izagu-101502.pdf

Jabir, I (2001) The shift in US oil demand and its impact on OPEC's market share. *Energy Economics* **23**: pp659–66.

Jantzen, J, Cofala, J and de Haan, B J (2000) Technical Report on Enlargement. RIVM report 481505022. RIVM website, http://www.europe.eu.int/comm/environment/enveco/priority-study/enlargement.pdf

Jenzen, H (1998) *Energiesteuern im nationalen und internationalen Recht: eine verfassungs-, europa- und welthandelsrechtliche Untersuchung*. Peter Lang, Frankfurt/M.

Jochem, E (1991) Long-term potentials of rational energy use – the unknown possibilities of reducing greenhouse gas emissions. *Energy & Environment* (2): pp31–44.

Jochem, E and Turkenburg, W (2003) Long-term potentials of energy and material efficiency. *Energy & Environment* (submitted).

Johansson, B and Ahman, M (2002) A Comparison of Technologies for Carbon-neutral Passenger Transport. Transportation Research Part D 7. Elsevier website, http://www.elseviercom/locate/trd

Johnson, S, McMillan, J and Woodruff, C (1999) *Property Rights, Finance and Entrepreneurship. EBRD Working Paper 43*. EBRD, London.

Jotzo, F and Michaelowa, A (2002) Estimating the CDM market under the Marrakech Accords. *Climate Policy* **2**(2-3): pp179–201.

Kämpf, K, Schulz, J, Walter, C, Benz, T, Steven, C and Hülser, W (2000) *Umweltwirkungen von Verkehrsinformations- und -leitsystemen im Straßenverkehr*. Umweltbundesamt (UBA), Berlin.

Kainer, A and Spielkamp, R (1999) *Gebündelte Macht. Entgleitet die Liberalisierung in Oligopolstrukturen?* Frankfurter Allgemeine Zeitung 2783/B3, Frankfurt/M.

Kaltschmitt, M, Reinhardt, G and Stelzer, T (1997) Life cycle analysis of biofuels under different environmental aspects. *Biomass and Bioenergy* **12**: pp121–34.

Kaltschmitt, M, Bauen, A and Heinz, A (1999) *Perspektiven einer energetischen Nutzung organischer Ernte- und Produktionsrückstände in Entwicklungsländern*. Universität Stuttgart. Institut für Energiewirtschaft und Rationelle Energieanwendung, Stuttgart.

Kaltschmitt, M, Merten, D, Fröhlich, N and Nill, M (2002) *Energiegewinnung aus Biomasse*. Expertise for the WBGU Report "World in Transition: Towards Sustainable Energy Systems". WBGU website, http://www.wbgu.de/wbgu_jg2003_ex04.pdf

Kashiwagi, T (1995) 'Chapter 20'. In Intergovernmental Panel on Climate Change (IPCC) (ed) *Second Assessment Report*. Cambridge University Press, Cambridge: pp649–77.

Kebede, B and Kedir, E (2001). *Energy Subsidies and the Urban Poor in Ethiopia: The Case of Kerosene and Electricity. Draft Report*. African Energy Policy Research Network (AFREPREN), Nairobi.

Keilhacker, M, Gibson, A, Gormezano, C, Lomas, P J, Thomas, P R, Watkins, M L, Andrew, P, Balet, B, Borba, D and JET Team (1999) High fusion performance from deuterium-tritium plasmas in JET. *Nuclear Fusion* **39**(2): pp209–35.

Keohane, R O (1996) 'Analyzing the effectiveness of international environmental institutions'. In Keohane, R O and Levy, M (eds) *Institutions for Environmental Aid: Pitfalls and Promise*. Cambridge University Press, Cambridge: pp3–27.

KfW (Kreditanstalt für Wiederaufbau) (ed) (2001) *Stellungnahme zu einem Fragekatalog der Enquete-Kommission Nachhaltige Energieversorgung. Thema: Neue Institutionen zur Bewältigung globaler Umwelt- und Energieprobleme und Probleme bei der Finanzierung von Projekten zur Energieversorgung in den Entwicklungs- und Transformationsländern*. KfW, Berlin.

Kheshgi, H S, Prince, R C and Marland, G (2000) The potential of biomass fuels in the context of global climate change: Focus on transportation fuels. *Annual Review of Energy and the Environment* **25**: pp199–244.

Klare, M (2001) The New Geography of Conflict. *Foreign Affairs* **80**(3): pp49–61.

Knutti, R, Stocker, T F, Joos, F and Plattner, G-K (2002) Constraint on radiative forcing and future climate change from observations and climate model ensembles. *Nature* **416**: pp719–23.

Kriegler, E and Bruckner, T (2003) Sensitivity analysis of emissions corridors for the 21st century. *Climatic Change* (submitted).

Kronshage, S and Trieb, F (2002) *Berechnung von Weltpotenzialkarten*. Expertise for WBGU. Mimeo.

Kulshreshtha, S N, Junkins, B and Desjardins, S (2000) Prioritizing greenhouse gas emission mitigation measures for agriculture. *Agricultural Systems* **66**: pp145–66.

Kutter, A (2002) Neue Herausforderungen an die GEF. *DNR EU-Rundschreiben* 09.02: pp10–11.

Leach, G (1986) *Energy and Growth*. Butterworth, London.

Lee, J J, Unander, F, Murtishaw, S and Ting, M (2001) Historical and future trends in aircraft performance, cost and emissions. *Annual Review of Energy and the Environment* **26**(11): pp167–200.

Leimbach, M and Bruckner, T (2001) Influence of economic constraints on the shape of emission corridors. *Computational Economics* **18**: pp173–91.

Lerer, L B and Scudder, T (1999) Health impacts of large dams. *Environmental Impact Assessment Review* **19**: pp113–23.

Levine, M, Koomey, J, McMahin, J, Sanstad, A and Hirst, E (1995) Energy efficiency policies and market failures. *Annual Review of Energy and the Environment* (20): pp535–55.

Lewandowski, I, Clifton-Brown, J C and Walsh, M (2000) Miscanthus: European experience with a novel energy crop. *Biomass and Bioenergy* **19**: pp209–27.

Lookman, A A and Rubin, E S (1998) Barriers to adopting least-cost particulate control strategies for Indian power plants. *Energy Policy* **26**(14): pp1053–63.

LSMS (Living Standard Measurement Study) (2002) LSMS Household Surveys. World Bank's LSMS website, http://www.worldbank.org/lsms/

Luther, J, Preiser, K and Willeke, G (2003) 'Solar modules and photovoltaic systems'. In Bubenzer, A and Luther, J (eds) *Photovoltaics Guidebook for Decision Makers*. Springer, Berlin, Heidelberg and New York: pp41–106.

Luukkanen, J and Kaivo, J (2002) *ASEAN Tigers and Sustainability of Energy Use – Decomposition Analysis of Energy and CO_2 Efficiency Dynamics*. University of Tampere. Department of Regional Studies and Environmental Policy, Tampere, Finland.

Lux-Steiner, M C and Willeke, G (2001) Strom von der Sonne. *Physikalische Blätter* (11): pp47–53.

Mahmood, K (1987) *Reservoir Sedimentation: Impact, Extent and Mitigation*. World Bank, Washington, DC.

Margolis, R M and Kammen, D W (1999) Underinvestment: the energy technology and R&D policy challenge. *Science* **285**: pp690–2.

Markels jr., M and Barber, R T (2001) *Sequestration of CO_2 by Ocean Fertilization. Poster Presentation at NETL Conference on Carbon Sequestration*. Conference Secretariat, Washington, DC.

Marrison, C I and Larson, E D (1996) A preliminary analysis of the biomass energy production potential in Africa in 2025 considering projected land needs for food production. *Biomass and Bioenergy* **10**: pp337–51.

Martin, J H, Coale, K H, Johnson, K S, Fitzwater, S E, Gordon, R M, Tanner, S J, Hunter, C N, Elrod, V A, Nowicki, J L, Coley, T L, Barber, R T, Lindley, S, Watson, A J, Van Scoy, K, Law, C S, Liddicoat, M I, Ling, R, Stanton, T, Stockel, J, Collins, C, Anderson, A, Bidigare, R, Ondrusek, M, Latasa, M, Millero, F J, Lee, K, Yao, W, Zhang, J Z, Fredrich, G, Sakamoto, C, Chavez, F, Buck, K, Kolber, Z, Green, R, Falkowski, P G, Chisholm, S W, Hoge, F, Swift, R, Yungle, J, Turner, S, Nightingale, P I, Hatton, A, Liss, P and Tindale, N W (1994) Testing the iron hypothesis in ecosystems of the equatorial Pacific. *Nature* **371**: pp123–9.

Maskus, K E (2000) *Intellectual Property Rights in the Global Economy*. Institute for International Economics, Washington, DC.

Matthes, F C (1999) Führen Stromexporte aus Osteuropa die Bemühungen um Klimaschutz und Atomausstieg ad absurdum? Öko-Institut website, http://www.oeko.de/bereiche/energie/documents/ostimport.pdf

Maurer, C and Bhandari, R (2000) The Climate of Export Credit Agencies. World Resources Institute (WRI), Washington, DC.

Maurer, C (2002) *The Transition from Fossil to Renewable Energy Systems: What Role for Export Credit Agencies?* Expertise for the WBGU Report "World in Transition: Towards Sustainable Energy Systems". WBGU website, http://www.wbgu.de/wbgu_jg2003_ex05.pdf

McAllister, D, Craig, J, Davidson, N, Murray, D and Seddon, M (2000) Biodiversity Impacts of Large Dams. Contributing Paper prepared for the Thematic Review II.1 of the World Commission on Dams. World Commission on Dams (WCD) website, http://www.damsreport.org/docs/kbase/contrib/env245.pdf

McCully, P (1996) *Silenced Rivers. The Ecology and Politics of Large Dams*. Zed Books, London and New Jersey.

Melchert, A (1998) *Möglichkeiten des Least-Cost Planning zur Energiekostenminimierung für den Haushaltssektor in der Elektrizitätswirtschaft. Dissertation*. Technische Universität, Berlin.

Messner, S and Schrattenholzer, L (2000) MESSAGE-MACRO: Linking an energy supply model with a macroeconomic module and solving it iteratively. *Energy* **25**: pp267–82.

Michaud, C M, Murray, C J L and Bloom, B R (2001) Burden of disease – implications for future research. *JAMA* **285**(5): pp535–9.

Milly, P C D, Wetherald, R T, Dunne, K A and Delworth, T L (2002) Increasing risk of great floods in a changing climate. *Nature* **415**: pp514–7.

Ministry of Trade and Industry (1999) Greenhouse Gas Reduction by Bioenergy Based on the Targets of the Action Plan. VTT Energy website, http://www.vtt.fi/ene/tuloksia/uusiutuvat/Ed-kalvot-engl.ppt

Mitra, S, Jain, M C, Kumar, S, Bandyopadhyay, S K and Kalra, N (1999) Effect of rice cultivars on methane emission. *Agriculture, Ecosystems and Environment* **73**: pp177–83.

Morita, T, Nakicenovic, N and Robinson, J (2000) Overview of mitigation scenarios for global climate stabilization based on new IPCC emission scenarios (SRES). *Environmental Economics and Policy Studies* **3**: pp65–88.

Morse, E L (1999) A new political economy of oil? *Journal of International Affairs* **53**(1): pp1–29.

Moss, D, Davies, C and Roy, D (1996) CORINE *Biotope Sites. Database Status and Perspectives 1995.* European Environment Agency (EEA), Copenhagen.

Muchlinski, P (1998) 'Towards a multilateral investment agreement (MAI)'. In Weiss, F. (ed) *International Economic Law With a Human Face.* Kluwer Law International, Den Haag: pp429–51.

Müller, F (2002) Energiepolitische Interessen in Zentralasien. *Das Parlament – Beilage "Aus Politik und Zeitgeschichte"* (B8-02): pp23–31.

Münchner Rückversicherung (2001) Topics 2001. Jahresrückblick Naturkatastrophen. Münchner Rückversicherung website, http://www.munichre.com/pdf/topics_2001_d.pdf

Murphy, J T (2001) Making the energy transition in rural East Africa: is leapfrogging an alternative? *Technological Forecasting & Social Change* **68**: pp173–93.

Murray, C and Lopez, A (eds) (1996) *Global Burden of Disease.* Harvard University Press, Cambridge, MA.

Nakashiki, N and Oshumi, T (1997) Dispersion of CO_2 injected into the ocean at the intermediate depth. *Energy Conversion and Mangement* **38S**: pp355–60.

Nakicenovic, N, Grübler, A and McDonald, A (1998) *Global Energy Perspectives.* Cambridge University Press, Cambridge and New York.

Nakicenovic, N and Riahi, K (2001) *An Assessment of Technological Change Across Selected Energy Scenarios.* International Institute for Applied Systems Analysis (IIASA), Laxenburg.

Nash, L (1993) 'Water quality and health'. In Gleick, P H (ed) *Water in Crisis – A Guide to the World's Fresh Water Resources.* Oxford University Press, New York, Oxford: pp25–36.

National Energy Policy Development Group (ed) (2001) *National Energy Policy Report.* National Energy Policy Development Group, Washington, DC.

Newman, P and Kenworthy, M (1990) *Cities and Automobile Dependence.* Gower, London.

Nitsch, J and Staiß, F (1997) 'Perspektiven eines solaren Energieverbundes für Europa und den Mittelmeerraum'. In Brauch, H G (ed) *Energiepolitik.* Springer, Berlin, Heidelberg and New York: pp473–86.

Nitsch, J (2002) *Potenziale der Wasserstoffwirtschaft.* Expertise for the WBGU Report "World in Transition: Towards Sustainable Energy Systems". WBGU website, http://www.wbgu.de/wbgu_jg2003_ex06.pdf

Nordhaus, W D and Boyer, J (2000) *Warming the World. Economic Models of Global Warming.* MIT Press, Cambridge, MA.

Odell, P (2001) Book Review. *Energy Policy* **29**: pp943–4.

Odum, E P (1969) The strategy of ecosystem development. *Science* **164**: pp262–70.

OECD (Organisation for Economic Co-operation and Development) (ed) (2000) Synthesis Report of the OECD Project on Environmentally Sustainable Development (EST). OECD website, http://www.oecd.org/env/ccst/est/estproj/estproj1.htm

OECD (Organisation for Economic Co-operation and Development) (ed) (2001) *Development Co-operation: 2000 Report: Efforts and Policies of the Development Assistance Committee.* OECD, Paris.

OECD (Organisation for Economic Co-operation and Development) (ed) (2002) *Entwicklungszusammenarbeit. Bericht 2001.* OECD, Paris.

Öko-Institut and DIW (Deutsches Institut für Wirtschaftsforschung) (2001) Analyse und Vergleich der flexiblen Instrumente des Kioto-Protokolls. Expertise of the Enquete Commission "Nachhaltige Energieversorgung unter den Bedingungen der Globalisierung und der Liberalisierung". Öko-Institut and DIW, Berlin.

OPEC (Organization of Petroleum Exporting Countries) (ed) (2001) *Annual Statistical Bulletin 2000.* OPEC, Geneva.

OTA (Office of Technology Assessment) (ed) (1991) *Energy in Developing Countries.* Congress of the United States, Washington, DC.

Ott, B (2002) *Vorlesungsbegleitende Unterlagen für Technik im Sachunterricht der Primärstufe SS 2002.* Universität Dortmund. Lehrstuhl Technik und ihre Didaktik I, Dortmund.

Paine, L K, Peterson, T L, Undersander, D J, Rineer, K C, Bartlet, G A, Temple, S A, Sample, D W and Klemme, R M (1996) Some ecological and socio-economic considerations for biomass energy crop production. *Biomass and Bioenergy* **10**: pp131–242.

Parson, E A and Keith, D W (1998) Fossil fuels without CO_2 emissions. *Science* **282**: pp1053–4.

Pearce, D (1991) The role of carbon taxes in adjusting to global warming. *The Economic Journal* **101**(407): pp938–48.

Pearce, F (1992) *The Damned. Rivers, Dams and the Coming World Water Crisis.* Bodley Head, London.

Pehnt, M (2002) *Ganzheitliche Bilanzierung von Brennstoffzellen in der Energie- und Verkehrstechnik.* Fortschritt-Berichte VDI, Reihe 6: Energietechnik, Nr. 476. VDI-Verlag, Düsseldorf.

Pehnt, M, Bubenzer, A and Räuber, A (2003) 'Life cycle assessment of photovoltaic systems 1. Trying to fight deep-seated prejudices'. In Bubenzer, A and Luther, J (eds) *Photovoltaics Guidebook for Decision Makers.* Springer, Berlin, Heidelberg, New York: pp179–214.

Percy, K E, Awmack, C S, Lindroth, R L, Kubiske, M E, Kopper, B J, Isebrands, J G, Pregitzer, K S, Hendrey, G R, Dickson, R E, Zak, D R, Oksanen, E, Sober, J, Harrington, R and Karnosky, D F (2002) Altered performance of forest pests under atmospheres enriched by CO_2 and O_3. *Nature* **420**: pp403–7.

Petschel-Held, G, Block, A, Cassel-Gintz, M, Kropp, J, Luedeke, M K B, Moldenhauer, O, Reusswig, F and Schellnhuber, H J (1999) Syndromes of Global Change: a qualitative modelling approach to assist global environmental management. *Environmental Modeling and Assessment* **4**: pp295–314.

Petsonk, A (1999) The Kyoto Protocol and the WTO: Integrating Greenhouse Gas Emissions Allowance Trading into the Global Marketplace. *Duke Environmental Law & Policy Forum* **10**: pp185–220.

Philips, M and Browne, B H (ny) Accelerating PV Markets in Developing Countries. Renewable Energy Policy Project website, http://www.repp.org/repp_pubs/articles/pv/7/7.html

Ploetz, C (2002) *Sequestrierung von CO_2 – Technologien, Potenziale, Kosten und Umweltauswirkungen.* Expertise for the WBGU Report "World in Transition: Towards Sustainable Energy Systems". WBGU website, http://www.wbgu.de/wbgu_jg2003_ex07.pdf

Quaschning, V (2000) *Systemtechnik einer klimaverträglichen Elektrizitätsversorgung in Deutschland für das 21. Jahrhundert. Fortschritt-Bericht VDI. Reihe 6: Energietechnik.* VDI Verlag, Düsseldorf.

Raeder, J, Cook, I, Morgenstern, F H, Salpietro, E, Bunde, R and Ebert, E (1995) *Safety and Environmental Assessment of Fusion Power (SEAFP).* European Commission DG XII, Brussels.

Ramsey, F P (1928) A mathematical theory of saving. *Economic Journal* **38**: pp543–59.

Raphals, P (2001) *Restructured Rivers: Hydropower in the Era of Competitive Markets.* International Rivers Network (IRN), Berkeley.

RECS (Renewable Energy Certificate System) (2002) Renewable Energy Certification System. RECS website, http://www.recs.org

Reddy, A K N (2002) 'Energy technologies and policies for rural development'. In Goldemberg, T B (ed) *Energy for Sustainable Development.* United Nations Development Programme (UNDP), New York: pp115–35.

Reichle, D, Houghton, J, Kane, B and Ekmann, J (1999) Carbon Sequestration Research and Development. Oak Ridge National Laboratory (ORNL) website, http://www.ornl.gov/carbon_sequestration/

Reusswig, F (1994) *Lebensstile und Ökologie.* Institut für sozial-ökologische Forschung, Frankfurt.

Reusswig, F, Gerlinger, K and Edenhofer, O (2002) *Lebensstile und globaler Energieverbrauch. Analyse und Strategieansätze zu einer nachhaltigen Energiestruktur.* Expertise for the WBGU Report "World in Transition: Towards Sustainable Energy Systems". WBGU website, http://www.wbgu.de/wbgu_jg2003_ex08.pdf

Revenga, C, Murray, S, Abramovitz, J and Hammond, A (1998) *Watersheds of the World: Ecological Value and Vulnerability.* World Resources Institute and Worldwatch Institute, Washington, DC.

Riahi, K and Roehrl, R A (2000) Greenhouse gas emissions in a dynamics-as-usual scenario of economic and energy development. *Technological Forecasting and Social Change* **63**: pp175–205.

Riahi, K (2002) *Data From Model Runs With Message.* Expertise for the WBGU Report "World in Transition: Towards Sustainable Energy Systems". Mimeo.

Roehrl, R A and Riahi, K (2000) Technology dynamics and greenhouse gas emissions mitigation: a cost assessment. *Technological Forecasting and Social Change* **63**: pp231–61.

Royal Commission on Environmental Pollution (ed) (2000) *Energy – The Changing Climate. Twenty-second Report.* Royal Commission, London.

Ruttan, V (2000) *Technology, Growth and Development: An Induced Innovation Perspective.* Oxford University Press, Oxford, New York.

Salameh, M G (2000) Global oil outlook: return to the absence of surplus and its implications. *Applied Energy* **65**: pp239–50.

Sanderson, M A, Reed, R L, Mclaughlin, R B, Conger, B V, Parrish, D J, Taliaferro, C, Hopkins, A A., Ocumpaugh, W R and Tischler, C R (1996) Switchgrass as a sustainable bioenergy crop. *Bioresource Technology* **56**: pp83–93.

Sankovski, A, Barbour, W and Pepper, W (2000) Quantification of the IS99 emission scenario storylines using the atmospheric stabilization framework. *Technological Forecasting and Social Change* **63**: pp263–87.

Sanstad, A H (2002) Information Technology and Aggregate Energy Use in the U.S.: Empirical and Theoretical Issues. Lawrence Berkeley National Laboratory, USA. IEA Workshop on Impact of Information & Communication Technologies on the Energy System, Paris, France, February 21, 2002. International Energy Agency (IEA) website, http://www.iea.org/effi/index.htm

Schellnhuber, H J (1998) 'Earth system analysis: The scope of the challenge'. In Schellnhuber, H J and Wenzel, V (eds) *Earth System Analysis.* Springer, Berlin, Heidelberg, New York: pp3–5.

Schneider, L C, Kinzig, A P, Larson, E D and Solorzano, L A (2001) Method for spatially explicit calculations of potential biomass yields and assessment of land availability for biomass energy production in Northeastern Brazil. *Agriculture, Ecosystems and Environment* **84**: pp207–26.

Scholz, F (2002) Energiesicherung oder neue politisch-globale Strategie? Vorder- und Mittelasien im Brennpunkt der US-Interessen. *Geographische Rundschau* **54**(12): pp53–8.

Schulze, E-D, Lloyd, J, Kelliher, F M, Wirth, C, Rebmann, C, Lühker, B, Mund, M, Knohl, A, Milyukova, I M, Schulze, W, Dore, S, Grigoriev, S, Kolle, O, Panfyorov, M, Tchebakova, N and Vygodskaya, N N (1999) Productivity of forests in the Eurosiberian boreal region and their potential to act as a carbon sink – a synthesis. *Global Change Biology* **5**: pp703–22.

Schulze, E-D, Valentini, R and Sanz, M-J (2002) The long way from Kyoto to Marrakesh: Implications of the Kyoto Protocol negotiations for global ecology. *Global Change Biology* **8**: pp505–18.

Schulze, E-D (2002) Die Bedeutung der naturnahen Forstwirtschaft für den globalen CO_2-Haushalt. *AFZ – Der Wald* **20**: pp1051–3.

Scott, M J, Sands, R D, Rosenberg, N J and Izaurralde, N C (2002) Future N_2O from US agriculture: projecting effects of changing land use, agricultural technology, and climate on N_2O emissions. *Global Environmental Change* **12**: pp105–15.

Seibel, B A and Walsh, P J (2001) Potential impacts of CO_2 injection on deep-sea biota. *Science* **294**: pp319–20.

Smith, K S, Corvalan, C F and Kjellström, T (1999) How much global ill health is attributable to environmental factors? *Epidemiology* **10**(5): pp573–84.

Smith, K R, Samet, J M, Romieu, I and Bruce, N (2000) Indoor air pollution in developing countries and acute lower respiratory infections in children. *Thorax* **55**: pp518–32.

Smith, K R (2000) National burden of disease in India from indoor air pollution. *PNAS* **97**(24): pp13286–93.

Smith, K R (2002) In Praise of Petroleum? *Science* **298**: p1847.

Snedacker, S C (1984) 'Mangroves: a summary of knowledge with emphasis on Pakistan'. In Haq, B U and Milliman, J D (eds) *Marine Geology and Oceanography of Arabian Sea and Coastal Pakistan*. Van Nostrand Reinhold, New York: p99.

Spreng, D and Semadeni, M (2001) *Energie, Umwelt und die 2000 Watt Gesellschaft*. Centre for Energy Policy and Economics (cepe), Zurich.

SRU (Rat von Sachverständigen für Umweltfragen) (ed) (1994) *Umweltgutachten 1994. Für eine dauerhaft umweltgerechte Entwicklung*. SRU, Wiesbaden.

SRU (Rat von Sachverständigen für Umweltfragen) (ed) (2002) *Umweltgutachten 2002. Für eine neue Vorreiterrolle*. SRU, Berlin.

SRW (Sachverständigenrat zur Begutachtung der gesamtwirtschaftlichen Entwicklung) (1998) *Vor weit reichenden Entscheidungen*. Kohlhammer, Stuttgart and Mainz.

Stahel, W R (1997) The service economy: wealth without resource consumption? *Philos T Roy Soc A* (355): pp1386–8.

Stanley, D G and Warne, A G (1993) Nile Delta: recent geological evolution and human impact. *Science* **260**: pp628–34.

Statistisches Bundesamt (2002) *Statistisches Jahrbuch 2002 für die Bundesrepublik Deutschland*. Statistisches Bundesamt, Wiesbaden.

Stucki, S, Palumbo, R, Baltensperger, U, Boulouchos, K, Haas, O, Scherer, G G, Siegwolf, R and Wokaun, A (2002) *The Role of Chemical Processes in the Transition to Sustainable Energy Systems*. Paul Scherrer Institut (PSI), Villingen, Switzerland.

Stüwe, M (1993) *Sonnige Zukunft: Energieversorgung jenseits von Öl und Uran*. Greenpeace International, Amsterdam.

Subedi, S P (1998) 'Foreign investment and sustainable development'. In Weiss, F. (ed) *International Economic Law With a Human Face*. Kluwer Law International, Den Haag: pp413–28.

Svensson, B (1999) *Greenhouse Gas Emissions from Hydroelectric Reservoirs – The Need of an Appraisal. Presentation Given at the COP-6 Meeting of the UNFCCC in The Hague, The Netherlands*. Mimeo. SwedPower AB, Stockholm.

Swiss Re (2001) Natural Catastrophes and Man-made Disasters in 2000: Fewer Insured Losses Despite Huge Floods. Swiss Re website, http://www.swissre.ch

Tahvanainen, L and Rytkönen, V-M (1999) Biomass production of *Salix viminalis* in southern Finland and the effect of soil properties and climate conditions on its production and survival. *Biomass and Bioenergy* **16**: pp103–17.

Tamburri, M N, Peltzer, E T, Friederich, G E, Aya, I, Yamane, K and Brewer, P (2000) A field study of the effects of CO_2 ocean disposal on mobile deep-sea animals. *Marine Chemistry* **72**: pp95–101.

Terivision (2002) Integrated Energy Solutions for Rural Families. Data Energy Research Institute website, http://www.teriin.org/news/terivsn/issue6/specrep.htm

Thuille, A, Buchmann, N and Schulze, E-D (2000) Carbon stocks and soil respiration rates during deforestation, grassland use and subsequent Norway spruce afforestation in the Southern Alps, Italy. *Tree Physiology* **20**: pp849–57.

Tol, R S J and Verheyen, R (2001) Liability and Compensation for Climate Change Damages – A Legal and Economic Assessment. Universität Hamburg website, http://www.uni-hamburg.de/Wiss/FB/15/Sustainability/Liability.pdf

Torp, T A (ed) (2000) *Final Report "Saline Aquifer CO_2 Storage" Project (SACS)*. Statoil, Bergen.

Toth, F L, Bruckner, T, Füßel, H M, Leimbach, M, Petschel-Held, G and Schellnhuber, H J (1997) 'The tolerable window approach to integrated assessments'. In CGER – Center for Global Environmental Research (ed) *Climate Change and Integrated Assessment Models – Bridging the Gap*. CGER, Ibaraki, Tokio: pp403–30.

UBA (Umweltbundesamt) (ed) (1996) *Manual on Methodologies and Criteria for Mapping Critical Levels/ Loads and Geographical Areas Where They are Exceeded. UN-ECE Convention on Long-range Transboundary Air Pollution. UBA Texte 71/96*. UBA, Berlin.

UBA (Umweltbundesamt) (ed) (1997) *Nachhaltiges Deutschland – Wege zu einer dauerhaft umweltgerechten Entwicklung*. Erich Schmidt, Berlin.

UN (United Nations) (1997) Resolution Adopted by the General Assembly for the Programme for the Further Implementation of Agenda 21. United Nations General Assembly, Nineteenth Special Session, New York, 23-27 June 1997. UN website, http://www.un.org/esa/sustdev/enr1.htm

UN-Ad Hoc Inter-Agency Task Force on Energy (2001) Briefing Paper on Energy Activities of the United Nations. UN website, http://www.un.org/esa/sustdev/csd9/briefing_iaenr.pdf

(United Nations Conference on Trade and Development) (2000) *Trade Agreements, Petroleum and Energy Policies.* , New York and Geneva.

(United Nations Conference on Trade and Development) (2001) *Energy Services in International Trade: Development Implications. A Note by the Secretariat.* , New York and Geneva.

(United Nations Conference on Trade and Development) (2002) *World Investment Report 2002. Transnational Corporations and Export Competitiveness.* , New York and Geneva.

UNDP (United Nations Development Programme) (ed) (1997) *Energy After Rio. Prospects and Challenges*. UNDP, New York.

UNDP (United Nations Development Programme) (ed) (2002a) *Human Development Report 2002. Deepening Democracy in a Fragmented World*. Oxford University Press Oxford, New York.

UNDP (United Nations Development Programme) (2002b) *Weltbevölkerungsbericht. Wege aus der Armut: Menschen, Chancen und Entwicklung*. UNDP, New York.

UNDP (United Nations Development Programme), UN-DESA (United Nations Department of Economic and Social Affairs) and WEC (World Energy Council) (eds) (2000) *World Energy Assessment. Energy and the Challenge of Sustainability*. UNDP, New York.

UNEP (United Nations Environment Programme) and IEA (International Energy Agency) (2001) Energy Subsidy Reform and Sustainable Development: Challenges for Policymakers. Synthesis Report. IEA website, http://www.iea.org/workshop/sustain/

UNEP (United Nations Environment Programme) (2002) Global Environmental Outlook. UNEP website, http://www.unep.org/geo/geo3/english/index.htm

UN-ECE (United Nations Economic Commission for Europe) and FAO (Food and Agricultural Organisation) (2000) *Forest Resources of Europe, CIS, North America, Australia, Japan and New Zealand*. UN-ECE and FAO, New York, Geneva and Rome.

UN-ECE (United Nations Economic Commission for Europe) (2001) Energy Efficiency and Energy Security in the CIS. ECE Energy Series No. 17 (Dokument ECE/ENERGY/44). UN website, http://www.unece.org/energy/nrghome.html

UN-ECOSOC (United Nations Economic and Social Council) (2001) Commission on Sustainable Development: Report on the Ninth Session. UN website, http://www.un.org/esa/sustdev/csd9/ecn172001-19e.htm#Decision%209/1

UNEP-CCEE (United Nations Environment Programm Collaborating Centre on Energy and Environment) (2002) Implementation of Renewable Energy Technology – Opportunities and Barriers. UNEP-CCEE website, http://www.uccee.org/RETs/SummaryCountryStudies.pdf

UNESCO (United Nations Educational Scientific and Cultural Organization) (2001) World Solar Programme 1996-2005. UNESCO website, http://www.unesco.org/science/wsp/background/full_report.htm#action

UNFCCC (United Nations Framework Convention on Climate Change) (2002) *Report on the Conference of the Parties on Its Seventh session, Held at Marrakesh from 29 October to 10 November 2001, FCCC/CP/2001/13/Add.1, 21 January 2002*. UNFCCC, New York.

US-DOE (U.S. Department of Energy) (1999) Carbon Sequestration – Research and Development. DOE website, www.fe.doe.gov/coal_power/sequestration/reports/rd/index.shtml

US-DOE (U.S. Department of Energy) (2002) World Production of Primary Energy by Selected Country Groups (Btu), 1991-2000. DOE website, http://www.eia.doe. gov/emeu/iea/table29.html

Van Beers, C and de Moor, A (2001) *Public Subsidies and Policy Failures: How Subsidies Distort the Natural Environment, Equity and Trade, and How to Reform Them*. Edward Elgar, Cheltenham, MA.

Van Vurren, D and Bakkes, J (1999) *GEO-2000 Alternative Policy Study for Europe and Central Asia – Energy-related environmental impacts of policy scenarios, 1990-2010*. UNEP and RIVM, Nairobi and Bilthoven.

Varma, C V J, Lafitte, R and Schultz, B (2000) Open Letter from ICOLD, IHA and ICID on the Final Report of the World Commission on Dams. UNEP website, http://www.unep-dams.org/print.php?doc_id=68

VENRO (Verband Entwicklungspolitik deutscher Nichtregierungsorganisationen) (2001) *Armut bekämpfen – Gerechtigkeit schaffen. Folgerungen aus der internationalen und nationalen Debatte über Armutsbekämpfung für die deutsche Entwicklungspolitik*. VENRO, Bonn.

Vesterdal, L, Ritter, E and Gundersen, P (2002) Change in soil organic carbon following afforestation of former arable land. *Forest Ecology and Management* **169**: pp137–47.

von Bieberstein Koch-Weser, M (2002) *Nachhaltigkeit von Wasserkraft*. Expertise for the WBGU Report "World in Transition: Towards Sustainable Energy Systems". WBGU website, http://www.wbgu.de/wbgu_jg2003_ex01.pdf

von Hirschhausen, C and Engerer, H (1998) Post-Soviet gas sector restructuring in the CIS: a political economy approach. *Energy Policy* **26**: pp1113–23.

Waelde, T, Bamberger, C and Linehan, J (2000) The Energy Charter Treaty in 2000. In A New Phase. CEPMLP-Journal 7-1. University of Dundee website, http://www.dundee.ac.uk/ecpmlp/journal/html/article7-1.html

Wancura, H, Keukeleere, D and Jensen, P (2001) Reaching the EC Alternative Fuels Objectives – A Multi-Path Mapping Analysis European Alternative Fuel Policy Status and Outlook. Enigmatic Network website, http://www.enigmatic-network.org/screening_workshop_paper_v1.4.pdf

Watson, A J, Bakker, D C E, Ridgwell, A J, Boyd, P W and Law, C S (2000) Effect of iron supply on Southern Ocean CO_2 uptake and implications for glacial atmospheric CO_2. *Nature* **407**: pp730–3.

WBCSD (World Business Council on Sustainable Development) (2001) Mobility 2001. World Mobility at the End of the Twentieth Century and its Sustainability. WBCSD website, http://www.wbcsdmobility.org

WBGU (German Advisory Council on Global Change) (1995) *Scenario for the Derivation of Global CO_2 Reduction Targets and Implementation Strategies. Statement on the Occasion of the First Conference of the Parties to the Framework Convention on Climate Change in Berlin. 1995 Special Report*. WBGU, Bremerhaven.

WBGU (German Advisory Council on Global Change) (1997a) *World in Transition: The Research Challenge. 1996 Report*. Springer Berlin, Heidelberg and New York.

WBGU (German Advisory Council on Global Change) (1997b) *Targets for Climate Protection 1997. A Study for the Third Conference of the Parties to the Framework Convention on Climate Change in Kyoto. Special Report 1997*. WBGU, Bremerhaven.

WBGU (German Advisory Council on Global Change) (2000) *World in Transition: Strategies for Managing Global Environmental Risks. 1998 Report*. Springer, Berlin, Heidelberg and New York.

WBGU (German Advisory Council on Global Change) (2001a) *World in Transition: Conservation and Sustainable Use of the Biosphere. 1999 Report*. Earthscan, London.

WBGU (German Advisory Council on Global Change) (2001b) *World in Transition: New Structures for Global Environmental Policy. 2000 Report*. Earthscan, London.

WBGU (German Advisory Council on Global Change) (2001c) *The Johannesburg Opportunity: Key Elements of a Negotiation Strategy. WBGU Policy Paper 1*. WBGU, Berlin.

WBGU (German Advisory Council on Global Change) (2002) *Charging the Use of Global Commons. Special Report 2002*. WBGU, Berlin.

WCD (World Commission on Dams) (2000) *Dams and Development. A New Framework for Decision-Making*. Earthscan, London.

WEC (World Energy Council) (2000) Energy for Tomorrow's World – Acting Now! WEC website, http://www.worldenergy.org/wecgeis/publications/reports/etwan/exec_summary/exec_summary.asp

WEED (World Economy Ecology and Development) (2002a) Stärkung von Umweltschutz bei Auslandsdirektinvestitionen. Kompromisspapier des vom BMU initiierten Dialogprozesses "Umwelt und Auslandsdirektinvestitionen" vom 23.05.2002. WEED website, http://www.weedbonn.org/unreform/bdidialog.htm

WEED (World Economy Ecology and Development) (2002b) Leitlinien zur Nachhaltigkeit von Auslandsdirektinvestitionen der deutschen Wirtschaft, WEED-Entwurf vom Januar 2002. WEED website, http://www.weedbonn.org/unreform/bdidialog.htm

Werksmann, J (2001) 'Greenhouse-gas emissions trading and the WTO'. In Chambers, W B (ed) *Inter-linkages. The Kyoto-Protocol and the International Trade and Investment Regimes*. United Nations University Press, Tokyo, New York and Paris: pp153–90.

WHO (World Health Organization) (ed) (1999) *Air Quality Guidelines*. WHO, Geneva.

WHO (World Health Organization) (ed) (2000) *Air Pollution. Fact Sheet N 187*. WHO, Geneva.

WHO (World Health Organization) (ed) (2002a) *An Anthology on Women, Health and Environment: Domestic Fuel Shortage and Indoor Pollution*. WHO, Geneva.

WHO (World Health Organization) (ed) (2002b) *World Health Report 2002. Reducing Risks, Promoting Healthy Life*. WHO, Geneva.

WHO (World Health Organization) (2002c) Women's Health and the Environment. WHO website, http://w3.whosea.org/women2/environment.htm

Wiemken, E, Beyer, H, Heydenreich, G W and Kiefer, K (2001) Power characteristics of PV ensembles: experiences from the combined power production of 100 grid connected PV systems distributed over the area of Germany. *Solar Energy* **70**(6): pp513–8.

Williams, R (2000) *Advanced Energy Supply Technologies. Chapter 8 in World Energy Assessment*. UNDP, New York.

Winter, C J and Nitsch, J (1989) *Wasserstoff als Energieträger. Technik, Systeme, Wirtschaft*. Springer, Berlin, Heidelberg and New York.

WISE (World Information Service on Energy) (2001) *Mögliche toxische Auswirkungen der Wiederaufbereitungsanlagen in Sellafield und La Hague*. WISE, Paris.

Wiser, G M (1999) The Clean Development Mechanism Versus the World Trade Organization: Can Free-Market Greenhouse Gas Emissions Abatement Survive Free Trade? *Georgetown International Environmental Law Review* **11**(3): pp531–97.

Wissenschaftlicher Beirat beim Bundesministerium der Finanzen (1997) *Umweltsteuern aus finanzwissenschaftlicher Sicht. Schriftenreihe des BMF Heft 63*. Stollfuß, Bonn.

World Bank (1991) *Environmental Assessment Sourcebook. Band III. Guidelines for Environmental Assessment of Energy and Industry Projects*. World Bank Environmental Department, Washington, DC.

World Bank (1993) The World Bank's Role in the Electric Power Sector: Policies for Effective Institutional, Regulatory and Financial Reform. World Bank Policy Paper. World Bank website, http://www.worldbank.org/energy/subenergy/policy_papers.htm

World Bank (2000) Energy Services for the World's Poor. Energy and Development Report 2000. World Bank website, http://www.worldbank.org/html/fpd/esmap/energy_report2000/

World Bank (2001a) *Reforming India's Energy Sector (1978-99)*. World Bank, Washington, DC.

World Bank (2001b) Topical Briefing to the Board of Directors on Energy. The World Bank Group's Energy Program. Poverty Alleviation, Sustainability, and Selectivity. World Bank website, http://www.worldbank.org/energy/pdfs/business_renewal.pdf

World Bank (2001c) *World Development Indicators 2001*. World Bank, Washington, DC.

World Bank (2002a) *World Development Indicators 2002*. World Bank, Washington, DC.

World Bank (2002b) *Global Development Finance*. World Bank, Washington, DC.

World Bank (2002c) *Energy and the Environment. Energy and Development Report 2001*. World Bank, Washington, DC.

World Tourism Organization (2002) Latest Data. World Tourism Organization website, http://www.world-tourism.org/market_research/facts&figures/menu.htm

WRI (World Resource Institute) (2001) *World Resources 2000–2001*. WRI, Washington, DC.

WRI (World Resource Institute) (2002) Energy Consumption by Economic Sector, Table ERC3. WRI website, http://earthtrends.wri.org/

WRI (World Resources Institute), UNEP (United Nations Environment Programme) and WBCSD (World Business Council for Sustainable Development) (eds) (2002) *Tomorrow's Markets. Global Trends and Their Implications for Business*. WRI, UNEP and WBCSD, Washington, DC.

WTO (World Trade Organisation/Council for Trade in Services) (1998) Energy Services – Background Note by the Secretariat (S/C/W/52) of 9 September 1998. WTO website, http://www.docsonline.wto.org/gen_search.asp

WTO (World Trade Organisation) (2001) *International Trade Statistics 2001*. WTO, Geneva.

Zan, C S, Fyles, J W, Girouard, P and Samson, R A (2001) Carbon sequestration in perennial bioenergy, annual corn and uncultivated systems in southern Quebec. *Agriculture, Ecosystems and Environment* **86**: pp135–44.

Zittel, W and Altmann, M (1996) Molecular Hydrogen and
Water Vapour Emissions in a Global Hydrogen Energy Economy. Proceedings 11th World Hydrogen Energy Conference.
Hydrogen Energy Conference website, http://www.hydro-
gen.org/Wissen/Vapour.htm

Glossary

1 joule = 1J
- is the energy needed by a bee to fly 120m,
- is the amount of electrical energy needed by a pocket calculator to carry out 50 multiplications.

1 kilojoule = 1kJ (10^3J)
- is the energy needed by a person to swim 1m, walk 5m, cycle 12m or go up 8 steps.

1 megajoule = 1MJ (10^6J)
- is enough to watch about 2 football matches on a colour TV,
- is consumed by a person doing nothing at all for 3.5 hours (basal metabolic rate).

1 gigajoule = 1GJ (10^9J)
- meets the washing and drying requirements of a 4-person household for 3 months, or lighting needs for 8 months.

1 terajoule = 1TJ (10^{12}J)
- is the content of 31,000l petrol, enough to travel 8 times around the world by car,
- is the amount wasted by a poorly insulated single-family house in 7 years.

1 petajoule = 1PJ (10^{15}J)
- represents a 6 m high pile of hard coal covering an entire football field.

1 exajoule = 1EJ (10^{18}J)
- is the amount of energy received by the Earth from the sun in 6 seconds,
- is the worldwide consumption of primary energy in 21 hours (in the year 2000).

Source: Ott, 2002

Annex B countries: Group of countries listed in Annex B to the ↑Kyoto Protocol that have committed themselves to limit or reduce their greenhouse gas emissions. These include all Annex I countries except Turkey and Belarus.

Annex I countries: Group of countries listed in Annex I to the ↑United Nations Framework Convention on Climate Change (UNFCCC). Includes all developed countries in the OECD, as well as countries undergoing the process of transition to a market economy in eastern Europe and the CIS states. All other countries are automatically termed non-Annex I countries. In Articles 4.2 (a) and 4.2 (b) UNFCCC, Annex I countries specifically commit themselves to return their greenhouse gas emissions to 1990 levels individually or jointly by the year 2000.

Annex II countries: Group of countries listed in Annex II to the ↑United Nations Framework Convention on Climate Change (UNFCCC). Comprises all developed countries in the OECD, being a subset of the ↑Annex I country group. Article 4.3 UNFCCC obligates these countries to provide financial resources in order to assist developing countries in meeting their commitments, such as the preparation of national reports. It also obligates Annex II countries to support the transfer of environmentally sound technologies to developing countries.

baseline scenario: A ↑scenario that characterizes the development of the economy and of other drivers of emissions (such as population, technologies) that is to be expected without policy interventions, in particular without explicit climate policy measures. It serves in, for instance, economic cost-benefit analysis as a basis on which to develop climate change mitigation scenarios that take climate policy measures into account.

biofuels: Liquid fuels such as biodiesel and bioethanol that result from the conversion of ↑biomass.

biogas: Generic term referring to gases from which energy can be recovered and that are created when ↑biomass decomposes under anaerobic conditions. In decomposition, about two-thirds methane (CH_4) and one-third carbon dioxide (CO_2) are released. The methane gas is the fraction of the biogas that can be utilized for energy recovery. Biogas has a high calorific value (25MJ per cubic metre).

biomass: The total organic mass of the biotic environment, either as living or dead biomass (e.g. fuelwood, charcoal and dung). Important conversion products of biomass include ↑biogas and ↑biofuel. In developing countries, ↑traditional biomass use is the dominant form.

capacity utilization: A measure of the difference between peak demand on the electricity market and overall installed power plant capacity.

capacity: Capacity is energy per unit of time. Electric capacity is measured in watts (W), kilowatts (kW), megawatts (MW), etc.

carbon dioxide (CO_2): A naturally occurring gas, and also a by-product of burning fossil fuels and biomass. CO_2 is also emitted as a result of deforestation and other land-use changes, as well as from industrial processes such as cement production.

carbon intensity: Carbon dioxide emissions per unit primary energy input.

Clean Development Mechanism (CDM): One of the ↑Kyoto mechanisms introduced by the ↑Kyoto Protocol, that permits an industrialized-country investor to carry out emissions-reducing projects in a developing or newly industrializing country. The greenhouse gas reduction attributable to the project is credited to the industrialized country.

climate change: A statistically significant variation in either the mean state of the climate or in its variability, persisting for an extended period (typically decades).

climate sensitivity: The °C rise in surface temperature that results if the pre-industrial CO_2 concentration in the atmosphere doubles from 280 to 560 ppm. The ↑IPCC states a range of climate sensitivity of 1.5–4.5°C, without stating a best estimate.

CO_2 storage (or sequestration): Storage, by human action, of atmospheric carbon in terrestrial ecosystems, geological formations or the oceans. For instance, through new technologies the ↑carbon dioxide resulting from combustion processes can be captured, possibly liquefied and then pumped into underground repositories such as depleted gas and oil fields. In addition, natural carbon storage takes place in vegetation, in which carbon dioxide is bound as ↑biomass over lengthier periods.

cogeneration (combined heat and power, CHP): Facilities with combined heat and power production not only generate electricity from the fuel consumed, but also make use of the waste heat at the same time. For instance, this heat can be used for space heating purposes (as in district heating systems). In industry, it can be used for heat-dependent production processes.

Commission on Sustainable Development (CSD): A commission of the Economic and Social Council (ECOSOC) of the United Nations established in 1992 as the key forum for the Rio follow-up process. It monitors and supports implementation of the AGENDA 21 adopted at the United Nations Conference on Environment and Development (UNCED) held in Rio de Janeiro in 1992. The annual meetings of the CSD are attended by governments and international organizations, but also by more than 1,000 non-governmental organizations.

contracting: A financing model under which the investment costs of energy-saving measures or new energy-generating plant are refinanced from the energy costs saved. The contractor, e.g. a private-sector company or a municipal energy supplier, implements measures that reduce the energy requirement of a building or installation. To do so, the contractor invests in new technology or, for instance, insulates a building. The host institution, such as a local authority that does not itself command over the necessary capital or knowledge, is guaranteed by contract a certain percentage of the energy savings and thus cost savings. The remaining savings are the profit that goes to the contractor. In energy equipment contracting arrangements, the contractor also operates and maintains the equipment.

Demand Side Management (DSM): A voluntary control and planning instrument to tap efficiency potentials on the demand side, by means of economic incentives (e.g. in the form of load management making use of variable tariff structures).

deregulation: Reduction of ↑regulation.

development partnerships (PPP: Public-Private Partnerships): In development partnerships, private companies cooperate with state development cooperation agencies in implementing projects that pursue development policy goals and at the same time yield micro-economic benefit for the participating companies. The advantage of this type of cooperation with industry from a development cooperation perspective is that the participation of private companies ensures that activities are cost-effective, efficient and have sustained impact.

Disability Adjusted Life Years (DALYs): An indicator of the overall disease burden of a population, integrating premature death, disease and disability.

efficiency: Measure of the effectiveness of an energy conversion process. The efficiency of a system is the ratio of energy output to energy input, and is stated as a percentage.

electricity-to-heat ratio: The ratio of electricity output to utilizable heat of a ↑cogeneration process.

emissions trading (or certificate trading): Economic instrument for the cost-efficient limitation or reduction of environmentally harmful emissions. Generators of emissions are subject to reduction goals that they must either meet themselves or can have met in part or in whole by other generators. To that end, reduction commitments can be traded among the participants in a trading system, which produces a cost-optimal allocation of the set overall reduction. The ↑Kyoto Protocol introduces this instrument as one of the ↑Kyoto mechanisms at state level for ↑Annex B countries. The Protocol also sets out the specific emissions reduction commitments. A transfer of reduction commitments from countries to companies is possible.

energy carriers: Substances or media which contain energy that can be utilized in a cost-effective manner. A distinction is made between e.g. fossil (coal, mineral oil, natural gas), renewable (biomass, geothermal, solar, wind, hydro) and nuclear (uranium) ↑primary energy carriers.

Energy Charter Treaty (ECT): The Treaty evolved from the 1991 European Energy Charter. It entered into force in 1998. 46 states, mainly in Europe and Central Asia, have ratified the Treaty (as at 11.09.2002). The aim of the ECT is to promote economic growth by liberalizing investment and trade. In addition, it establishes minimum standards with respect to foreign investment and energy transport. The environmental aspects of energy policy have been set out in greater detail in a legally non-binding protocol (Protocol on Energy Efficiency and Related Environmental Aspects, PEEREA).

energy efficiency: The technical efficiency of end-use equipment (such as household appliances) or systems (such as power plants), usually quantified in terms of their efficiency in converting energy.

energy input: This corresponds to the frequently used term 'energy consumption', the latter being, strictly speaking, incorrect, and means the amount of energy deployed.

energy intensity: The ratio of energy input to ↑gross domestic product generated by that input (inverse of ↑energy productivity).

energy mix: Combination of different energy carriers for energy supply.

energy poverty: Energy poverty refers to the lack of sufficient access to energy services, in order to meet basic needs, that are affordable, reliable, high-quality and safe, and cause no undue health or environmental impacts. Countries where energy poverty is widespread are generally characterized by major development problems. Energy poverty affects some 2,400 million people, who are dependent upon traditional biomass use. 1,600 million people have no access to electricity.

energy productivity: The ratio of ↑gross domestic product to ↑energy input required to produce that product. In contrast to ↑energy efficiency, energy productivity can be improved not only through technological efficiency, but also through structural changes in energy systems (such as a transition from coal power plants to more efficient gas-fired plants), economic structural changes towards less energy-intensive products and services, altered patterns of energy use or changes in lifestyles.

energy service: The actual utility gained by using ↑useful energy: a brightly illuminated working space, refrigerated food, clean laundry, transportation of goods from one place to another, etc. The quantity of energy used is irrelevant to the value of the energy service (e.g. the quality of lighting is important, not the electricity consumed, transportation to the destination is decisive, not the petrol consumed).

energy: Energy is the capacity of a system to do work. A distinction is made between e.g. chemical, mechanical and electrical energy, as well as heat.

final energy: Energy that is available in a utilizable form after conversion of ↑primary energy to ↑secondary energy and after transportation and distribution to the final consumer (e.g. briquettes, electricity from the socket, or petrol at the petrol pump). Final energy is the third stage in the energy flow chain from ↑primary over ↑secondary to ↑useful energy.

fossil fuels: Carbon-based fuels from fossil carbon deposits, including coal, oil and natural gas. Their combustion releases carbon dioxide, which is the main driver of human-induced global warming.

fuel cell: Fuel cells convert chemical energy directly into electrical energy, i.e. without an intermediate thermal phase. The ideal energy carrier is hydrogen, which can be produced from e.g. green electricity or by reforming fossil energy carriers, but also from biomass. Electric efficiencies of 30–50 per cent are currently achieved. Efficiencies of up to 60 per cent are anticipated for the future. Water vapour is the sole emission. There are various designs of fuel cells, e.g. polymer electrolyte membrane fuel cells (PEMFC) operating in the low-temperature range of 50–100°C, which are particularly suitable for mobile applications, molten carbonate fuel cells (MCFC) for the medium-temperature range around 650°C and solid oxide fuel cells (SOFC) which operate in the high-temperature range of 800–1000°C.

geothermal energy: Heat from the Earth's core that reaches the Earth's surface or can be utilized there.

global change: Refers to the interlinkages among global environmental changes, economic globalization and cultural transformation.

Global Environment Facility (GEF): A multilateral financing mechanism established in 1991. The GEF is implemented jointly by UNDP, UNEP and the World Bank. It provides grants and low-interest loans to developing countries and eastern European transition countries to help them carry out projects and measures to relieve pressures on global ecosystems. The focal areas are climate protection, biodiversity conservation, ozone layer protection and the protection of international waters. Soil conservation measures in arid zones and forest conservation measures also receive support if they have links to one of the four focal areas.

global radiation: Direct and diffuse solar irradiation incident upon a horizontal surface. The level of global radiation depends upon the position of the sun (in turn dependent upon latitude and time of year) and atmospheric conditions (cloud cover, atmospheric particles).

green electricity: A common term for electricity produced from ↑renewable sources. It also embraces power produced through ↑cogeneration.

Green Energy Certificates: This is a further development of flexible, tradable ↑quotas. Producers of ↑green electricity receive these certificates from a state-controlled issuing body as verification that they have produced a certain amount of electricity (such as 1MWh). Besides energy suppliers and generators, final consumers can also participate in a system of tradable Green Energy Certificates.

greenhouse gases: Greenhouse gases are those gaseous constituents of the atmosphere that, due to their selective absorption of thermal radiation, cause a warming of the lower atmosphere. The primary anthropogenic greenhouse gases are ↑carbon dioxide, ↑nitrous oxide and ↑methane. They also include industrial gases such as hydrofluorocarbons (HFCs), perfluorocarbons (PFCs) and sulphur hexafluoride (SF_6).

gross domestic product (GDP): Equals the total of all income from employment plus unearned income in the reporting period generated in the course of production in a country, to which are added depreciation, and levies on production and imports (reduced by subsidies).

guard rail: Guard rails demarcate the domain of free action for the people-environment system from those domains which represent undesirable or even catastrophic developments and which therefore must be avoided. Pathways for sustainable development run within the corridor defined by these guard rails.

heat pump: Heat pumps raise heat from a low temperature level to a higher, more useful temperature level, using external energy input. All heat pumps need external energy to drive the process; it is essential to take this into account in an energy-balance assessment.

hydrogen economy: An economy based upon hydrogen as energy storage medium. In this technology system, at first the ↑primary energy of solar radiation or of wind is converted into electricity. Electricity from e.g. ↑renewable sources can be used to produce hydrogen through electrolytic separation from water. Alternatively, hydrogen can also be produced from ↑biomass or ↑fossil fuels. It can serve as a distributed storage medium for electricity, for instance in periods of surplus power production, or can be used to operate ↑fuel cells in buildings and vehicles.

IAEA: The International Atomic Energy Agency was founded in 1957 as a specialized agency of the United Nations with seat in Vienna, and has 123 member states. Its tasks include reviewing compliance with the Non-Proliferation Treaty and the worldwide monitoring of nuclear facilities. It promotes the civilian use of nuclear energy, cooperation in nuclear engineering and research, and the exchange of scientific-technical experience through support programmes.

IEA: The International Energy Agency was established in 1974 as an autonomous organization within the OECD framework in Paris, with the aim of ensuring the security of primary energy supply. The 26 member states have agreed to reduce dependency upon oil imports through promoting the use of fossil and also renewable energies, to exchange key energy information, to coordinate their energy policies and to collaborate in programmes for the efficient use of energy. The IEA regularly publishes the World Energy Outlook, the most important source worldwide for energy statistics and analyses of the energy sector.

Intergovernmental Panel on Climate Change (IPCC): The Panel was founded in 1988 and is one of the most influential international scientific institutions for climate policy. The IPCC laid the scientific foundation for negotiations on the ↑United Nations Framework Convention on Climate Change, and publishes regular assessment reports on global climate change. The Third

Assessment Report (TAR), published in 2001, was authored by more than 3,000 scientists from all around the world.

international regimes: International regimes are institutions by which a group of states addresses a transboundary problem – in the environmental sphere e.g. climate change. They comprise principles, norms, rules and decision-making procedures, and are based upon formal or informal intergovernmental agreements. A conference of the parties meeting periodically generally forms the core of the decision-making process of a regime. Although they do often have small secretariats, regimes are not autonomous actors, in contrast to international organizations.

Joint Implementation (JI): One of the flexible mechanisms of the ↑Kyoto Protocol (↑Kyoto mechanisms), that permits developed countries (↑Annex I countries) to carry out climate change mitigation projects jointly with another Annex I country. The project (e.g. the erection of a wind turbine) is carried out in country A, but financed by country B. The emissions thus prevented in country A can be emitted additionally by country B within the commitment period, or country B can have them credited to its account. A corresponding quantity of emission rights is deducted from country A's account.

joule (J): Unit of energy.
$1J = 1Nm = 1Ws = 1kg\ m^2\ s^{-2}$
$1kWh = 3,600,000J$

kilowatt peak (kWp): Output of a photovoltaic module under standard test conditions, i.e. global radiation 1000 W/m^2, cell temperature 25°C and the spectrum of sunlight in central Europe.

kilowatt-hour (kWh): Commonly used measure of energy. For larger installations, energy is often stated in megawatt-hours (MWh) per year.

Kyoto mechanisms: 'Flexible' mechanisms envisaged in the ↑Kyoto Protocol, such as ↑emissions trading, ↑Clean Development Mechanism (CDM) and ↑Joint Implementation (JI). These permit the crediting of emissions reductions achieved outside of a country that has adopted commitments.

Kyoto Protocol to the United Nations Framework Convention on Climate Change: Agreement under international law that sets out greenhouse gas emissions reduction targets for developed countries, as well as key implementation modalities. The Protocol was adopted in 1997 at the 3rd Conference of the Parties to the ↑United Nations Framework Convention on Climate Change (UNFCCC) in Kyoto, Japan. It commits ↑Annex B countries to reduce emissions of certain ↑greenhouse gases by around 5 per cent from the base year 1990 in the commitment period 2008–2012.

The Kyoto Protocol has not yet entered into force because Russia, which has announced its intention to ratify, has not yet actually done so. The USA declared in March 2003 that it does not intend to ratify the Protocol.

learning curve: The drop in specific production costs in step with growing cumulative production.

liberalization: A general term referring to the dissolution of formerly monopolistic structures and the introduction of market-based conditions, i.e. competition. In Germany, the Energy Industry Act (Energiewirtschaftsgesetz) adopted in April 1998 led to the removal of the territorial monopolies previously held by the power suppliers. Since then, consumers are free to choose their power supplier.

methane (CH_4): Greenhouse gas emitted above all from rice cultivation, livestock breeding, the combustion of biomass and the extraction and combustion of fossil fuels.

microgrid: Is a closed, spatially discrete power supply network that is not connected to further (including public) networks.

modern biomass use: ↑Biomass from which energy can be recovered (e.g. agricultural residues, forestry residues and small-diameter wood, industrial wood waste and discarded timber, as well as annual or perennial energy crops cultivated specifically for energy production) which, giving due regard to ecological and health restrictions, is used to produce heat and/or power, as well as biogas (cf. ↑traditional biomass use).

net metering: Fixed rates received by electricity generators for power sold to the public grid, produced from e.g. ↑renewables. The rates paid for power sold to the grid are a key determinant of the cost-effectiveness of power-generating installations. They generally taper over time.

new renewables: These include those ↑renewables that still have a major expansion potential because their use is presently in the initial phase of technology development, e.g. solar, wind and modern biomass. They do not include hydropower.

nitrous oxide (N_2O): A persistent greenhouse gas released above all through the use of nitrogen fertilizers in agriculture, and through the combustion of biomass and fossil fuels.

OPEC: The Organization of Petroleum Exporting Countries was founded in 1960 by Saudi Arabia, Venezuela, Iraq, Iran and Kuwait. Qatar (1961), Indonesia (1962), Libya (1962), the United Arab Emirates (1967), Algeria (1969) and Nigeria (1971) joined later. Today OPEC is a powerful alliance of newly industrializing economies operating on the international energy market.

overall efficiency: In contrast to momentaneous ↑efficiency, which is the ratio of output to input under defined momentaneous conditions, overall efficiency is this ratio across a certain period.

path dependency (or lock-in effects): The limitations imposed upon policy options to steer technologies, as a result of historical developments (lock-in). For instance, that a technology prevails over another on the market is not necessarily due to its superiority, but can be the outcome of chance historical events and a self-amplifying process: The costs of the 'traditional' technology are low compared to the initial investment associated with a new technology, as learning effects have been used in the traditional technology's application and compatible techniques and standards are available.

portfolio approach: The requirement upon actors subject to quota commitments to use a certain proportion of renewables for electricity production.

primary energy: The energy content of natural energy carriers such as coal, oil, natural gas or natural uranium. It is the input parameter of energy flows, which characterize energy use by humankind. Primary energy is the first link in the energy flow chain and is converted, e.g. in power plants, into ↑secondary energy.

Public-Private Partnership (PPP): Longer-term cooperation between the state and the private sector to pursue a common goal (such as development partnerships). Both sides bring their own specific resources (assistance measures, know-how, etc.) to the cooperation.

Purchasing Power Parity (PPP): A special measure of ↑gross domestic product permitting comparisons of purchasing power among countries. PPP relates to a 'shopping basket' in which current exchange rates have no influence. Thus one kilo of rice receives the same value in Japan as in Indonesia, even though its dollar value is about 7 times higher. PPP tends to reduce the difference in GDP per capita between industrialized and developing countries.

quantitative approach: Generic term applying to the promotion of renewables through quantitative stipulations set by the state (minimum quantity or minimum proportion of renewable energies that must be implemented within a certain period). Quantitative instruments include the various forms of ↑quotas, as well as ↑tendering procedures.

quotas: An instrument to promote the deployment of renewable energy sources in energy supply. A quantitative goal is set at the policy level – usually a national-level minimum target for energy pro-

duction from renewable sources within a certain period and/or for specific technology realms. The overall quota is broken down into sub-quotas for energy producers or suppliers. To improve flexibility and thus economic efficiency, it is conceivable to establish a quota trading system and to further develop the approach towards ↑Green Energy Certificates.

regime: ↑international regime

regulation: Statutory standards established by a state, regulating markets. The intervention in market processes that these standards produce can be more or less deep. The energy sector generally requires a certain degree of regulation because transmission and distribution networks represent natural monopolies.

renewables: These include the energy of the sun, water, wind, tides, modern biomass and geothermal energy. Their overall potential is in principle unlimited or renewable, and is CO_2-free or -neutral.

re-regulation: The renewed ↑regulation of markets that were previously liberalized (deregulated).

scenario: A plausible description of how the future may develop, based on analysis of a coherent and internally consistent set of assumptions, trends, relationships and driving forces.

secondary energy: Readily storable and/or transportable forms of energy (e.g. electricity, fuels, hydrogen), produced from ↑primary energy carriers in e.g. power plants or refineries. Secondary energy is the second stage in the energy flow chain that begins with ↑primary energy. It is transported through e.g. the electricity grid to consumers, where it is available as ↑final energy.

sequestration: ↑CO_2 storage.

shadow subsidies: Non-accounting of the costs of external effects, for instance of conventional energy production, whose exact quantitative calculation is extremely difficult.

solar thermal power plants: In solar thermal power plants, direct sunlight is concentrated by means of optical elements onto an absorber. The radiation energy thus absorbed heats a heat transfer medium. This heat energy is subsequently used to drive largely conventional prime movers, such as steam turbines or ↑Stirling engines.

Stirling engine: Is a cyclically operating thermodynamic machine that converts external heat inputs into mechanical energy.

sustainable development: This term is mostly understood as a concept of environment and development policy that was formulated by the Brundtland Report and further refined at the United Nations Conference on Environment and Development in Rio de Janeiro in 1992. The concept implies that democratic decision-making and implementation processes should promote development that is ecologically, economically and socially sustainable, and should take into account the needs of future generations.

technology transfer: The set of processes relating to the exchange of knowledge, money and goods among different stakeholders that leads to the spreading of use of technologies, e.g. for mitigating climate change or securing sustainable energy development. Transfer often has two meanings: Diffusion of technologies and technological cooperation across and within countries.

tendering procedure: An approach towards providing support for renewables. In this procedure, the state calls for tenders for precisely stipulated generation capacities or grid feed-in quantities from certain energy sources. Contracts are generally awarded to those investors who make the cheapest bid.

traditional biomass use: Form of energy production from ↑biomass, such as wood, dung, harvest residues, etc., above all for cooking and heating. Worldwide about 2,400 million people, predominantly in developing countries, are dependent upon traditional biomass and are thus often exposed to emission-related health hazards due to inadequate combustion technologies.

United Nations Framework Convention on Climate Change (UNFCCC): The Convention was adopted in 1992 and entered into force in 1994. Its ultimate objective is "Stabilization of greenhouse gas concentrations in the atmosphere at a level that would prevent dangerous anthropogenic interference with the climate system. Such a level should be achieved within a time-frame sufficient to allow ecosystems to adapt naturally to climate change, to ensure that food production is not threatened and to enable economic development to proceed in a sustainable manner." The ↑Kyoto Protocol, adopted in 1997, sets out binding commitments to reduce greenhouse gas emissions.

United Nations Millennium Declaration: In the UN Millennium Declaration adopted in 2000, the signing states committed themselves to contribute to overcoming extreme poverty. To this end, they agreed eight international development goals, most of which are to be attained by the year 2015:
1. Eradicate extreme poverty and hunger
2. Achieve universal primary education
3. Promote gender equality and empower women
4. Reduce child mortality
5. Improve maternal health
6. Combat HIV/AIDS, malaria and other diseases
7. Ensure environmental sustainability
8. Develop a global partnership for development

useful energy: Is the energy actually utilized to perform a certain task. It is the last link in the energy flow chain that begins with ↑primary energy.

watt (W): Unit of energy output.

World Bank: Founded in 1944, the World Bank is today the largest source of development assistance finance. The objective of the Bank is to reduce poverty in developing countries. The Bank grants loans and provides policy advice, technical assistance and, increasingly, services for knowledge exchange. The priorities in granting loans are: Health and education, environmental protection, supporting private-sector economic development, strengthening the capability of governments to provide services efficiently and transparently, supporting reforms to attain stable economic conditions, and social development and poverty reduction.

World Energy Council (WEC): The World Energy Council was founded in 1924 with seat in London. Its work is supported by member committees in 90 countries. The objective of these non-governmental organizations is to promote sustainable energy policy by means of research, analysis, policy advice and cooperation. WEC published a World Energy Assessment in 2000 together with the United Nations.

Index

spatial planning 93
standards 34, 126, 154, 159, 164, 187, 190, 209
 – CDM standard 163
stoves; *cf* cookers
subsidies 18, 27, 35, 144, 146, 148, 159, 161, 184, 193; *see also*
 incentive systems
 – removing subsidies 146, 152, 166, 176, 183, 216
 – report on subsidies 147
 – target-group specific subsidies 163, 213
sufficiency 85, 200; *see also* lifestyles
supply strategies 76-77, 79
supply systems/networks 75, 86, 203, 212; *see also* power
 supply
sustainable development 85, 92, 101, 107, 115, 143, 168, 193

T

tariffs 35, 158, 186, 190
taxation 144-145, 164; *see also* financing
technology risks; *cf* risks
technology transfer 38, 173, 178, 183, 187, 209; *see also*
 knowledge transfer
terrorism 51, 53, 124, 192, 211
thermal insulation 86, 206
thermohaline circulation 110; *see also* climate change
tourism 23, 110
trade 29, 35, 40, 158, 186, 188-189
Trade-Related Aspects of Intellectual Property Rights
 (TRIPS) 35, 186, 188
transformation path; *cf* exemplary transformation path
transformation strategy 139, 143-144, 166, 207; *see also*
 exemplary transformation path
transition countries 26, 42, 111, 146, 151, 154, 157, 163
transport 14, 16-17, 25-28, 92, 120, 164, 206; *see also*
 mobility
information systems 93, 165
turbines 14, 52, 64, 73-74, 154

U

United Nations (UN) 32-33, 37, 41, 201
United Nations Conference on Environment and
 Development (UNCED) 32, 176, 200
United Nations Development Programme (UNDP) 32, 37,
 122, 135, 171, 173
United Nations Educational Scientific and Cultural
 Organization (UNESCO) 37, 172
United Nations Environment Programme (UNEP) 32, 34,
 37, 170, 173, 185, 215
United Nations Framework Convention on Climate
 Change (UNFCCC) 35, 38, 41, 48, 109, 166, 178, 182; *see*
 also climate policy, agreements
 – Conference of the Parties 35, 37, 178
United Nations Industrial Development Organization
 (UNIDO) 37, 180
urbanization 16, 23, 42, 87, 100
USA 21, 30, 34, 112, 114, 138, 170, 178, 198, 201
useful energy 43, 83, 85, 185; *see also* energy

user charges 183, 216; *see also* financing

V

vegetation 90, 198, 207

W

wind energy 63-64, 68, 94, 127, 194, 203; *see also* energy
 carriers
 – offshore systems 63-65, 116, 203
women 22, 61-62, 117, 162
World Bank 38-39, 54, 173, 175, 181-182, 212
World Commission on Dams (WCD) 54, 115, 159, 204; *see*
 also dams
World Energy Assessment (WEA) 32, 128, 135, 167, 168
World Energy Charter 109, 167, 169-170, 172, 184, 207, 215
World Energy Council (WEC) 32, 118-119
World Energy Outlook (WEO) 32, 199; *see also*
 International Energy Agency (IEA)
World Energy Research Coordination Programme
 (WERCP) 168, 174, 215
World Health Organization (WHO) 22, 37, 47, 124
World Meteorological Organization (WMO) 32, 37
World Solar Programme 37; *see also* solar energy
World Summit on Sustainable Development (WSSD) 34,
 169, 183, 213, 214
World Trade Organization (WTO) 28, 34, 146, 185